Combined Proceedings International Plant Propagators' Society 2023

IPPS Volume 73

Editors:

Robert L. Geneve
Ranjith Pathirana
William Barnes Jr.
Masanori Tomita
Jill Reader
Elsa du Toit
Fred T. Davies Jr.
Ryan Stewart

ISBN: 978-1-7138-9909-9

Printed from e-media with permission by:

Curran Associates, Inc.
57 Morehouse Lane
Red Hook, NY 12571

Printed with permission by Curran Associates, Inc. (2025)

International Plant Propagators' Society (IPPS)
174 Crestview Drive
Bellefonte, PA 16823-8516
USA

Phone: +01 814-355-2467

secretary@ipps.org

Additional copies of this publication are available from:

Curran Associates, Inc.
57 Morehouse Lane
Red Hook, NY 12571 USA
Phone: 845-758-0400
Fax: 845-758-2633
Email: curran@proceedings.com
Web: www.proceedings.com

THE INTERNATIONAL PLANT PROPAGATORS' SOCIETY

Combined Proceedings

Volume 73

2023

Dr. Robert L. Geneve, International Editor

Dr. Ranjith Pathirana, Editor – Australian Region

William Barnes, Jr., Editor – Eastern Region, North America

Dr. Masanori Tomita, Editor – Japan Region

Jill Reader, Editor – New Zealand Region

Dr. Elsa du Toit, Editor – Southern Africa Region

Dr. Fred T. Davies, Jr., Editor – Southern Region, North America

Dr. Ryan Stewart, Editor – Western Region, North America

IPPS

Sharing Plant Production Knowledge Globally

Published 2023

Combined Proceedings International Plant Propagators' Society (ISSN 0538-9143) is a yearly publication of the International Plant Propagators' Society (IPPS) headquartered in Pennsylvania, U.S.A.
Email: Secretary@ipps.org
Website: www.ipps.org

Business Office. Memberships are processed at the regional level, with dues being allocated to both international and regional expenses. Contact the regional secretary / treasurer in your area of the world for membership information and applications.

Table of Contents

Australian Region

Eastern Region - USA

Combined Proceedings IPPS 73:1-9. 2023.

South Africa Exchange – Presentation Summary

Stephanie Hastie

TAFE, South Australia (+61) 434 377 530

Stephanie.hastie@tafesa.edu.au

Keywords: industry experience, Australia, IPPS exchange

Summary

Earlier this year I was fortunate to be given the opportunity by the IPPS do the 3-week exchange in South Africa and give a talk at the 2023 Conference. The purpose of this paper is to provide a brief introduction to some of the Australian native plants that are common in the different regions of Adelaide, South Australia, and beyond.

INTRODUCTION

I am from Adelaide, the capital city of South Australia. The is the home of the Kaurna people, the original indigenous custodians and First Nations people of the Adelaide Plains, who have lived in Australia and cared for the land for over 50,000 years. Adelaide is a relatively small capital city compared with that of Sydney, Melbourne and Brisbane – and less well known as a result. It's a beautiful place to live, with the Adelaide Plains wedged between the southern coastline adjoining the Gulf of St Vincent and the Mt Lofty ranges – also referred to as the Adelaide Hills.

I am a lecturer at an institution called TAFESA (Technical and Further Education

South Australia). TAFESA focuses on hands on vocational training. Within our program, we cover a wide range of topics that cover the foundational knowledge needed to work in the nursery and gardens industry. We give students an introduction to botany, soils, and plant nutrition. We teach them about common pests, diseases and weeds, and integrated pest management approaches that can be used to control them. We also offer courses in irrigation, hydroponics, propagation, pruning, turf, machinery, and several other electives.

Adelaide Coastline

The Adelaide Coastline boasts spectacular views and pristine beaches. It is, however, subject to harsh winds, salt spray and erosion, which leads to a need for plants that can handle these conditions. Some examples of plants native to this region include *Allocasuarina verticillata*, *Carpobrotus glaucescens*, and *Leucophyta brownii.*

Allocasuarina verticillata (also known as the Dropping Sheoak) is a nitrogen fixing tree native to southeastern Australia (**Fig. 1**). The Drooping Sheoak has a soft look to it, and it whispers in the breeze. One of the indigenous names for the Sheoak can be translated as 'hair tree'. The European scientific name comes from the Malay word of Cassowary, which is Kasuari. More recently a variation of the Greek word Allos, was added to the start of the name, to indicate that it was 'different to' the now separate genus Casuarina. Verticillata refers to the vertical position of the leaf scales on stems, otherwise known as cladodes. I like showing my students this plant and getting them to pull apart the different sections of the brachlets to look at the scale-like leaves, which form a crown shape at the top of the cladodes. The first nations people soaked the cones of allocasuation in their drinking water to give it a lemon flavour and to add vitamin C (Murphy, 2001). When water is scarce, it can be chewed on to increase saliva flow and reduce the need to drink on long journeys. It is said that you do not get snakes under Sheoaks, because the leaf scales get under the scales on the belly of a snake, so the snakes avoid this area.

Figure 1. *Allocasuarina verticillata.*

Carpobrotus glaucescens (known as angular sea-fig or pigface) is a prostrate, creeping succulent which long trailing stems to 2m in length (**Fig. 2**). It is very useful as a soil stabliser along sand dunes. The botanical name refers to the edible fruits. It comes from the Greek 'karpos' meaning 'fruit' and 'brota' meaning 'edible'.

Figure 2 - *Carpobrotus glaucescens*. Image used under license from Shutterstock.com.

The common name refers to the flower, which is said to resemble a pig's face – although I would say this stretches the imagination almost to breaking point. Both the leaves and the fruits are edible. It can be made into a pickle or a jam, although I have not tried it myself. I have tried the leaves on their own, and they have a slightly astringent, tangy salty flavour.

Leucophyta brownii is a rounded shrub endemic to Australia. It has silver-coloured stems and tiny silver-coloured leaves. (**Fig. 3**). To my mind, it has a coral like appearance. It has the common name of cushion bush because the first nations people used it as a cushion to sit on. It has a lot of ornamental value – it can be trimmed into tight rounded shapes, and the silver of the stems and leaves offer a wonderful colour contrast with other plants.

Figure 3. *Leucophyta brownii.*

Situated between the Gulf St Vincent coast to the west and the Mount Lofty Ranges to the east, the Adelaide Plains covers a large area including Adelaide City and surrounds. As the name suggests, the area is reasonably flat and receives less rainfall than much of the Adelaide Hills area. The intermediary between the coast and the hills, the plains are home to a wide selection of native plants that can be found not only in the plains but also in surrounding areas.

Some examples of plants native to this region include *Eremophila nivea*, *Maireana sedifolia*, and *Santalum acuminatum*.

Eremophila nivea, commonly known as Silky Eremophila, is shrub with lovely soft grey foliage (**Fig. 4**). It is a good attractor for nectar eating birds and insects and produces a mass of purple flowers during Spring and Summer. It is also considered a fire-retardant plant. Some species of *Eremophila* were used as traditional medicines, although I'm not aware of any such uses for the Silky Eremophila.

Figure 4. *Eremophila nivea*. Images used under license from Shutterstock.com.

Maireana sedifolia, also known as the bluebush or pearl bluebush, is a shrub endemic to many parts of Australia (**Fig. 5**). This striking plant offers a mass of soft grey succulent foliage all year round, which is also used in the floristry industry. It is a hardy plant that grows happily in the plains and tolerates coastal exposure. The species name sedifolia tells us that it has leaves like Sedum.

Figure 5. *Maireana sedifolia.*

Santalum acuminatum, also known as Quandong or Native peach or desert peach, is a hemi parasitic tree that produces bright red fruits that are a popular traditional bush food (**Fig. 6**). It was a valuable food source for the First Nations people; the entire drupe including the flesh but even more so the brain like nut inside. It is a popular plant, but it is difficult to propagate and often you have to go on a waiting list to get the plant from the State Flora nursery in South Australia. It's often grafted, as the fruits can be variable when propagating using seed.

Figure 6. *Santalum acuminatum*. Image used under license from Shutterstock.com.

Adelaide Hills

The Adelaide Hills covers a large area overlooking the Adelaide plains. The Adelaide Hills are known to reach temperatures cool enough for snow in some months, whilst suffering from high risk of bushfires in the summer. Only about 10% of the original native vegetation of the Adelaide Hills remains, and so it is important to ensure that endemic species are continued to be planted in the area (Adelaide Hills Council, 2023). Some examples of plants native to this region include *Arthropodium strictum*, *Calostemma purpureum*, and *Kennedia prostrata*.

Arthropodium strictum is a herbaceous perennial plant (**Fig. 6**). The scent of the flowers resemble chocolate,caramel or vanilla, which give it it's common name of Chocolate Lily (Tucker, 2020). The tubers, which are juicy and slightly bitter, were a traditional food source. The flowers of *A. strictum* are purple and held atop of slender, drooping stalks.

Figure 6. *Arthropodium strictum*. Image used under license from Shutterstock.com.

Calostemma purpureum, or Garland lily, is a perennial flowering herb from the Amaryllidaceae family (**Fig. 7**). 'The family is not well represented in Australia and *Calostemma* is the only wholly endemic genus.' (Australian Native Plants Society, 2023). The leaves die off it spring, and (much like South African Nerines) it often flowers in summer without any leaves. It is highly ornamental and can be used in rockery plantings.

Figure 7. *Calostemma purpureum* featured on the cover of Focus on Flora: Native Plants of the Adelaide Hills and Barossa.

Kennedia prostrata, which has the whimsical common name of Running postman, is a perennial scrambling ground cover (**Fig. 8**). It has bright red pea-shape/papilionate flowers and undulate trifoliate leaves.

Figure 8. *Kennedia prostrata*. Image used under license from Shutterstock.com.

Botanical Gardens of South Australia

Adelaide is home to three botanical gardens, although I'm going to mention four.

First, the Adelaide Botanic Gardens, is located within the city centre and home to 50 hectares of gardens, which includes the last museum of economic botany in the world. It displays a permanent collection of about 3,500 objects exhibiting the practical, medicinal and economic use of plant materials. The pomological collection is perhaps the highlight of the museum, which features 129 pears and 192 apples (including some varieties that no longer exist) expertly made out of papier- mâché by artist Heinrich Arnoldi. The models were acquired for the museum in the late 1800s by its director, Dr. Richard Schomburgk, a German immigrant. There is another similar collection of 210 papier-mâché models showcasing fungi in different stages of growth.

The second, the Mount Lofty Botanic Gardens is dedicated to the cultivation and display of cooler climate plants and is well known for its rich display of autumn colours. It is made up of 97-hectare of gardens, which include the Heritage Rose Garden, the Woodland Garden (which is home to a variety of Northern Hemisphere tree species), Fern Gully, Rhododendron Gully, South American Gully, Southeast Asian Gully (where Camellias are the dominant feature), and the West Asian Gully (which boast a collection of Viburnum species).

The third, and perhaps most interesting to those with an interest in Australian plants is Wittunga Botanic Gardens – which includes 13-hectare of gardens comprised of water-wise plants from both Australia and South Africa, with a focus on the *Proteaceae* family. It has brilliant displays of *Ericas* and *Proteas* in spring (**Fig. 9**). There is a Fynbos section which includes approximately 50 species and cultivars of *Ericas* ("heaths"). Australian *Hakeas* and *Banksias* can also be found throughout Wittunga.

Figure 9. Protea in bloom at Wittunga Botanic Gardens. Photograph by Stephanie Hastie.

Last but not least, is the Arid Lands Botanic Gardens. Although these particular gardens are remote – approximately 3.5 hours away from Adelaide - the arid lands are home to thousands of highly evolved plant communities that are specially adapted to thrive in an environment where temperatures are extreme and drought can last for decades. A testament to the vastness of the outback, these gardens are the largest of the four – comprising 250-hectare of gardens. I have only visited these gardens once, it was my first stop on a 4-day drive to Darwin, in the northern territory, and it was during a sweltering heatwave in 2019 and on a day when the temperature at the Gardens was about 46°C, which meant we only spent about 15 minutes walking around the garden before having to seek the respite of shade. Nevertheless, the Arid Lands Botanic Gardens fulfil the important role of educating us about the importance, value and beauty of these fragile and precious ecosystems. Likewise, all four botanic gardens help to bridge the connection between plants and people, that has been mostly lost in the modern world – the concrete jungle – that many of us live in today.

Eucalypts

South Australia is also home to the Currency Creek Arboretum of eucalypts. It holds the largest collection of eucalypt species in the world. The eucalypt collection, which was established in 1992, now includes 1,000 species and subspecies in the genera *Eucalyptus*, *Angophora* and *Corymbia* (which we refer to generally as 'eucalypts'). It is a privately managed arboretum and it used for scientific research purposes, although it is open to the public over two weekends each year (in Spring and in Autumn). In South Africa you are probably familiar with *Eucalyptus grandis*, native to NSW and Queensland, and Eucalyptus camaldulensis, which is native to many parts of Australia including my home state, and a significant weed here in South Africa.

However, with over 700 *Eucalyptus* species, 9 species of Angophora and about 100 species of *Corymbia* - there is significant variation among the Eucalypts, and many different species to fall in love with, and so I wanted to talk briefly about a few of my favourite species.

Being a horticulturist, I'm mostly looking at plants from an ornamental perspective. In Australia there is, sadly, a growing trend of larger houses and smaller gardens. And, even in a larger garden, fitting in large eucalypt species can be difficult or undesirable. Hence, my favourite species mostly fall into the mallee category – which tend to have an open and wispy habit, grow to a reasonable height of usually no more than 10 meters, and play well with others (that is, you can plant an understory of other species underneath them).

Having said that, the first species I want to mention is *Angophora subvelutina. Angophora subvelutina*, commonly known as the broad-leaved apple, is a species of woodland tree that is endemic to eastern Australia. The bark is rough, fibrous to flaky, and grey in colour. It has opposite, usually sessile, leaves – which have a lovely somewhat lime-y green colour, sometimes more blueish grey, and illuminate beautifully in the presence of sunlight. The species is summer flowering, with white to creamy white flowers. The branches do not tend to get too large, which means they pose less of a safety risk compared to the larger eucalypt species in terms of falling branches. The branches also tend to have an interesting slightly gnarly and bent appearance, giving an impression of wizened age. It is a medium-sized tree to 20 meters tall, so not overly suited to small backyards, however it can be trained as a mallee once established by frequent cutting and pruning of main stems. It is a nice landscaping tree and, back home, I hope to see it used more frequently as a street tree.

Eucalyptus caesia is another on my list. It goes by the common names of Silver Princess or Gungurru (**Fig. 10**). *Eucalyptus caesia* is a mallee species endemic to Western Australia. It is an incredibly attractive, highly ornamental species. It grows up to 15m tall, has an open slightly weeping habit. It flowers with red- pink filaments and yellow anthers, on ghostly white pedicles and peduncles, in winter. It has lovely reddish-brown peeling minni richi bark on the trunk and branches. The juvenile foliage is glossy green, with a rounded cordate shape. The adult leaves take on the classic lanceolate to falcate shape mostly commonly associated with species of Eucalypt, and they are usually a duller grey-green colour.

It is the mix of the stunning large hot pink flowers, the powdery white to glossy red of the new stems and the reddish brown of the mature trunk and branches, that I love. And the sparse habit makes it workable in a smaller space.

Figures 10. *Eucalyptus caesia*. Photographs by Stephanie Hastie.

Eucalyptus sepulcralis (known commonly as the Weeping Mallee) is a mallee that is another species that is endemic to Western Australia. It is a mallee that grows to 7 m tall. The stems are slender, sparse, and pendulous. It gives it a lovely weeping, wispy habit that moves elegantly in the breeze. It has smooth bark, glossy green leaves, and pale-yellow flowers in summer.

Perhaps my favourite species of eucalypt is *Eucalyptus minniritchi*, which may be a synonym of *Eucalyptus orbifolia*, commonly known as the Round-Leaved Mallee. Both names tell you a bit about the features of the plant. Like the *Eucalyptus caesia* it has the lovely reddish-brown peeling minnirichi bark on the trunk and branches. Its flowers are yellow and appear anywhere from late autumn to winter to early spring. The greyish green foliage keeps what it characteristically juvenile growth for a eucalypt – it remains rounded to heart shaped. It is a mallee that grows to 5 m tall, but with canopy often reaching to ground level – giving it more of a shrubby look. Like the other eucalypts I've mentioned, it forms a lignotuber and it can be coppiced to keep it smaller and to refresh the plant.

I really enjoy the leaf litter that falls from the *Eucalyptus minniritchi* – because it turns a lovely golden brown and the rounded shape almost looks like a coin or large heart shaped confetti (**Fig. 11**). It is native to Western Australia and central Australia.

Figure 11. *Eucalyptus minniritchi*. Photograph by Stephanie Hastie.

Conclusion

I would like to finish by saying, having been fortunate enough to attend both the 2023 IPPS conferences in Durban and in Geelong, I would highly recommend the South Africa exchange to any young propagators (from South Africa or Australia alike). My heartfelt thanks go to the IPPS and all the people who contributed to make it happen. It was a pleasure to meet so many passionate plants people.

LITERATURE CITED

Adelaide Hills Council. (2023). *Biodiversity in the Hills*, Adelaide Hills Council, viewed 1 February 2023, <https://www.ahc.sa.gov.au/environment/biodiversity-in-the-hills>.

Australian Native Plants Society (Australia) (2023). *Callostemma purpureum*, Australian Native Plants Society Australia, viewed 1 February 2023, <https://www.anbg.gov.au/gnp/gnp7/callostemma- purpureum.html>.

Australian Native Plants Society (Australia). (2023). *Kennedia prostrata*, Australian Native Plants Society Australia, viewed 1 February 2023, <https://anpsa.org.au/plant_profiles/kennedia- prostrata/>.

Tucker, B. (2020). Chocolate Lily - *Arthropodium strictum*, Bush Tucker, viewed 1 February 2023, <https://tuckerbush.com.au/chocolate-lily-arthropodium-strictum/>.

Kersbrook Landcare Group. (2014). Focus on Flora: Native Plants of the Adelaide Hills and Barossa. Axiom Distributors Pty Ltd, South Australia.

Murphy, S. (2021). Drooping Sheoaks - connecting with nature one plant at a time. Recreating the Country. <https://www.recreatingthecountry.com.au/blog/drooping-sheoaks-connecting-with- nature-one-plant-at-a-time>.

Combined Proceedings IPPS 73:10-14. 2023.

Horticultural Lighting in Nurseries from the Ground UP

Athon Human

Energywise System, Durban, South Africa

Earth@energywise.co.za

Keywords: plant growth, PAR, PPFD, light quality, light spectrum

Summary

This paper describes how supplemental lighting can extend the growing season increase yields and promote healthy plant growth. It includes descriptions of light's properties and how to measure those properties.

INTRODUCTION

Light is an electromagnetic wave. Humans see light between 380 nm and 750 nm (Fig. 1). Plants use light wavelengths between 400 nm to 700 nm (Fig. 1).

Plants use light wavelengths in the photosynthetic active radiation region (PAR) to produce fit itself and grow.

IPPS Vol. 73 – 2023

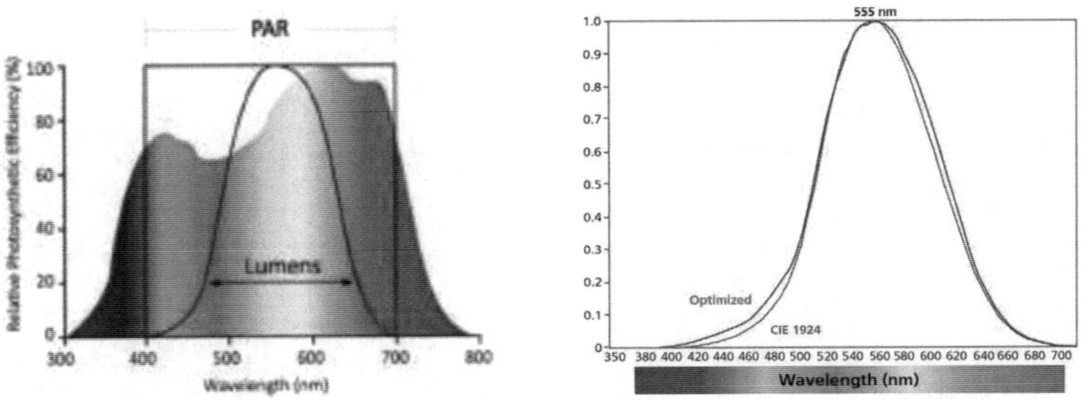

Figure 1. Light wavelength sensitivity and use by humans and plants.

Supplemental horticultural sources may deliver monochromatic or full spectrum light (**Fig. 2**). Monochromatic is an energy efficient supplemental light option but it can be more costly compared to full spectrum sources. Full spectrum lighting is more similar to sunlight and may be used as a supplemental light source or as the sole lighting source for indoor growth systems.

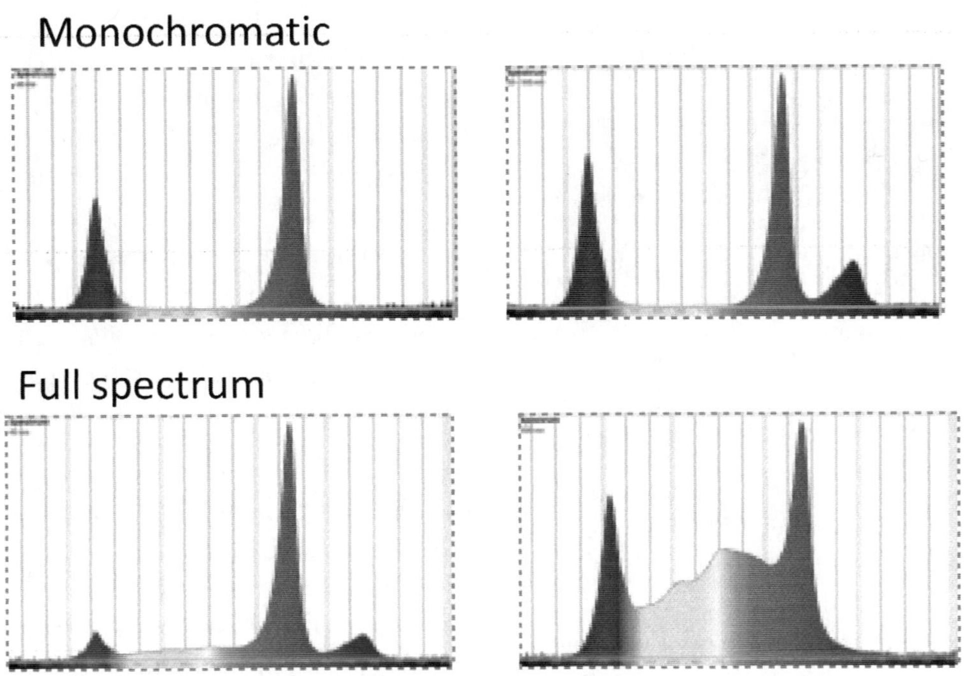

Figure 2. Monochromatic and full spectrum light wavelengths.

Supplemental light influences plant growth depending on the light spectra. Light sources rich in blue light promotes seedling root and vegetative leaf growth as well as shorter internodes (**Fig. 3**).

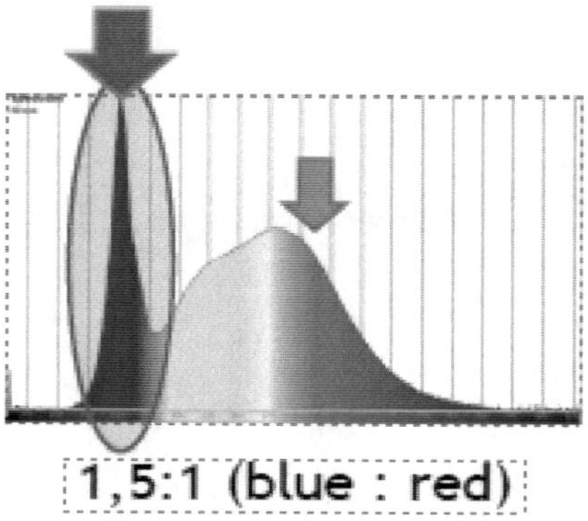

Figure 3. Light spectra enriched for blue light.

Supplemental light with more red wavelengths tends to promote stem growth, flowering and fruit production. This is accompanied by increased photosynthesis, leaf counts and subsequent yield (**Fig. 4**).

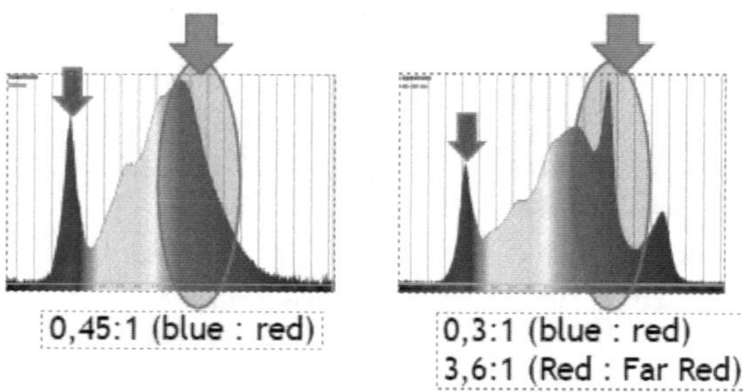

Figure 4. Light spectra enriched for red light.

When trying to identify which light spectra to specify for your plant needs, it is often related to red, blue and far-red ratios. In general, consider more blue light for propagation and vegetative growth and red/far red for flowering and fruit development (**Fig. 5**).

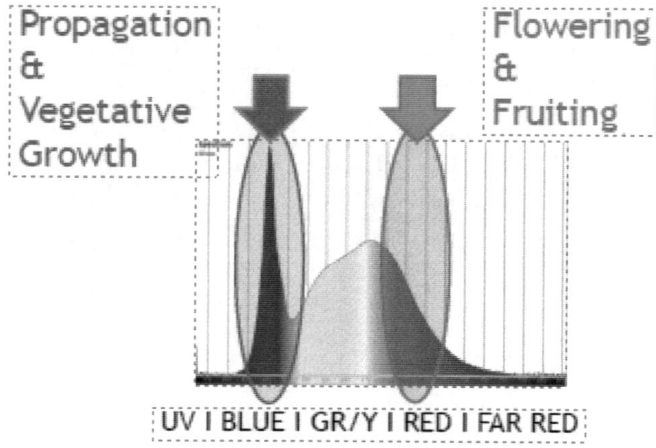

Figure 5. Light spectra related to plant growth.

Photons of light can be expressed in a number of different terms including lumens, lux and moles. All can be useful, but expression in moles can best describe light in the photosynthetic range for plants as photosynthetic photon flux density (PPFD). Instantaneous light intensity (flux density) is measured as the micromoles of light per meter squared per second (μmol/m^2/sec). The accumulated light over a specific day is expressed as a daily light integral (DLI) expressed as moles per day. The daily light integral can be increased by using supplemental lighting including extending the day length.

PPFD is an intensity value required to make up a particular DLI over the photoperiod. PPFD maps (often referred to as PAR maps) show the range of μmol/m^2/sec (light intensity) over a particular plant grow space – easily generated through simulation software (**Fig. 6**).

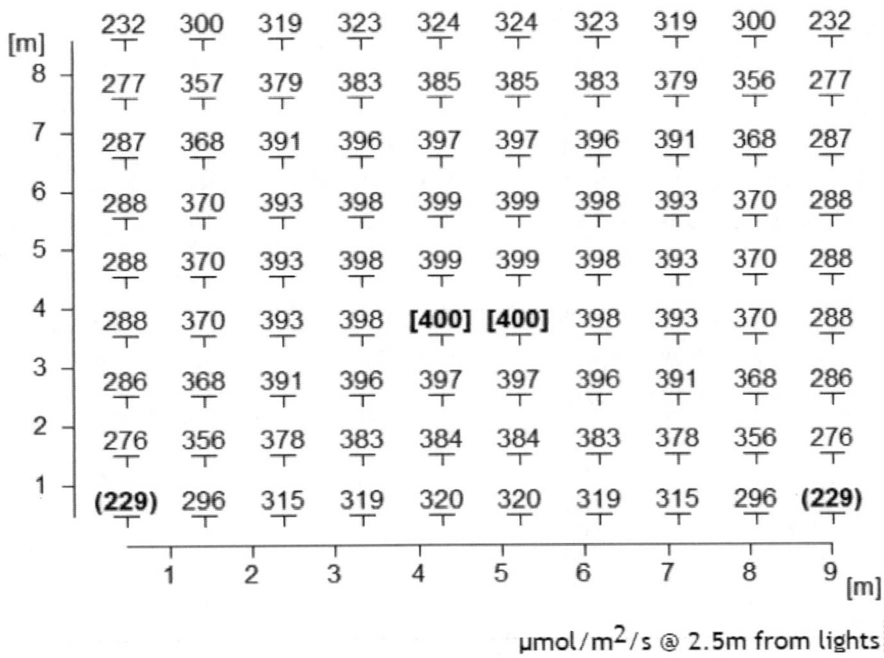

Figure 6. This PPFD the map shows intensity values over the grow space at a particular distance away from the light source/sec.

Plant daily light integral (DLI) should reflect the optimal amount of light per day for a given plant growth cycle. It can be calculated as:

DLI = PPFD (intensity) x light hours per day (photoperiod) x (3600/1000000)

Photoperiod is the number of light hours a plant receives in one day. Put simply, this is the number of hours the plant is "awake" to feed and grow.

The optimal photoperiod is given per plant, per growth cycle/ stage. So, you may find propagation and vegetative growth cycles photoperiods are similar and the flowering photoperiod something else. Often simulating seasons that induce /stress the plants into different growth cycles/ stages. Remember that under the sun's capabilities, the photoperiod will have dawn and dusk times that do not reach the optimal light intensity.

Research your crop/s on the internet to find DLI's and/ or photoperiods for the stages of growth you are concerned with. After some simple calculations you can produce the "magic numbers" for an intensity for a particular number of hours. There is a lot of information on the internet on how to do trials to find out optimal DLI, intensity and photoperiod – lead your field!

Now that you know what your crop can optimally handle, it's time to understand what you are giving it at the moment, and then throughout the year? Today in Port Edward we have just under 13 hours of daylight with fewer hours at any optimum intensity value, the first of January had just over 14 hours daylight and 21 June bottoming out at just over 10 hours daylight. Let's just assume, for low intensity requirements, that the first and last hour of each day equate to 1 full hour at optimal intensity – best case as I see it –so that's 13 optimal intensity hours on 1 January and 9 hours mid-Winter. What if your seedlings/ transplants could handle that intensity for 18 hours? 28% more time at optimum growth conditions mid-Summer and 100% more time at optimum growth conditions mid-Winter.

An advantage can be key to success in a competitive market. Knowledge is power that allows you to create your own advantage. You need to be able to measure environmental variables like light. These can include installed and personal device sensors that are readily available in the market. Personal devices can measure intensity at a moment and installed devices can measure DLI over a day.

Run trials, record data... Agriculture 4.0, it has been given a name.

Combined Proceedings IPPS 73:15-19. 2023.

Effect of Light Quality on In Vitro Growth and Flowering in *Perilla frutescens*

Masamichi Torii, Arisa Noguchi, Hiroko Nakatsuka and Wakanori Amaki

Department of Agriculture, Tokyo University of Agriculture, 1737 Funako, Atsugi, Kanagawa 243-0034, Japan

96233003@nodai.ac.jp

Keywords: tissue culture, plant evaluation, medicinal plant

Summary

This study investigated the effects of light quality on in vitro growth and flowering in *Perilla frutescens* (Crispa type). In this experiment, three cultivars, 'Hoko Ao-shiso', 'Hoko Aka-shiso' and 'Hoko Uraaka-shiso' were used, and each seedling of them was raised in a test tube dispensed the medium (1/2 MS medium, 30g/L sucrose, 8g/L agar, pH 5.8) from the seed. Then they were cultured under different light quality conditions (Mixed White (W): Red + Green + Blue, Blue (B): $\lambda = 470$ nm, Green (G): $\lambda = 525$ nm, Red (R): $\lambda = 660$ nm). Seeds were sterilized with 70 % ethanol for 5 minutes followed by 5% sodium hypochlorite (NaClO) for 5 minutes. Four light treatments (W, B, G, R, PPFD = 100μmol $m^{-2}s^{-1}$, 16 hr-Light/8 hr-Dark) started after appearing two unfolded true-leaves under the cool-white fluorescent lamps (PPFD = 130μmol $m^{-2}s^{-1}$, 16 hr-Light/8 hr-Dark). After the 3 months of culture, measured the growth and flowering responses. 'Hoko Ao-shiso' were flowered only under the G light treatment. 'Hoko Aka-shiso' and 'Hoko Uraaka-shiso' were flowered under the W, G and R light treatments respectively. Flowering was not occurred under the B light treatment. This is the first report, that *Perilla frutescens* in vitro flowering depending on light quality under non-inductive photoperiod (long day conditions). Hence, it is possible to analyze the flowering reaction of perilla by light quality, excluding stress-induced flowering by in vitro experimental system.

INTRODUCTION

Flowering is a critical event in the plant life cycle, referring the transition from vegetative to reproductive growth. The process is regulated by a complex mechanism, which varies between plant species (Amaki and Watanabe, 2016). Research on the control of flowering has important implications for the horticulture industry, including improving yield and achieving early flowering.

Perilla frutescens L. is an annual herb in the Lamiaceae family, with its leaves, flowers, and seeds commonly used for food, medicine, and dyeing. It is a qualitative short-day plant, meaning that it grows vegetatively under long day condition and transitions to reproductive growth under short day condition. The critical day length of the perilla is 14 hours and 15 minutes (Takimoto and Ikeda, 1961). Due to its high responsiveness, perilla plant has been used to investigate the florigen and the control mechanisms of flowering.

Light quality is an important factor in plant morphogenesis, serving as one of the environmental signals. Previous studies have shown that blue light inhibits perilla flowering, regardless of the cultivars, while flowering occurs under green light in long day conditions (Kawana, 2010; Tadokoro, 2014; Ayata, 2021). However, perilla has been observed the flowering under long day condition by various environmental stresses, including nitrogen deficiency (Wada and Totsuka, 1982) and low light intensity (Wada et al., 2010). Therefore, it was necessary to confirm the flowering response to light quality by constructing an experimental system that excluded those environmental stress factors.

In vitro experiments provide a stable experimental system, which has been used to study the gene functions in various plants by clarifying the effects of changes in gene expression under constant cultivation conditions. In vitro flowering has been confirmed in perilla by adding plant growth regulators to the tissue culture medium (Zhang, 2007). However, there is no report on the effects of light quality on flowering under the in vitro experimental system. Hence, the purpose of this study was to investigate the effects of light quality on the growth and flowering of perilla under the in vitro experimental system.

MATERIALS AND METHODS

Seeds of *Perilla frutescens* L. 'Hoko Aoshiso' ("Ao"), 'Hoko Aka-shiso' ("Aka") and 'Hoko Uraaka-shiso' ("Ura") obtained from Nakahara Seed Corporation. The perilla seeds were sterilized with 70% ethanol for 5 minutes, followed by 5% sodium hypochlorite (NaClO) for 5 minutes. The sterilized seeds were then washed three times for 1 minute with sterilized water. The half strength of Murashige and Skoog (MS) medium (Murashige and Skoog, 1962), which contained 30 g/L sucrose, 8 g/L agar, and was adjusted to pH 5.8. Twenty mL of the medium was dispensed into each of a flat-bottomed test tube (ø 40×130 mm) and sealed with aluminum foil (8×8cm). The test tubes were then autoclaved at 121°C for 15 minutes. The sterilized seeds were sowed on the surface of medium and cultured under fluorescent lamps (FLR40S·EX-N/M-H, TOSHIBA Lighting & Technology Corp.) provided a photosynthetic photon flux density (PPFD) of 130 μmol $m^{-2}s^{-1}$ for 16 hours of light and 8 hours of darkness at 24 ± 1°C until the seedling stage with two unfolded true leaves. The seedlings were cultured under different light qualities using light emitting diode (LED) panel light source (ISL series, CCS Inc.) including Mixed White (W: Red + Green + Blue), and the 3 types of monochromatic LED, Blue (B) (λ = 470 nm), Green (G) (λ = 525 nm) and Red (R) (λ = 660 nm). Each LED provided 100 μmol $m^{-2}s^{-1}$ PPFD. After three months of culture, the growth and flowering responses were measured. All the data were analyzed for the statistical significance using analysis of variance (ANOVA) with mean separation by Tukey's multiple range test. Statistical analyses were performed using RStudio (R ver.4.1.2).

RESULT AND DISCUSSION

The results of this experiment were shown in **Figure 1**. Plant height tended to be higher under blue light (B) treatment regardless of cultivars. This tendency for B treatment to promote stem elongation was consistent with the results of previous studies (Kawana, 2010; Tadokoro, 2014; Ayata, 2021). The numbers of stem node and alive leaf showed the highest value in "Aka" under B treatment. No significant differences were observed between other treatments. The reason for the cultivar difference was thought to be that the growth rate of "Aka" was accelerated by B treatment compared to other cultivars (**Fig. 1**).

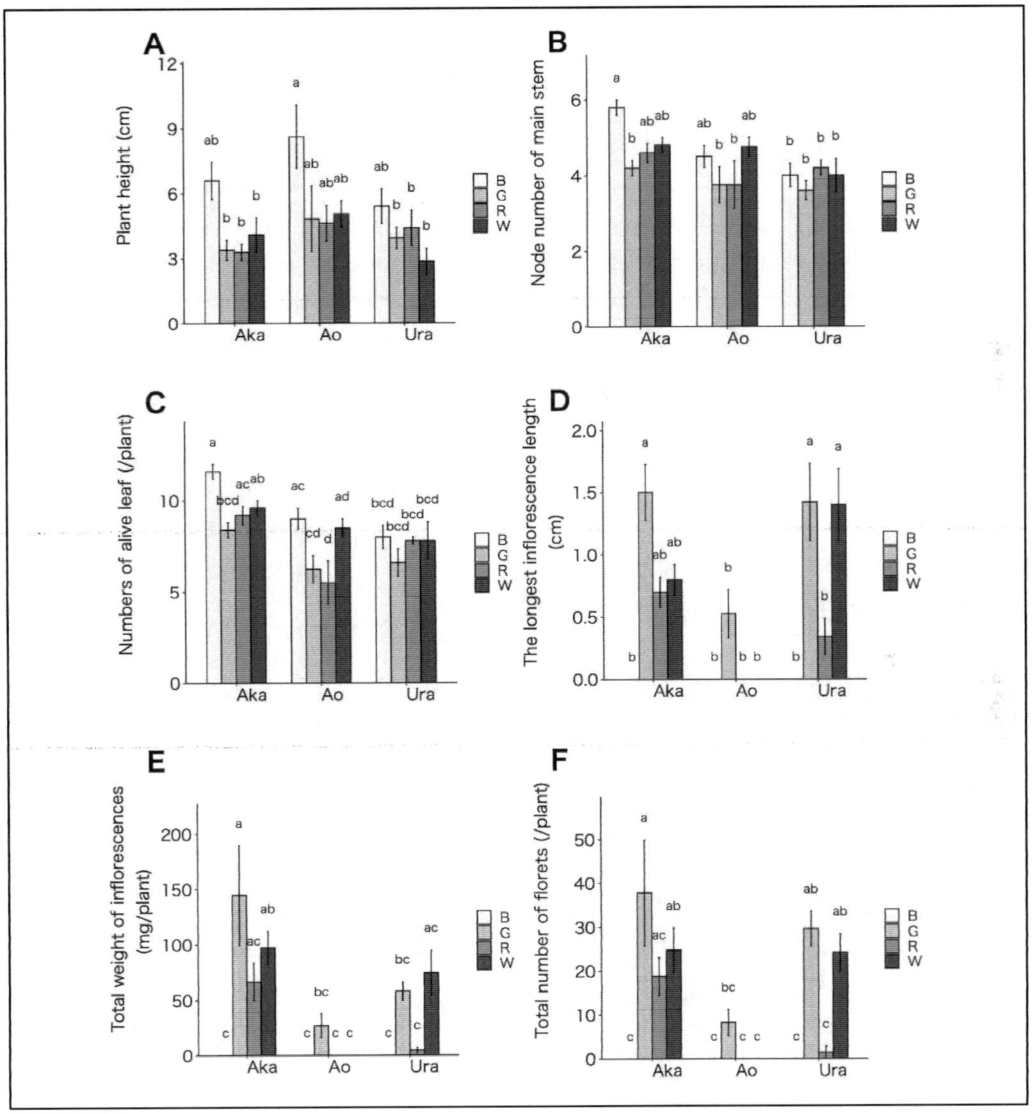

Figure 1. Effect of light quality on the growth and flowering of perilla under in vitro environment. Plant height (A), number of stem nodes (B), number of alive leaves (C), inflorescence length (D), total inflorescence weight (E), total number of floletss (F). "Aka" indicates 'Hoko Akashiso' (n=5), "Ao" indicates 'Hoko Aoshiso' (n=4), "Ura" indicates 'Hoko Uraakashiso' (n=5). Tukey's multiple range test showed a significant difference at the 5% level between different alphabets (error bars are standard error).

All perilla cultivars used in this experiment showed a tendency to increase under green light (G) treatment for inflorescence length, total weight of inflorescence and total number of florets. Flowering reactions were observed in "Aka" and "Ura" under mixed white light (W), G and red light (R) treatments, but flowering reactions were observed in "Ao" only under G treatment. From these results, in cultivar "Ao", the lowering was strongly suppressed under all light quality conditions under long-day (non-inductive flowering) conditions compared to the other two cultivars. Therefore, it was revealed that cultivar "Ao" exhibits the strongest flowering suppression response among the three cultivars used under the long-day conditions. On the other hand, cultivar "Aka" exhibited a stronger flowering response than cultivar "Ura" in the G and R treatments, indicating that "Aka" is a cultivar that is less susceptible to flowering suppression due to the long day conditions.

Seed formation was observed in all flowering perilla plants. The size of the seeds was comparable to commercially available seeds, and it appeared possible to propagate them through auto-self-pollination under the in vitro conditions (**Fig. 2**).

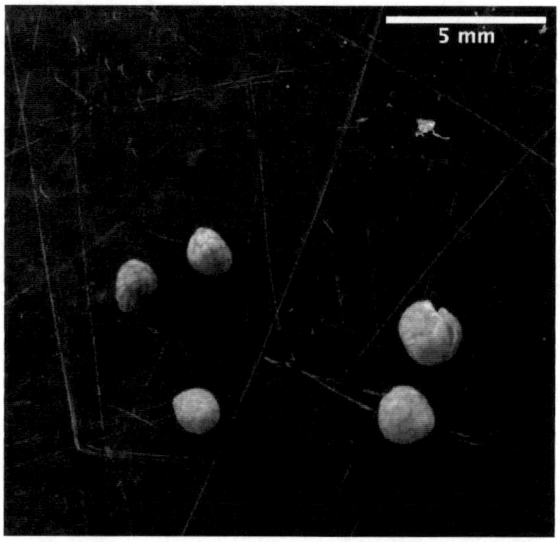

Figure 2. Comparison of 'Hoko Aoshiso' seeds collected by in vitro auto-self-pollination and commercially available seeds. Left: Commercially available seeds. Right: Seeds collected by in vitro auto-self-pollination.

In this study, in line with findings from prior research (Ayata, 2021; Nguyen and Oh, 2021), blue light was found to promote vegetative growth and suppressing flowering responses, both in vitro and in potted cultivation. Furthermore, under the monochromatic green light, flowering responses were clearly promoted compared to other treatments of light quality. These findings suggest the possibility of replicating perilla's photomorphogenesis within an in vitro environment, with the potential to eliminate environmental stress-induced flowering reactions through precise control of the in vitro conditions. Additionally, since successful seed formation through auto-self-pollination in vitro was achieved, it is anticipated that differences in responses to among lots of perilla seed can be effectively mitigated.

In conclusion, the use of in vitro experimental system offers the opportunity to investigate perilla's flowering response in greater detail by eliminating environmental stress-induced flowering and reducing the influence of seed lot variations.

LITERATURE CITED

Amaki, W. and Watanabe, H.. (2016). Effect of light quality on flowering response under different photoperiod conditions. Horticultural Research *15* (2): 446. (In Japanese)

Ayata, M. (2021). Effects of day length, light intensity, and light quality on the growth and flowering of perilla. Master thesis, Tokyo University of Agriculture. (In Japanese)

Kawana, T. (2010). Effects of green light irradiation on flowering response and growth of perilla. Bachelor thesis, Faculty of Agriculture, Tokyo University of Agriculture (In Japanese)

Murashige, T. and Skoog, F. (1962). A revised medium for rapid growth and bioassays with tobacco tissue cultures. Physiol. Plant. *15* (3): 473–497.

Nguyen, T. K. L. and Oh. M. M. (2021). Physiological and biochemical responses of green and red perilla to LED-based light. Journal of the Science of Food and Agriculture *101*(1):240-252. https://doi.org/10.1002/jsfa.10636

Tadokoro, H. (2014). Effect of light quality on the flowering reaction of perilla during cultivation, night interruption and supplemental lighting. Bachelor thesis, Faculty of Agriculture, Tokyo University of Agriculture (In Japanese)

Takimoto, A. and Ikeda, K. (1961). Effect of twilight on photoperiodic induction in some short-day plants. Plant and Cell Physiology. *2:*213-219.

Wada, K. and Totsuka, T. (1982). Long-day flowering of perilla plants cultured in nitrogen-poor media. Plant and Cell Physiology *23*(6):977-985. https://doi.org/10.1093/oxfordjournals.pcp.a076451

Wada, K.C., Kondo, H. and Takeno, K. (2010). Obligatory short-day plant, *Perilla frutescens* var. *crispa* can flower in response to low-intensity light stress under long-day conditions. Physiologia Plantarum *138*: 339–345. https://doi.org/10.1111/j.1399-3054.2009.01337.x

Zhang, T. (2007). In vitro flowering of *Perilla frutescens*. In Vitro Cellular & Developmental Biology-Plant *43*(2): 91-94. https://doi.org/10.1007/s11627-007-9038-5

Combined Proceedings IPPS 73:20-25. 2023.

Effects of Blue Light and Cultivar in the Internode Elongation in Sweet Basil

Arisa Noguchi and Hiroko Nakatsuka

Department of Agriculture, Tokyo University of Agriculture, 1773 Funako, Atsugi City, Kanagawa Prefecture, 243-0034 Japan.

a3noguch@nodai.ac.jp

Keywords: LED, *Ocimum*, plant height, light quality

Summary

Two sweet basil F_1 cultivars, 'Campione', a tall Genovese type, and 'TSGI-156', a short Italian large-leaf type, were used to investigate the effect of light quality on the internode elongation using a mixed white LED light (blue: 470 nm, green: 530 nm, red: 630 nm, 1:1:1 mix ratio) and a monochromatic blue LED light (peak wavelength: 470 nm). Plant height and internodal length above the second node were higher and longer under the blue LED for both cultivars. The effect on internode elongation was greater for 'TSGI-156' than for 'Campione'. Cortex cells were larger under the blue LED, and in 'TSGI-156', the long axis length of cells under the blue LED was about 60% longer than that of the white LED. Blue light has been reported to have an inhibitory effect on plant stem elongation, but in sweet basil, it was found to have a promotive effect.

INTRODUCTION

Plants need light not only for photosynthesis, but for regulation of their growth and development. In recent years, the effects of light quality have been studied for effective plant production with artificial light (Brian, 2006; Massa et al., 2008). Stem elongation is one of the most important plant characteristics in horticultural production, used in the production of graft and compact seedlings. Previous studies have indicated that blue light was more effective than red light in suppressing stem elongation in many species (Laskowski and Briggs, 1989; Shimizu et al., 2006). However, its effects have also been reported to vary with species, light intensity, and blue/red ratio (Hirai et al., 2006; Nanya et al., 2012). In sweet basil, the lower internodes of seedlings grown under blue light elongated (Amaki et al., 2012.). However, basil seedlings tested in Amaki's experiment showed variation in plant height immediately after germination and the inherited character was not fixed.

In this experiment, two sweet basil F_1 cultivars, 'Campione', a Genovese type with tall stem, shallowed cupped, egg-shaped leaves, and 'TSGI-156', an Italian large-leaf type with short stem, pointed serration, wrinkled wide leaves, were used to investigate their response to blue light.

MATERIALS AND METHODS

Sweet basil 'Campione' (Kobayashi Seed Co., Ltd., Japan) and 'TSGI-156' (Tokita Seed Co., Ltd., Japan) were sown in 6 cm diameter plastic pots. Seedlings were grown in an incubator (LI-200, Nippon Medical & Chemical Instruments Co., Ltd., Japan) with white fluorescence lamps, photosynthetic photon flux density (PPFD) of 120 $\mu mol \cdot m^{-2} \cdot s^{-1}$ at 16 hours light/8 hours darkness, and day temperature 25/night 20°C until 4 true leaves developed. Light quality treatment was performed in a temperature-controlled room at 24±2°C. The monochromatic blue LED light (blue LED had peak wavelengths of 470 nm) and a mixed white LED light (blue: 470 nm, green: 530 nm, red: 630 nm, 1:1:1 mix ratio) put on 16-hr photoperiod. Each of treatments was screened with black cloth to preclude extraneous lights. The PPFD of each LED was adjusted to 120 $\mu mol \cdot m^{-2} \cdot s^{-1}$ at the top of plants. Eight plants were placed in each light quality treatment area for 14 days. Plant height and internode length were measured every 7 days, and the measurement sites were shown in **Figure 1**.

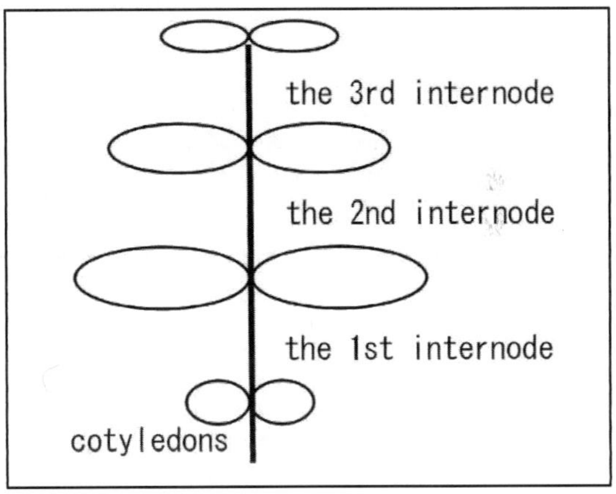

Figure 1. An explanatory diagram of the internodal positions targeted of sweet basil for investigation.

At the end of treatment, the third internode was collected and immediately fixed with formaldehyde-alcohol-acetic acid (FAA) (70% ethanol: formaldehyde: acetic acid, 90:5:5) solution and subsequently dehydrated with ethanol series solutions. All treated samples were embedded in paraffin, and sections for microscopic observation were stained with 0.05% toluidine blue solution. Slides were observed using a digital microscope (VHX-1000, Keyence Corporation, Japan), The size of cortex cells was measured using the ImageJ software program.

RESULTS AND DISCUSSION

Plant height and internode length above the second node were higher and longer under the blue LED for both cultivars (**Fig. 2**). In Figure 2, each internodal length is shown in a different filled pattern, and the plant height is shown by stacking these internodal lengths. It should be noted that measurement was possible up to the 5th internode.

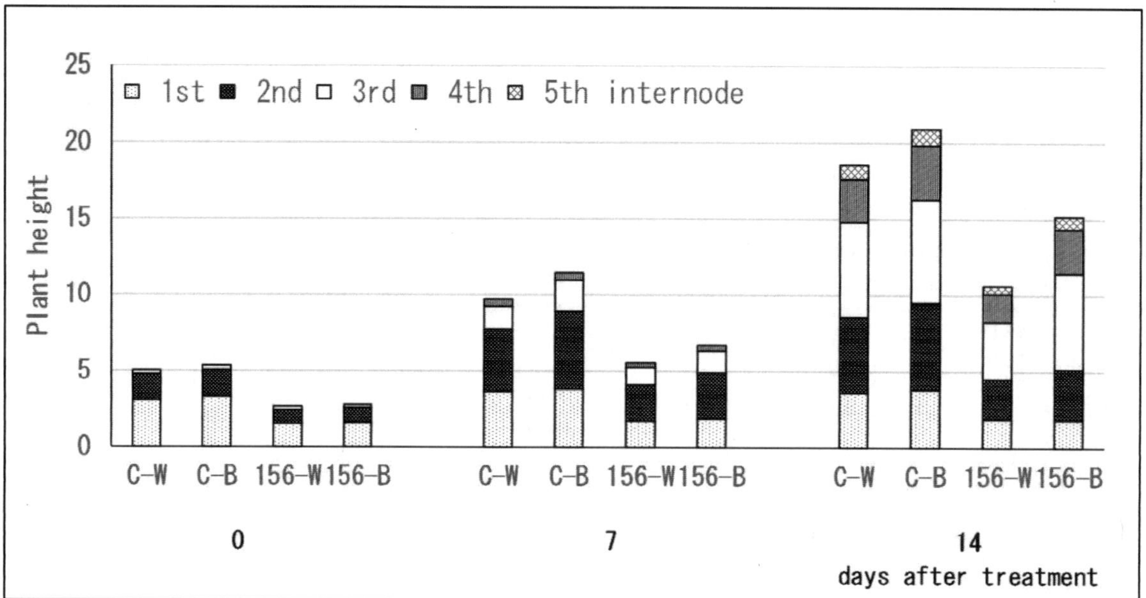

Figure 2. Plant height and individual internode length of sweet basil under white or blue light (n=8) C: 'Campione', 156: 'TSGI-156', -W: white LED, -B: blue LED.

The internode lengths at the second to fifth under the blue LED was 15% longer in 'Campione' and 32% longer in 'TSGI-156' than those under the white LED. The elongation rate was particularly high above the third internode of 'TSGI-156' under the blue LED. This may be due to the influence of light quality from the formation of the leaf primordium. The size of the cortex cells was not significantly different in 'Campione', although the short axis length of cells under the blue LED was about 4 μm longer than that under white LED. In 'TSGI-156', the long axis length of cells under the blue LED was about 60% longer than that under the white LED, a significant difference at the 1% level (**Table 1, Fig. 3**).

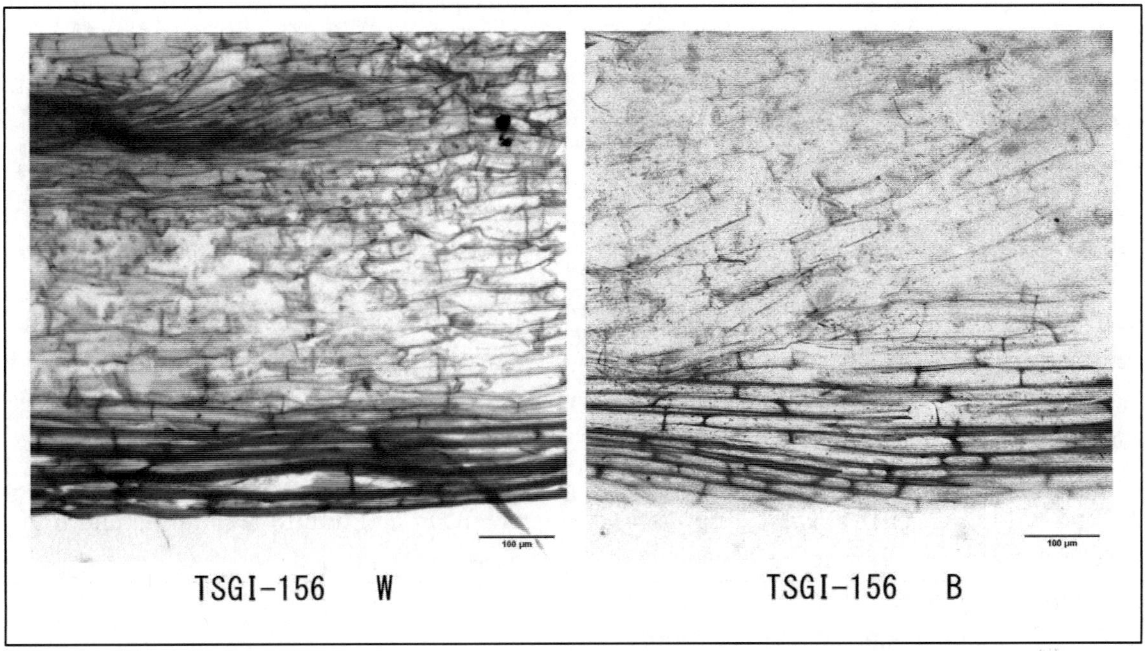

Figure 3. Longitudinal section of tissue the third internode of sweet basil under white or blue light (The lower side is the epidermis.

Table 1. Effect of light quality on the cell size of cortex of the third internode (μm) in sweet basil.

Cultivar name	Light quality of irradiation light	Cortex cell length (μm)	
		Long axis	Short axis
Campione	White	157.6 ± 31.3	29.9 ± 5.0
	Blue	157.3 ± 32.8	34.0 ± 9.3
TSGI-156	White	137.3 ± 47.0	28.6 ± 6.6
	Blue	218.6 ± 47.2	28.7 ± 6.1

n=30. average±standard deviation.

Blue light has been reported to have an inhibitory effect on plant stem elongation (Laskowski and Briggs, 1989; Shimizu et al., 2006). In this experiment, however, blue light promoted internodal elongation of sweet basil, the same result as previously reported by Amaki et al. (2012). In 'TSGI-156', the cortex cells were significantly longer in the stem axial direction, which increased internode length and plant height. On the other hand, in 'Campione', which has a larger plant height than 'TSGI-156', showed less elongation effect by blue light. In sweet basil, the higher percentage of blue light irradiated resulted in increased plant height and stem thickening, as well as reduced dry matter weight and leaf area (Larsen et al., 2020). Under 100% (monochromatic) blue light, a reduction in biomass and leaf dry mass unit area and phytochrome activity has been reported in arugula, mustard and petunia (Johnson et al., 2020; Kong et al., 2018). These responses were like shade-avoidance responses induced by increased red light percentage and less light (Franklin and Whitelam, 2005; Ballaré and Pierik, 2017).

Sweet basil has greater transpiration than other plants (Patiño et al., 2018), and blue light promotes transpiration through stomata opening (Assmann and Shimazaki, 1999). The 'TSGI-156' used in this experiment has larger leaves than 'Campione'. We think that the attempt to make the leaves smaller to reduce transpiration by opening the stomata under blue light resulted in increased distribution of photosynthetic to the stem and internode than in 'Campione'. To determine the effects of blue light on sweet basil morphogenesis, measurements of leaf area, dry matter weight, and phytohormone content in addition to respiration and transpiration are needed.

LITERATURE CITED

Amaki, W., Takahashi, A. and Watanabe, H. (2012). Effect of blue light irradiation on internode elongation of sweet basil. Horticultural Research *11*(Supplement.1): 386. (In Japanese).

Assmann, S.M. and Shimazaki, K. (1999). The multisensory guard cell. Stomatal responses to blue light and abscisic acid. Plant Physiology *119*:809–816. https://doi.org/10.1104/pp.119.3.809

Ballaré, C.L. and Pierik, R. (2017). The shade-avoidance syndrome: multiple signals and ecological consequences. Plant, Cell & Environment *40*:2530-2543. https://doi.org/10.1111/pce.12914

Brian, T. (2006). Light signals and flowering. Journal of Experimental Botany *57*:3387-3393. https://doi.org/10.1093/jxb/erl071

Franklin, K.A. and Whitelam, G.C. (2005). Phytochromes and shade-avoidance responses in plants. Annals of Botany *96*:169–175. https://doi.org/10.1093/aob/mci165

Hirai, T., Amaki, W. and Watanabe, H. (2006). Action of blue or red monochromatic light on stem internodal growth depends on plant species. Acta Horticulturae *711*:345-349. https://doi.org/10.17660/Acta-Hortic.2006.711.47

Johnson, R.E., Kong, Y. and Zheng, Y. (2020). Elongation growth mediated by blue light varies with light intensities and plant species: a comparison with red light in arugula and mustard seedlings. Environ. Exp. Bot. *169*:103898. https://doi.org/ 10.1016/j.envexpbot.2019.103898

Kong, Y., Stasiak, M., Dixon, M.A. and Zheng, Y. (2018). Blue light associated with low phytochrome activity can promote elongation growth as shade-avoidance response: a comparison with red light in four bedding plant species. Environ. Exp. Bot. *155*:345–359.

https://doi.org/10.1016/j.envexpbot.2018.07.021

Larsen, D.H., Woltering, E.J., Celine, C.S.N. and Marcelis, L.F.M. (2020). Response of basil growth and morphology to light intensity and spectrum in a vertical farm. Frontiers in Plant Science *11*:597906. https://doi.org/10.3389/fpls.2020.597906

Laskowski, M. J. and Briggs ,W.R. (1989). Regulation of pea epicotyl elongation by blue light 1: Fluence-response relationships and growth distribution. Plant Physiology *89*:293–298.
https://doi.org/10.1104/pp.89.1.293

Massa, G. D., Kim, H-H., Wheeler, R.M. and Mitchell, C. A. (2008). Plant productivity in response to LED lighting. HortScience *43*:1951–1956.
https://doi.org/10.21273/HORTSCI.43.7.1951

Nanya, K., Ishigami, Y., Hikosaka, S. and Goto, E. (2012). Effects of blue and red light on stem elongation and flowering of tomato seedlings. Acta Horticulturae *956*: 261-266.
https://doi.org/10.17660/Acta-Hortic.2012.956.29

Patiño, L. S., Diego, A. C. and Aníbal, O. H. (2018). Influence of 1-MCP and modified atmosphere packaging in the quality and preservation of fresh basil. Postharvest Biology and Technology *136*:57-65.
https://doi.org/10.1016/j.postharvbio.2017.10.010

Shimizu, H., Ma, Z., Tazawa, S., Douzono, M., Runkle, E.S. and Heins, R.D. (2006). Blue light inhibits stem elongation of chrysanthemum. Acta Horticulturae *711*:363-367.
https://doi.org/10.17660/Acta-Hortic.2006.711.50

Combined Proceedings IPPS 73:26-31. 2023.

In Vitro Flowering Response in *Iberis umbellata*

Syunya Shimazaki*, Hiroko Nakatsuka, Arisa Noguchi and Wakanori Amaki

Department of Agriculture, Tokyo University of Agriculture, 1737 Funako, Atsugi, Kanagawa 243-0034, Japan; *Currently affiliated with ZEN-NOH PEARL RICE Co., Ltd

amaki@nodai.ac.jp

Keywords: tissue culture, micropropagation, bedding plant

Summary

Iberis species are perennial or annual herbs of the genus *Iberis* in the Brassicaceae family, and is native to the Mediterranean coast, North Africa, and southwestern Asia. In vitro flowering is considered a convenient system to study the entire mechanism of reproductive processes. There have been no reports of in vitro flowering in *Iberis*, although there have been several reports on tissue culture for other species, *I. amara* and *I. semperflorens*. *I. umbellata* used in this research has never been used in in vitro research. Therefore, we investigated the effects of sucrose concentration and explant collected position on the in vitro flowering response. As a result, the addition of sucrose to the medium was essential for flowering, and judging from the flowering reaction, 30 g/L seemed to be optimal concentration. In addition, as a result of examining the explants from different positions, it became clear that the explants containing the apical bud were most likely to flowering.

INTRODUCTION

Iberis species are perennial or annual herbs of the genus *Iberis* in the Brassicaceae family, and is native to the Mediterranean coast, North Africa, and southwestern Asia. The flowering period in Japan is from April to June, and it has the characteristics of long-day plant. It is called "Candytuft" in English because the flowers look like swollen sugar candy (The Royal Horticultural Society, 1992). In Japan, it is called "Magaribana" (meaning a flower that looks crooked) because of its unbalanced flower shape, of which the inner two petals are small and the outer two petals are large. *I. umbellata* L., which is the test material for this research, has small flowers but a wide variety of flower colors, ranging mainly from rose to pink, but also including red, purple, lavender, and white. It has been improved for use in flower beds and is highly branched, reaching a height of about 25 cm when grown outdoors (Tsuda, 1988).

To date, it has been shown that flowering can be induced in various species in vitro (Scorza, 1982; Van Staden and Dicken, 1991; Murthy, 2012). There are various purposes for in vitro flowering. For example, one purpose is to analyze flowering physiology. In vitro flowering is considered a convenient system to study the entire mechanism of reproductive processes, including flower development, flower organ formation, and flower maturation. Another purpose, plants that bloom in vitro are called "in vitro flowers," and they can create new value that is different from conventional flower usage. To date, there have been no reports of in vitro flowering in this genus, *Iberis*, although there have been several reports on tissue culture for other species, *Iberis amara* L. and *Iberis semperflorens* L. (Mudgal et al., 1981; Iapichino and Bertolino, 2009). *Iberis umbellata* used in this research has never been used in in vitro research. Therefore, we investigated the in vitro flowering response of *Iberis umbellata*.

MATERIALS AND METHODS

Cultivation of Plant Material

Seeds of *Iberis umbellata* Candytuft (TAKII & Co., Ltd.) were sterilized in 70 % ethanol for 1 minute and in a 1 % sodium hypochlorite aqueous solution containing 1 drop of Tween 20 for 15 minutes. It was then washed three times with sterile water. Two to three seeds were sown in each test tube (φ40 x 30 mm). A double layer of aluminium foil was used to close the test tube. At the stage when the cotyledons were unfolded, the seedlings were replaced with aluminium plugs with holes of 8 mm in diameter and an air ventilation membrane (MilliSeal, Millipore Co., Ltd.) attached to the hole to improve the ventilation rate, and one good seedling was left and the others were thinned out to serve as test material.

The basal medium used was 1/3 strength of Murashige and Skoog (1962) medium supplemented with 30 g/L sucrose and 8 g/L agar. The pH of the medium was adjusted to 5.8. Twenty mL of the medium was dispensed into each of test tubes, and the tubes were capped with aluminium foil and were autoclaved at 121°C for 15 minutes.

A culture incubator (CL-301, Tomy Seiko Co., Ltd.) was used, and the culture conditions were 20°C, white fluorescence light (FLR40S·EX-N/M-H, Toshiba Lighting and Technology Co., Ltd.) 24-hour lighting, and photosynthetic photon flux density (PPFD) was 120 μmol · m^{-2} · s^{-1} (When seeds of Brassicaceae plants are sown under 8 or 16 hours of light per day, the hypocotyls tend to elongate unnecessarily. Under 24-hour lighting, this elongation phenomenon is suppressed, so in this experiment, 24-hour continuous lighting was used).

Effects of sucrose concentration on the flowering (Exp.1)

Approximately 1 cm long apical segments cut out from in vitro seedlings (29 days after sowing) were used as explants. Sucrose concentration of the basal medium was changed to 0, 15, 30 and 60 g/L, and the flowering response was observed after culturing for 26 days. Other culture procedures were performed in accordance with the in vitro seedling growth described above.

Effect of stem nodal position on flowering response (Exp.2)

Seedlings (30 days after sowing) were divided into four explants from the top (apical segment 1 cm long) to the base (3 of one-node stem segments, 15 to 20 mm long) and used as explants.Other culture procedures were performed in accordance with the in vitro seedling growth described above. Observations were made on the 45th day after the start of culture.

RESULTS AND DISCUSSION

Effects of sucrose concentration on the flowering

When the sucrose concentration was 0 g/L, the flowering rate was 0 %. In addition, the shoot length was significantly shorter than in other treatments, and no callus was formed. Furthermore, most of the plants were malformed. On the other hand, Flowering was observed in all media containing sucrose, regardless of its concentration (**Table 1**).

Table 1. Effects of sucrose concentration on the in vitro growth and flowering of apical stem explants in *Iberis umbellata*.

Sucrose concentration (g/L)	Number of alive explants	Flowering rate (%)	Shoot length (mm)	Callus formation (%)
0	7	0	15.3b*	0
15	6	33.3	55.5ab	100
30	7	71.4	59.8a	100
60	7	71.4	34.0ab	100

* There are significant differences (5% level) between different letters by Tukey's multiple range test.

These results suggest that sucrose plays an important role in shoot growth and flowering reactions. In addition, although the number of florets was greater at 60 g/L sucrose concentration than at other concentrations, the plant height was small and the stems turned red (**Table 1** and **Fig. 1**), suggesting that growth inhibition was caused by osmotic stress. At 30g/L, flowering started earlier than at other concentrations and the final flowering rate was the highest. From the above, it is considered that adding 30 g/L of sucrose is suitable for growth and flowering in *Iberis umbellata*.

Figure 1. Growth and flowering of apical explants at different sucrose concentrations in Iberis umbellate (Photographed on the 26th day after the start of culture). From the left: 0, 15, 30, 60g/L sucrose.

Effects of stem nodal position on flowering response

Flowering was observed only in one of the apical explants (**Table 2** and **Fig. 2**). There was no significant difference in shoot length depending on the location of explant collection. Although there was no statistically significant difference in the number of lateral buds formed, there was a tendency for the number of lateral buds to increase in lower explants. The existence of such a gradient of physiological responses from the top to the base of the stem has been reported in tobacco, torenia, perilla, gentian, etc. (Tran Than Van et al., 1974; Tanimoto and Harada, 1979; Tanimoto and Harada, 1980; Zhang and Leung, 2002). In gentian, the shoot formation rate and flower bud formation rate tend to gradually decrease as the node position decreases from the top to the base (Zhang and Leung, 2002). As mentioned above, in tissue culture, when the explant collected position is changed from the top to the base of the stem, there are differences in reactions such as lateral bud formation, flower bud formation and adventitious bud formation depending on the collected position. This suggests that a physiological gradient clearly exists in stem. However, the trends are different between *Iberis* and gentian.

At least, when trying to induce in vitro flowering in *Iberis*, it is better to use the explants contained the apical bud.

Table 2. Effects of stem node position collected explant on the in vitro growth and flowering in *Iberis umbellata* (n = 6).

Stem node position of explant (from top to base)	Flowering rate (%)	Longest shoot length (mm)	Callus formation (%)	Callus formation (%)
Apical	16.7	128a*	15.3b*	0
Second	0	97a	55.5ab	100
Third	0	107a	59.8a	100
Fourth	0	97a	34.0ab	100

* There are significant differences (5% level) between different letters by Tukey's multiple range test.

Figure 2. Growth and flowering of explants collected apical explants at different stem position (Photographed on the 26th day after the start of culture). From the left: Apical explant, Second node, Third node, Fourth node.

LITERATURE CITED

Iapichino, G. and Bertolino, M. (2009). Propagation techniques for *Iberis semperflorens* L. Acta Hortic. *813*:427-434.

Mudgal, A.K., Goel, S., Gupta, S.C. and Chopra, R.N. (1981). Regeneration of *Iberis amara* plants from in vitro cultured leaf and stem explants. Z. Pflanzenphysiol. *101*:179-182.

Murashige, T. and Skoog, F. (1962). A revised medium for rapid growth and bioassays with tobacco tissue cultures. Physiol. Plant. *15*:473-497.

Murty, K.S.R., Kondamudi, R., Rao, C.P.V. and Pullaiah, T. (2012). In vitro flowering - A review. J. Agric. Tech. *8*:1517-1536.

Scorza R. (1982). In vitro flowering: a review. HortScience *4*:106-127.

Tanimoto, S. and Harada, H. (1979). Influence of environmental and physiological conditions on floral bud formation of *Torenia* stem segments cultured *in vitro*. Z. Pflanzenphysiol. *95*:33-41.

Tanimoto, S. and Harada, H. (1980). Hormonal control of morphogenesis in leaf explants of *Perilla frutescens* Britton var. *crispa*, Decaisnef., *viridi-crispa* Makino. Ann. Bot. *45*:321-327.

The Royal Horticultural Society. (1992). *Iberis*. p.643-644. In: A. Huxley (Chief ed.) The New Royal Horticultural Society Dictionary of Gardening. MacMillan Press, London.

Tran Than Van, M., Dien, N.T. and Chlyah, A. (1974). Regulation of organogenesis in small explants of superficial tissue of *Nicotiana tabacum* L. Planta *119*:149-159.

Tsuda, H. (1988). *Iberis*. p.234-235. In: T. Aiga (ed.) The grand dictionary of horticulture. Vol. 1. Shogakukan, Tokyo.

Van Staden, J. and. Dickens, C.W.S. (1991). In vitro induction of flowering and relevance to micropropagation. p.85-115. In: Y.P.S. Bajaj (ed.), Biotechnology in agriculture and forestry, Vol.17. High-tech and micropropagation I. Springer-Verlag, Berlin, Heidelberg.

Zhang, Z. and. Leung, D.W. (2002). Factors influencing the growth of micropropagated shoots and in vitro flowering of gentian. Plant Growth Regulation *36*: 245-251.

Combined Proceedings IPPS 73:32-35. 2023.

A Medicinal plant, *Eucommia ulmoides*: Possibility of In Vitro Propagation under Several Tissue Culture Conditions

Manami Inoue, Hiroko Nakatsuka, Arisa Noguchi, Wakanori Amaki and [1]Chieko Yasuma

Graduate School of Agriculture, Tokyo University of Agriculture, 1737 Funako, Atsugi, Kanagawa 243-0034, Japan: [1]Hekizanen Limited Company, 1438-5 Hanbara, Aikawa, Aiko, Kanagawa 243-0307, Japan

43322003@nodai.ac.jp

Keywords: micropropagation, hardy rubbertree, cytokinin, auxin inhibition

Summary

Eucommia ulmoides Oliver (eucommia) is a deciduous and dioecious woody plant that is native to China. In China, the bark has been used as a Chinese herbal medicine since ancient times, and its medicinal properties include lowering blood pressure, analgesia, and diuresis. Recently, it has been discovered that the leaves have similar effects. There have been several reports of propagation by tissue culture in the past using in vitro seedlings, but it cannot be said to be established. In the future it will be necessary to conduct line selection and propagation of eucommia. For this purpose, the following experiments were conducted using axillary buds of adult trees. We added 6-benzylaminopurine (BA) and 2,3,5-triiodobenzoic acid (TIBA) in combination and investigated their effects. As a result, by combining 2 mg/l BA and 1 mg/l TIBA and increasing the culture temperature to 28°C, adventitious shoots could be induced.

INTRODUCTION

Eucommia ulmoides Oliver (eucommia) is a deciduous woody plant that is native to China and belongs to the family Eucommiaceae, order Eucommiales. *Eucommia* belongs to one family, one genus, and one species, and is dioecious. In China, the bark has been used as a Chinese herbal medicine since ancient times, and its medicinal properties include lowering blood pressure, analgesia, and diuresis. Recently, it has been discovered that the leaves have similar effects (Harashima, 2012). There is currently a lot of fallow land in Aikawa Town, Aiko District, Kanagawa Prefecture, and local companies are using the fallow land to cultivate eucommia. At the same time, they are already producing and selling eucommia tea using eucommia leaves. The city has begun to revitalize through the sixth industrialization. However, production of eucommia leaves, the raw material for the product, has not kept up with demand.

Currently, seeds are mainly used to propagate eucommia, but only female trees set seeds, and the germination rate of seeds is extremely low when using normal sowing methods. There have been several reports of propagation by tissue culture in the past (Chen et al., 1995; Chen et al., 2007; 2008), but it cannot be said to be established. Furthermore, previous tissue culture propagation studies have prepared explants from hypocotyls of sterile seedlings. When we look at the field-grown eucommia, they are genetically mixed, perhaps because they are propagated by seedlings. In the future it will be necessary to conduct line selection and propagation of eucommia. For this purpose, it is necessary to induce multiple shoots from explants prepared from adult trees. The following experiments were conducted using axillary buds of adult trees.

MATERIALS AND METHODS

Seedlings provided by Hekizanen were planted, and branches were collected from a 6-year-old adult trees grown at the Atsugi campus of Tokyo University of Agriculture. The axillary buds prepared from collected branches were used. We conducted some preliminary experiments and found that the contamination rate was extremely high when using general sterilization methods. Therefore, we investigated various sterilization methods to reduce the contamination rate, and finally decided to sterilize the axially buds using the following method. The collected axillary buds were dipped in a solution of 1.25 g/L of benomyl (GF Benlate hydrating agent, Sumitomo Chemical Garden Products Inc.) and 1 drop of Tween 20, and then treated with an ultrasonic cleaner (ASU-6, As One Co., Ltd.) for 10 minutes. The axillary buds were sterilized by immersion in 70% ethanol for 5 minutes and then 2% sodium hypochlorite (NaClO) containing one drop of Tween 20 for 10 minutes in a clean bench. Thereafter, the axially buds were washed three times with sterilized water, the scale leaves of axially bud were removed with a scalpel, and the center of axillary buds were cut out to form explants.

The basic medium was a 1/2 strength Murashige and Skoog (1962) medium to which 30 g/l sucrose and 8 g/l agar were added, and the pH was adjusted to 5.8. A flat-bottomed glass test tube (φ40 x 130 mm) was used as the culture vessel, and after dispensing 40 ml of the medium, the tube was closed with aluminium foil and sterilized in an autoclave machine (LSX-500, TOMY SEIKO Co., Ltd) at 120°Cfor 15 minutes.

Primarily, the effects of plant growth regulators on axillary bud development were investigated using 2 mg/L 6-benzylaminopurine (BA), 2 mg/L 1-naphthaleneacetic acid (NAA), and 1 mg/L 2,3,5-triiodobenzoic acid (TIBA). The culture environment was 23 ± 1°C, white fluorescence light 20 μmol m^{-2} s^{-1} photosynthetic photon flux density PPFD, 16 hours light/8 hours dark. Although axillary bud development was observed in this experiment, there was a lot of white friable callus formation, and subsequent bud growth could not be observed. Therefore, we next used an incubator (CFH-305, TOMY SEIKO Co., Ltd.) for culture, changed the culture temperature to 28 ± 4.5°C, and investigated the effects of mixed 2 mg/L BA and 1 mg/L TIBA on shoot formation, which BA had been shown to be effective on axillary bud development in the previous experiment.

RESULTS AND DISCUSSION

There are few research reports on tissue culture of eucommia, and even when we conducted follow-up tests using previously reported methods using in vitro seedlings (Chen et al., 1995; Chen et al., 2007; 2008), there was almost no response. In this experiment, the addition of 2 mg/l BA was found to have some effect on promoting the axillary bud development (**Table 1**) but did not lead to adventitious shoot formation.

Table 1. Effects of TIBA, BA, and NAA on the development of axillary bud explants in *Eucommia ulmoides*.

Plant growth regulator (mg/L)			Survival rate (%)	Contamination rate (%)	Development of axially bud (%)
TIBA	BA	NAA			
0	0	0	65	0	0
1	0	0	60	0	0
0	2	0	75	0	15
0	0	2	20	0	0

*Culture temperature is 23±1°C; 16-hr light (PPFD at 20 μmol·m^{-2}·s^{-1}) / 8-hr dark. n=20)

Although not shown in the data, addition of 1 mg/L TIBA tended to suppress callus formation and proliferation. Therefore, we added BA and TIBA in combination and investigated their effects. As a result, by combining 2 mg/l BA and 1 mg/l TIBA and increasing the culture temperature to 28°C, adventitious shoots could be induced (**Table 2** and **Fig. 1**).

Table 2. Effects of TIBA and BA on the morphogenesis of axillary bud explants in *Eucommia ulmoides*.

Plant growth regulator (mg/L)		Survival rate	Contamination rate	Callus formation	Development of axially bud	Adventitious shoot formation
TIBA	BA	(%)	(%)	(%)	(%)	(%)
0	0	5	25	5	0	0
1	2	5	15	5	5	0
1	2	15	20	5	5	5

*Culture temperature is 23±1°C; 16-hr light (PPFD at 20 $\mu mol \cdot m^{-2} \cdot s^{-1}$) / 8-hr dark. n=20)

Figure 1. Adventitious shoot formation of axillary bud explants from adult tree on 1/2MS + 2 mg/L + 1 mg/L TIBA.

The fact that adventitious buds could be induced from the axillary buds of adult trees indicates that eucommia is dioecious, and that it is possible to select and breed superior female plants. The results of a series of experiments on eucommia suggest that the timing of collecting axillary buds, the ripeness of the branches, and the position of the axillary buds on a branch affect the results. Furthermore, the types and concentrations of plant growth regulators used need to be further optimized, and further research is required.

LITERATURE CITED

Chen, L.J., Hu, T.W. and Huang, L.C. (1995). A protocol toward multiplication of the medicinal tree, *Eucommia ulmoides* Oliver. In vitro Cell. Dev. Biol. Plant *31*:193-198.

Chen, R., Namimatsu, S., Nakadozono, Y., Nakazawa, Y. and Gyokusen, K. (2008). Efficient regeneration of the *Eucommia ulmoides* Oliver plant from hypocotyl explant. International symposium on *Eucommia ulmoides 1*:75-77.

Chen, R., Namimatsu, S., Nakadozono, Y., Bamba, T., Nakazawa, Y.and Gyokusen, K. (2007). Efficient regeneration of *Eucommia ulmoides* from hypocotyl explant. Biologia Plantarum *52*:713-717.

Harasima, H. (2012). Syoyakutan - Pharmacognosy-. Revised edition. p.118-119. NTS inc., Tokyo (In Japanese)

Murashige, T. and Skoog, F. (1962). A revised medium for rapid growth and bioassays with tobacco tissue cultures. Physiol. Plant. *15*:473-49.

Combined Proceedings IPPS 73:36-42. 2023.

Establishment of Species-Specific DNA Markers to Identify Interspecific Hybrids of *Hibiscus*

Masaki Ochiai, Kensuke Nakagomi, and Hirokazu Fukui

Faculty of Applied Biological Science, Gifu University 1-1, Yanagido, Gifu, Gifu, 501-1193, Japan

ochiai.masaki.h3@f.gifu-u.ac.jp

Keywords: DNA fingerprinting, plant breeding, PCR

Summary

Confirmation of hybridization is important for interspecific crosses. This study established species-specific DNA markers to select hybrids from interspecific crosses among 4 *Hibiscus* species: *H. cannabinus* L., *H. acetosella* Welw. ex Fic. 'Black King', *H. coccineus* Walt., and *H. mutabilis* L.. Start codon targeted (SCoT) primers were used to establish species-specific DNA markers. Species identification was performed based on the electrophoresis band patterns of PCR products using SCoT primers. The results show that some SCoT primers could be used as species-specific DNA markers among the 4 *Hibiscus* species: SCoT7 primer could distinguish *H. cannabinus* L. (blue flower type) or *H. coccineus* Walt. from the other species, a combination of SCoT2 and ScoT3 primers could distinguish *H. cannabinus* L. (white flower type) or *H. mutabilis* L. from the other species, and a combination of SCoT3 and SCoT7 primers could distinguish *H. acetosella* Welw. ex Fic. 'Black King' from the other species. Accordingly, these DNA markers were applied to detect the pollen parents of interspecific crosses, and the results were consistent with the results from morphological evaluation.

INTRODUCTION

Interspecific crossing is an important breeding method that is used to develop new characteristics in horticultural plants. However, because interspecific crosses generally have low fertilization rates, the presence of self-pollinated individuals among the individuals obtained by crossbreeding is problematic. Morphological evaluation, DNA marker analysis, and ploidy analysis can distinguish between cross- and self-pollinated individuals. However, a suitable method must be selected according to the plant materials in question. *Hibiscus*, such as *Hibiscus rosa-sinensis* L. (a tropical plant), are well-known ornamental plants. Temperate *Hibiscus* plants, such as *H. mutabilis* L. and *H. syriacus* L., are also renowned in East Asia. The flowers of well-known *Hibiscus* are red, pink, orange, yellow, or white, whereas only a few cultivars and species have blue or purple flowers. Therefore, to breed new cultivars with blue or purple flowers, our group made interspecific crosses among four *Hibiscus* species: *H. cannabinus* L., *H. acetosella* Welw. ex Fic. 'Black King', *H. coccineus* Walt., and *H. mutabilis* L. Accordingly, this study aimed to establish species-specific DNA markers to identify pollen parents of progenies obtained by crossing.

MATERIALS AND METHODS

Plant Materials

Tetraploids of *H. cannabinus* L. (blue flower type), tetraploids of *H. cannabinus* L. (white flower type), *H. acetosella* Welw. ex Fic. 'Black King' (tetraploid cultivar), tetraploids of *H. coccineus* walt., tetraploids of *H. mutabilis* L., and their cross populations were grown in a greenhouse at the Gifu Field Science Center, Gifu University.

Selection of DNA extraction method

The following methods were evaluated to find a suitable DNA extraction method for *Hibiscus* species: a sodium dodecyl sulfate (SDS)-based method (Edwards et al., 1991), a cetyltrimethylammonium bromide (CTAB)-based method (Escaravage et al., 1998), a modified CTAB method using polyvinylpolypyrrolidone (PVPP) with a high-pH condition (Kasajima et al., 2013), DNA sui sui-VS (Rhizo, Japan), a Blood & Cell Culture DNA Mini Kit (QIAGEN, Germany), and a DNeasy Plant Mini Kit (QIAGEN, Germany). Each DNA extraction method was performed using young leaves of *H. cannabinus* L. (blue flower type). The presence of DNA and its purity were checked by electrophoresis.

Establishment of species-specific DNA markers

Start codon targeted (SCoT) primers were screened to identify species-specific DNA markers that can distinguish species based on the specificity of PCR and electrophoresis bands (Collard and Mackill, 2009). MightyAmp DNA Polymerase Ver.3 (Takara Bio, Japan) was used for PCR with the following program: 98°C for 2 min followed by 30 cycles of 98°C for 10 s, 55°C for 15 s, and 68°C for 2 min. SCoT primers were evaluated alone or in combinations of 2 in PCR.

Evaluation of hybridization in progenies

The species-specific DNA markers identified above were applied to detect the pollen parents of the progenies obtained by crossing. Progenies were also subjected to morphological evaluation. Length of terminal leaflet, width of terminal leaflet, length of leaf, and attached angle of bottom lateral leaflets were measured in mature leaves. Flower color was subjectively evaluated in crosses between *H. cannabinus* L. (blue flower type) and *H. cannabinus* L. (white flower type).

RESULTS AND DISCUSSION

Selection of DNA extraction method

Hibiscus plants contain large quantities of polysaccharides, which interfere with DNA extraction. Therefore, 6 DNA extraction methods were evaluated to select a suitable method for *Hibiscus* plants. Extracts derived using the SDS-based method or DNA sui sui-VS did not show any DNA bands **(Fig. 1)**.

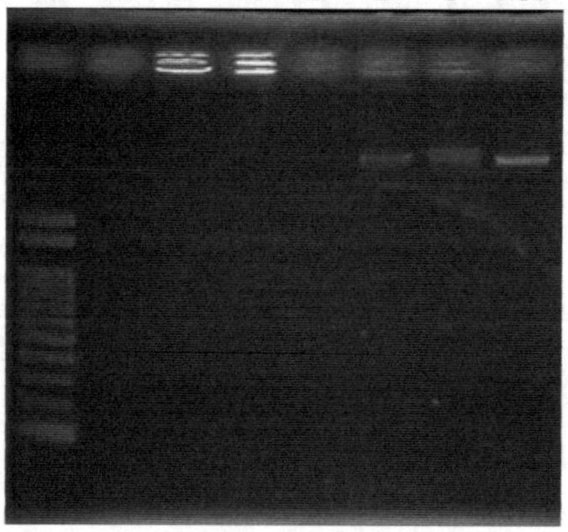

Figure 1. Electrophoresis of extracts derived using each DNA extraction method M: 100-bp DNA size marker, A: SDS-based method, B: CTAB-based method, C: modified CTAB method using PVPP and high-pH condition, D: DNA sui sui-VS, E: Blood and Cell culture DNA mini kit, F: DNeasy Plant mini kit, λ10 : λ-DNA as a positive control.

DNA could not be separated from unnecessary components in the intermediate step. Fluorescence was observed in the wells of agarose gels in the electrophoresis results of extracts derived using the CTAB-based method or modified CTAB-based method with PVPP and high pH. This phenomenon is observed when DNA extracts contain many high-molecular-weight compounds. Although these methods could extract DNA, they did not remove enough polysaccharides. Meanwhile, clear DNA binding was observed in the results electrophoresis results of extracts derived using the Blood & Cell Culture DNA Mini Kit and DNeasy Plant Mini Kit, suggesting their suitability for DNA extraction from *Hibiscus* plants. Accordingly, the DNeasy Plant Mini Kit was chosen for the following experiments owing to its simplicity of operation.

Establishment of species-specific DNA markers

PCR products with SCoT7 primer showed a band specific to *H. cannabinus* (blue flower type) between 1 and 2 kbp as well as a band specific to *H. coccineus* over 2 kbp (**Fig. 2**). SCoT7 primer distinguished *H. cannabinus* (blue flower type) or *H. coccineus* from the other samples. PCR products obtained using a combination of SCoT2 and SCoT3 primers showed an obvious band specific to *H. cannabinus* (white flower type) over 2 kbp and one for *H. mutabilis* over 2 kbp. The combination of SCoT2 and SCoT3 primers could distinguish *H. cannabinus* (white flower type) or *H. mutabilis* from the other species. PCR products obtained using a combination of SCoT3 and SCoT7 primers showed an obvious band specific to *H. acetosella* 'Black King' between 1 and 2 kbp.

Furthermore, the combination of SCoT3 and SCoT7 primers distinguished *H. acetosella* 'Black King' from the other species.

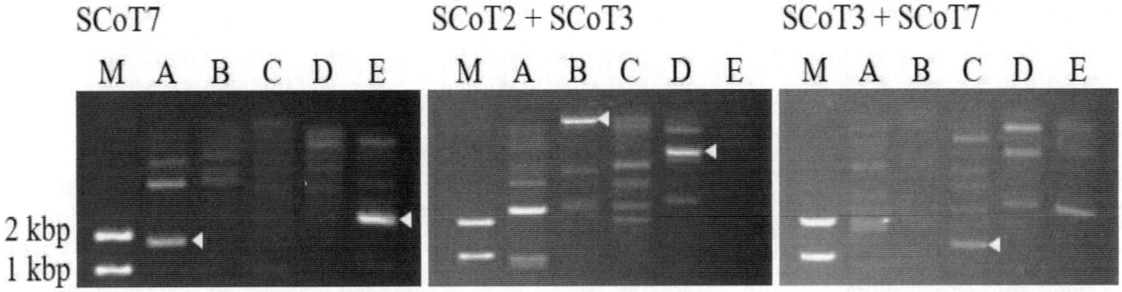

Figure 2. Electrophoresis of PCR products using SCoT primers. M: 100-bp DNA size marker, A: *H. cannabinus* (blue flower type), B: *H. cannabinus* (white flower type), C: *H. acetosella* 'Black King, D: *H. mutabilis* , E: *H. coccineus*. White arrowheads indicate deduced species-specific band.

Evaluation of hybridization in progenies

The species-specific DNA markers established above were applied to distinguish the progenies of interspecific crosses. Given that the embryonic parents are clear in the crossing populations, the species-specific DNA markers were used to identify pollen parents. Eleven progenies obtained by crossing between *H. cannabinus* (blue flower type) and *H. cannabinus* (white flower type) were examined using the marker specific to *H. cannabinus* (blue flower type) (**Fig. 3**).

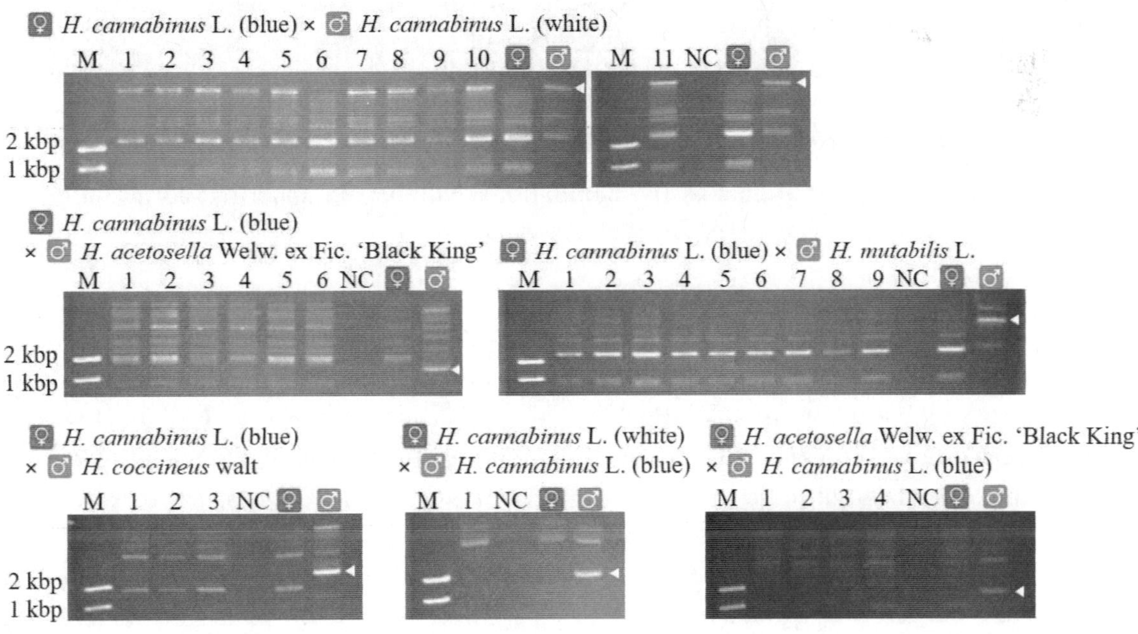

Figure 3. Detection of DNA markers in the crossing populations. M: 100-bp DNA size marker. Numbers indicate individual numbers of progenies in each crossing. NC: non-template control. White arrowheads indicate species-specific band of pollen parent.

Ten progenies exhibited a band specific to *H. cannabinus* (blue flower type), whereas 1 progeny did not. This suggests that the 10 progenies are hybrids between *H. cannabinus* (blue flower type) and *H. cannabinus* (white flower type); meanwhile, the remaining progeny is likely a self-crossing of *H. cannabinus* (blue flower type). Regarding flower color, *H. cannabinus* (blue flower type) and *H. cannabinus* L. (white flower type) exhibit dark purple and white petals, respectively. The 10 progenies that exhibited the band specific to *H. cannabinus* (blue flower type) had pale purple petals, whereas the 1 progeny that exhibited the band specific to *H. cannabinus* (white flower type) had white petals (**Fig. 4**). Thus, these morphological characteristics are consistent with the DNA marker results.

Figure 4. Flower morphology in crossing populations. Numbers in upper left corners represent individual numbers of progenies in each crossing.

Six progenies obtained by crossing *H. cannabinus* (blue flower type) and *H. acetosella* 'Black King' were examined using the marker specific to *H. acetosella* 'Black King' (**Fig. 3**). However, none of the progenies exhibited the band specific to *H. acetosella* 'Black King', suggesting that they are self-crossings of *H. cannabinus* L (blue flower type). The leaf shape of these progenies resembled that of *H. cannabinus* (blue flower type) (**Fig. 5**), which is consistent with the DNA marker results.

Nine progenies obtained by crossing *H. cannabinus* (blue flower type) and *H. mutabilis*. were examined using the marker specific to *H. mutabilis*. (**Fig. 3**). However, none of them exhibited the band specific to *H. mutabilis*, suggesting that they are self-crossings of *H. cannabinus* (blue flower type). The leaf shape of these progenies resembled that of *H. cannabinus* (blue flower type) (**Fig. 5**), which is consistent with the DNA marker results.

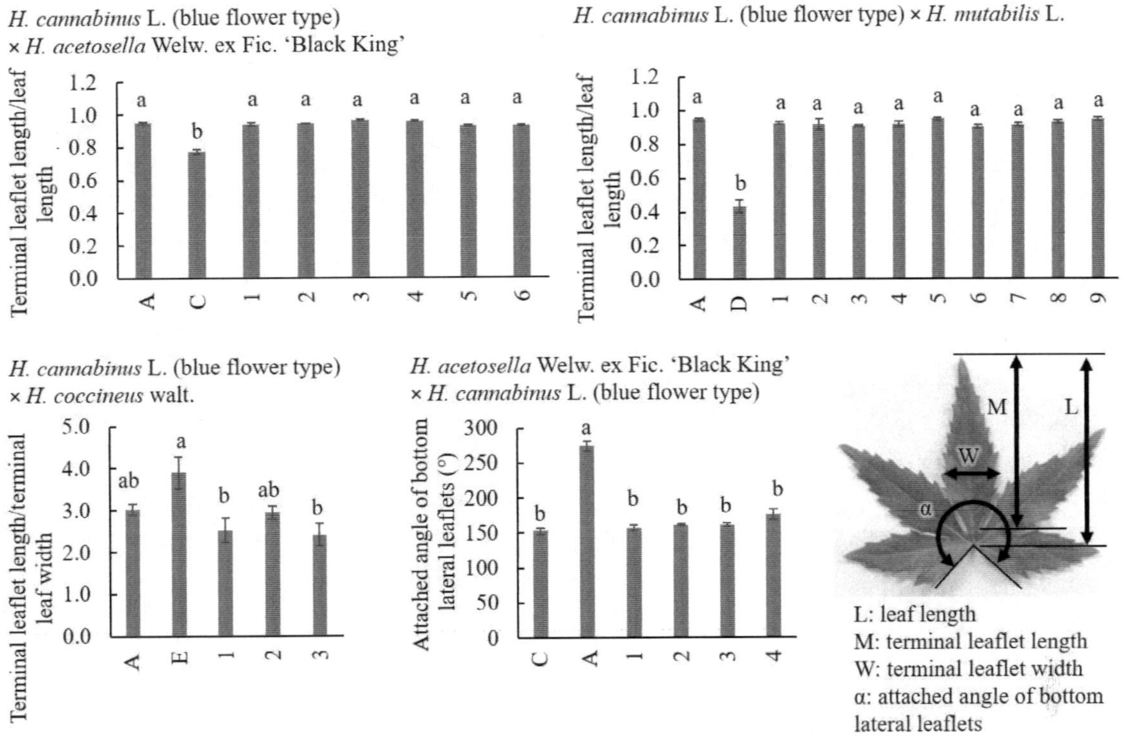

Figure 5. Leaf shape in crossing populations. A: *H. cannabinus* (blue flower type), C*: H. acetosella* 'Black King, D: *H. mutabilis*, E: *H. coccineus*. Numbers indicate individual numbers of progenies in each crossing. Error bars indicate standard deviation. Different lower-case letters in the same graph indicate a significant difference at P < 0.05 according to the Tukey–Kramer test.

Three progenies obtained by crossing *H. cannabinus*. (blue flower type) and *H. coccineus* were examined using the marker specific to *H. coccineus*. **(Fig. 3**). However, none of them exhibited the band specific to *H. mutabilis*, suggesting that they are self-crossings of *H. cannabinus* (blue flower type). The leaf shape of these progenies resembled that of *H. cannabinus* L. (blue flower type) (**Fig. 5**), which is consistent with the DNA marker results.

Four progenies obtained by crossing *H. acetosella* 'Black King' and *H. cannabinus* (blue flower type) were examined using the marker specific to *H. cannabinus* (blue flower type) **(Fig. 3**). However, none of them exhibited the band specific to *H. cannabinus* (blue flower type), suggesting that they are self-crossings of *H. acetosella*. 'Black King'. The leaf shape of these progenies resembled that of *H. acetosella* 'Black King' (Fig. 5), which is consistent with the DNA marker results.

In summary, this study established species-specific DNA markers for *H. cannabinus* (blue flower type), *H. cannabinus* (white flower type), *H. acetosella*. 'Black King', *H. coccineus* walt., and *H. mutabilis* The ability of these species-specific DNA markers to identify species was also confirmed by identifying the pollen parents of the progenies of crosses among the four *Hibiscus* species.

LITERATURE CITED

Collard, B. C. Y. and Mackill, D. J. (2009). Start Codon Targeted (SCoT) polymorphism: A simple, novel DNA marker technique for generating gene-targeted markers in plants. Plant Mol. Biol. Rep. *27*: 86-93.

Edwards, K., Johnstone, C. and Thompson, C. (1991). A simple and rapid method for the preparation of plant genomic DNA for PCR analysis. Nucleic Acids Res. *19*(6): 1349.

Escaravage, N., Questiau, S., Pornon, A., Doche, B. and Taberlet, P. (1998). Clonal diversity in a *Rhododendron ferrugineum* L. (Ericaceae) population inferred from AFLP markers. Mol. Ecol. *7*: 975-982.

Kasajima, I., Sasaki, K., Tanaka, Y., Terakawa, T. and Ohtsubo, N. (2013). Large-scale extraction of pure DNA from mature leaves of *Cyclamen persicum* Mill. and other recalcitrant plants with alkaline polyvinylpolypyrrolidone (PVPP). Sci. Hortic. *164*:65-72.

Combined Proceedings IPPS 73:43-45. 2023.

Pawpaw (*Asimina triloba*) Floral Differentiation Period in Miyazaki Prefecture, Japan

Takaaki Maeda, Shota Miyahara, and [1]Tatsuro Murata

Graduate School of Horticultural and Food Science, Minami Kyushu University 3764-1, Tatenocho, Miyakonojo-shi, Miyazaki, 885-0035, Japan; [1]Faculty of Agriculture, Tokai University 9-1-1, Toroku, Kumamoto Higashi-ku, Kumamoto, 862-8652, Japan

tmaeda@nankyudai.ac.jp

Keywords: fruit crop, yield, flowering

Summary

Pawpaw (*Asimina triloba*) is a North American fruit crop with potential in as a novel crop in Japan. This paper describes the flowering phenology of pawpaw under local conditions in Miyazaki Prefecture.

INTRODUCTION

Pawpaw (*Asimina triloba* (L.) Dunal) is a temperate fruit tree and can be cultivated over a wide area in Japan. The flower bud differentiation period of the pawpaw in Japan was in mid-May for plants cultivated in Osaka and Kyoto Prefectures (Hirai, 1954; Sobajima, 1955). With global warming, the flower bud differentiation period may differ depending on the cultivation area. Therefore, this study investigated the flower bud differentiation period of pawpaw grown in Miyazaki Prefecture, located south of the Kinki region, Japan.

MATERIALS AND METHODS

One 11-year-old grafted pawpaw and one 15-year-old grafted pawpaw, 'Rebecca

Gold', grown outdoors in Nichinan, Miyazaki Prefecture, Japan, were studied. Survey materials were collected on June 8, July 9, August 7, September 4, October 4, November 1, and December 1, 2017, January 5, February 8, and March 8, 2018, and May 16, May 28, and June 14, 2021. Each time, we selected three ca. 15-cm non-fruit-bearing branches grown in the current year at a near-horizontal angle. Five buds were collected randomly from each branch, each time, and immediately fixed in formalin acetic acid alcohol solution (50% ethanol : formaldehyde solution : acetic acid = 90 : 5 : 5). Pawpaw flower buds are covered with a hard brown outer skin, so after removing the outer skin, the flower buds were observed using a stereomicroscope. Some of the buds were sectioned longitudinally to a thickness of 15 μm using paraffin sectioning, and double-stained with Safranin O and Fast green. Under an optical microscope, we made morphological observations of flower bud differentiation.

RESULTS AND DISCUSSION

The flower buds on May 16, 2021 were in the early stages of differentiation, with the shoot apex thickening and beginning to change into a flat or dome-like shape (**Fig. 1A**). In the bud on May 28, 2021, the primordia of sepals, petals, and stamens were observed (**Fig. 1B**). In the buds on June 14, 2021, pistil primordia formation, and ovules, ovaries, and anthers were also observed (**Fig. 1C, D**). Many stamens were also seen in the buds observed under a stereomicroscope on June 8, 2017. These results confirmed that stamen primordia had already formed in late May. As shown in Figure 1, the pistil continued to develop until September, and the transverse diameter of the pistil on September 4, 2017, was 549.2 μm. No significant differences were observed in the development of flower buds from October until February of the following year (data omitted). However, in March 2018, one month before flowering, the flower buds enlarged rapidly, and the horizontal diameter of the flower buds was 5,048.4 μm (**Fig. 2C**). According to Miyahara et al. (2018), pawpaw trees enter a dormant period in mid-October and awaken from dormancy in early March, so it is thought that flower bud development may have stopped during that time.

Based on these results, the flower bud differentiation period of pawpaw in Miyazaki Prefecture, Japan, is in mid-May. The results were similar to those obtained by Hirai (1954) and Sobajima (1955) when they investigated pawpaw trees grown in Osaka and Kyoto Prefectures 68 to 69 years ago. Recently, global warming has progressed. Osaka/Kyoto Prefectures and Miyazaki Prefectures are at different latitudes, with Miyazaki Prefecture located far to the south; nevertheless, the flower bud differentiation period was similar. For many deciduous fruit trees, flower bud differentiation occurs from June to August, but for pawpaw, it occurs much earlier, in mid-May.

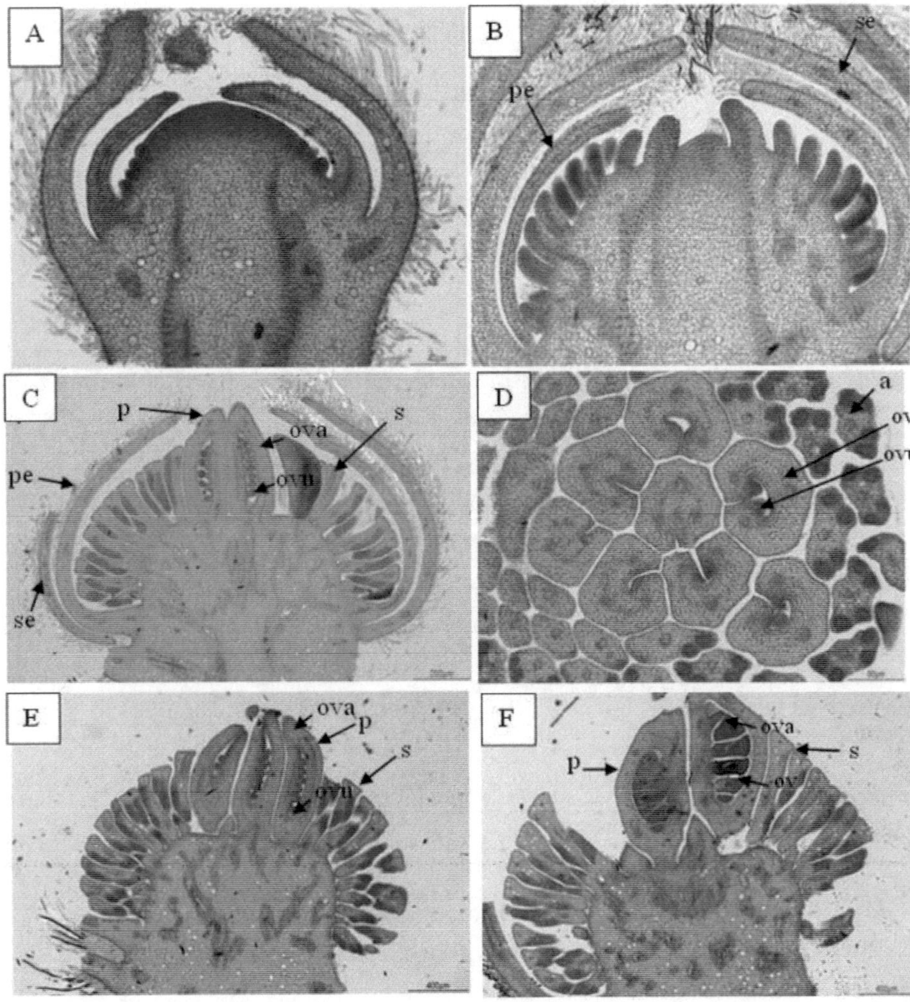

Figure 1. Observations of the flower bud formation process of pawpaw using an optical microscope. Collection dates: A) 2021.5.16, B) 2021.5.28, C D) 2021.6.14, E) 2017.8.9, and F) 2017.9.4 (A, B, C, E, and F: longitudinal sections, D: cross-section) s, stamen; p, pistil; pe, petal; se, sepal; ovu, ovule; ova, ovary; a, anther.

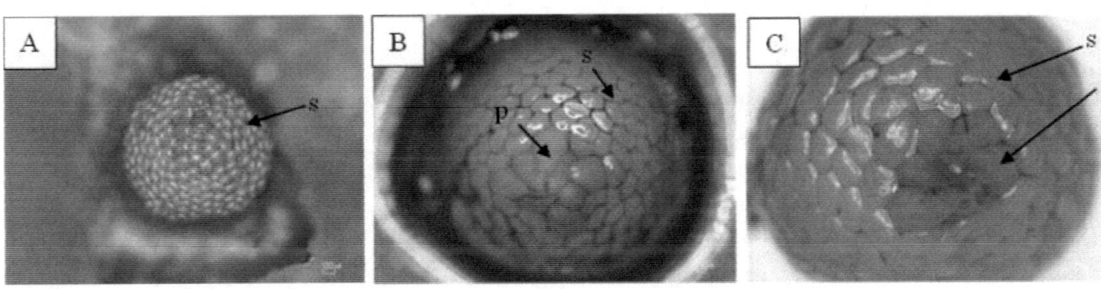

Figure 2. Stereomicroscope observations of pawpaw flower bud formation. Collection dates: A) 2017.6.8, B) 2017.7.9, and C) 2018.3.8. s, stamen; p, pistil.

Combined Proceedings IPPS 73:46-66. 2023.

Propagation – Essential to Life on Earth: 51st Conference of the International Plant Propagators' Society – Australia Region

Ranjith Pathirana[a] and Zoe Williams[b]

[a] School of Agriculture, Food and Wine, Waite Road, University of Adelaide, Urrbrae, Adelaide, SA 5064, Australia; [b] Waterworth's nursery, 1106 Steve Irvin Way, Glasshouse Mountains, QLD 4518, Australia

ranjith.pathirana@adelaide.edu.au

Keywords: IPPS, program, tours, awards, scholarships

Summary

The 51st Annual Meeting of the International Plant Propagators' Society-Australia Region convened on 25 May 2023 at the Novotel Hotel, Geelong, Victoria, Australia with President Bruce Higgs presiding.

INTRODUCTION

"Propagation – Essential to Life on Earth" was the theme of the 51st Conference of the Australian chapter of the International Plant Propagators' Society (IPPS) held in Novotel, Geelong, Victoria during 25 – 27 May 2023. The conference was organised by

Clive Larkman and Dermot Molloy, supported by the dedicated Secretary Pam Berryman. It was packed not only with interesting presentations from a variety of scientists, academics, hands-on nursery managers, company representatives and students, but also with pre- and post-conference tours, the traditional golf competition and a variety of trade displays. This editorial is meant to cover the activities associated with the conference as well as giving a brief rundown of the Conference Proceedings and introducing this year's IPPS Australia award winners.

PRE-CONFERENCE TOURS

Pre-conference Tour – Day 1

The state of Victoria represents a meagre 3% of Australia's territory (GA, 2023), but contributes a whopping 37% of the national gross value of nurseries and floriculture produce of AUD 1.9 billion (AV, 2023), the largest of any of the Australian States/Territories. Thus, this state has much to offer for those seeking knowledge in nursery and floriculture industries, and the Pre-Conference Tour offered participants a firsthand glimpse into the evolving landscape of plant propagation and a unique opportunity to delve into Melbourne's (Capital city of Victoria) diverse horticultural landscape, discover innovative propagation techniques, and engage in enriching discussions with fellow professionals.

First Stop: Rivers of Yarrambat

Rivers of Yarrambat where the tour began is a renowned garden center and leisure destination with a rich history (**Fig. 1**). What began as Plenty River Nurseries in 1981, Rivers of Yarrambat has transformed into a multifaceted retail hub, showcasing an array of plants, home decor, and culinary delights (Riversofyarrambat, 2023). The evolution of Rivers of Yarrambat stands as a testament to the resilience and adaptability of horticultural enterprises in meeting the diverse needs of consumers.

Figure 1. Rivers of Yarrambat visit by participants of the IPPS conference.

Second Stop: Alowyn Gardens

Nestled amidst picturesque landscapes of the Yarra Valley, Alowyn Gardens welcomed participants with warm hospitality and captivating stories of its inception (**Fig. 2**). Established in 1997, the gardens serve as a testament to the transformative power of passion and perseverance. As stewards of biodiversity, the creators of Alowyn Gardens John and Prue Van de Linde have crafted an immersive experience that celebrates the beauty of nature while imparting invaluable insights into gardening and environmental stewardship. Alowyn Gardens is designed along strong symmetrical lines that become blurred by the softness of the spaces and the grace of the trees. Within an area of approximately seven acres (~ three hectares), there are seven clearly defined areas including a perennial border, a silver birch forest, an inspiring edible garden, a parterre garden and another few display gardens. The formal parterre garden leads you through to a series of smaller courtyards and display gardens. Through all these areas runs a wisteria and rose covered archway of about 100 m with a sunken garden and a classical fountain as a centrepiece Visitors exit the gardens through an extensive plant nursery full of great plants, many of which feature in the gardens. The nursery tends to favour heat tolerant and frost hardy plants as they recognise the climate is changing towards extremes of temperatures (Alowyn Gardens, 2023).

Figure 2. Some of the features of Alowyn Gardens that the participants visited before they were treated to lunch.

Third Stop: Larkman Nurseries

At Larkman Nurseries, participants had the privilege of exploring a family-owned propagation nursery steeped in tradition and innovation (**Fig. 3**). Since its establishment in 1984, Larkman Nurseries has been at the forefront of plant propagation, embracing advancements such as post-entry quarantine and facility modernization. It is one of Australia's leading tubestock nurseries with a 2 million turnover of plants annually. Due to the wide variety of native and exotic genera being propagated (400,000 tubes and 300,000 cuttings at a time), the participants had a glimpse of several propagation facilities such as fog, mist, cold frame, mist/hot bed, tissue culture and straight heat (LarkmanNurseries, 2023). The nursery is actively involved in the famous Herb and Chilli Festival of Victoria. It was a heartwarming experience for many participants as the owners Clive and Di Larkman are active members of the IPPS; Clive is a Board Member and organizer of the Geelong Conference. The tour provided a glimpse into the meticulous processes and dedication that underpin the success of this esteemed establishment.

Figure 3. Visit to Larkman Nurseries in Lilydale near Melbourne, VIC by the conference participants during the pre-conference tour.

Fourth Stop: Kuranga Native Nursery

The first day of tour culminated with a visit to Kuranga Native Nursery (**Fig. 4**), a sanctuary of rare and indigenous flora nestled in the heart of Mt Evelyn (Kuranga, 2024). The business of the Nursery keeps growing as natives become increasingly popular for both home and commercial gardens across Australia as the effects of drought and hot summers are taking a toll on the exotics and the younger Australians realize the role of natives and indigenous species in environmental protection. With a passion for sustainability, the nursery uses solar energy and at the same time saves some AUD 13,000 in energy costs (ChoiceEnergy, 2024). The Paperback Café of the Kuranga Native Nursery has curated a menu that celebrates the flavours and nutritious qualities of Australin bush foods. The participants were treated to a sensory journey through Australia's botanical heritage, culminating in a delightful culinary experience infused with native ingredients sourced from the nursery and the Yarra Valley. Kuranga Native Nursery exemplifies a harmonious fusion of conservation, commerce, and culinary arts, inviting visitors to explore the rich tapestry of Australia's natural heritage.

Figure 4. Kuranga Native Nursery in Mount Evelyn was the last stop of the first day of IPPS pre-conference tour of 2023.

Pre-conference Tour – Day 2

The IPPS Australia Pre-Conference Tour continued for a second day in its exploration of Melbourne's horticultural landscape, uncovering new insights and innovations in plant propagation.

First Stop: Norwood Industries

Norwood Industries, Australia's leading manufacturer of horticultural labels, welcomed participants to its state-of-the-art facilities (**Fig. 5**). Since the 1960s, Norwood has been at the forefront of label printing technology, revolutionizing plant marketing with its patented tag display methods (Norwood, 2024). With sustainable practices in the forefront of their operations, Norwood has committed to making sure that 100% of its polypropylene waste is recycled in partnership with Garden City Plastics. The tour provided a behind-the-scenes glimpse at Norwood's commitment to quality and innovation, showcasing the company's contributions to the global horticultural industry.

Figure 5. Participants had an opportunity to see the operations at Norwood Nursery on the second day of pre-conference tour.

Second Stop: Garden City Plastics

Participants were immersed in the history and innovation of Garden City Plastics (GCP), a pioneering manufacturer of horticultural containers (**Fig. 6**). From its humble beginnings in 1975 to its recent move to a larger manufacturing headquarters, GCP has remained dedicated to sustainability and environmental stewardship and a large proportion of their plastic pots and containers are made from recycled plastic and 100% of their products are recyclable. The tour highlighted the company's efforts to promote recycling and reduce plastic waste, underscoring its commitment to a greener future for horticulture. Innovation is the key to the success of GCP. Equipped with over 100 injection molding machines, grinders and robots operated using advanced software, GCP provides integrated solutions effectively and timely to its customers (GCP, 2024). More about the plastic recycling in nursery and horticulture industry in Australia can be found in the paper by Mathew Mills in these Proceedings. GCP is one of the major sponsors of IPPS Australia conferences for many years and sponsors the 6 pack – a group of selected youth from the industry.

Figure 6. The pre-conference tour participants were able to witness the sustainable practices at Garden City Plastics where they utilize closed loop recycling of plastics in horticulture.

Third Stop: Ball Australia

Ball Australia welcomed participants to its 40-acre (~16 ha) production facility in Skye, 40 km from Melbourne, offering insights into its mission to lead the research, breeding, production, and marketing of ornamental crops (**Fig. 7**). With a focus on quality, innovation, and sustainability, Ball Australia has introduced numerous groundbreaking varieties to the Australian market. The tour showcased the company's dedication to excellence and its ongoing contributions to the horticultural community. With environmentally controlled glasshouses, an automated production facility and state of the art dispatch systems, Ball Australia has the flexibility of growing high-quality plants to supply the growers with plugs all year round. The three parts of the vegetative propagation system developed and used at Ball Australia's Melbourne facility is described in detail by Ian van Zanten in these Proceedings.

Figure 7. Some of the facilities visited by delegates at the Ball Australia nursery near Melbourne, Victoria.

Fourth Stop: The Diggers Foundation

In 1978 Penny and Clive Blazey set up Diggers Club in Melbourne, a commercial business to provide gardeners with all their gardening needs (**Fig. 8**). Although the Diggers Club was originally a commercial business, preservation and conservation always sat at the core of the Blazey's vision for Diggers. In 2011, The Diggers Foundation was registered as a non-profit charity and all proceeds are distributed to the foundation to deliver its charitable mission. With the theme of 'grow with purpose' as its vision, this charity supports rescuing heirloom seeds through "Support a Seed" initiative (Diggers, 2024), assist communities with "Seed the Future" program (over 1.2 million seeds have been donated to community groups since 2023), promotes sustainable organic gardening practices, provides education programmes for gardeners as well as preserves heritage buildings they own. The visit to The Diggers Foundation provided a glimpse into its rich heritage and enduring commitment to preservation and conservation. From historic gardens to championing heirloom seeds and advocating for environmental stewardship, The Diggers Foundation has remained at the forefront of promoting biodiversity and sustainability. The tour celebrated the foundation's multifaceted contributions to horticulture and its unwavering dedication to preserving Australia's botanical heritage.

Figure 8. At the Diggers Foundation, the participants of the pre-conference tour had the opportunity to take a glimpse of the charitable work being conducted by the foundation and its commitment to sustainability and conservation.

Fifth Stop: Van Loon's Garden Centre

The IPPS Pre-Conference Tour concluded with a visit to Van Loon's Garden Centre, a hub of inspiration and innovation for gardening enthusiasts (**Fig. 9**). With its diverse range of products and services, Van Loons Garden Centre embodies the spirit of community and creativity, inviting visitors to explore the wonders of gardening and horticulture.

Figure 9. Van Loon's Garden Centre was the last stop of the pre-conference tour. This nursery attracts lot of crowds for the wide range of plants and the popular café.

AWARDS

IPPS Australia has a suite of awards for people who excel in propagation-related activities or have contributed to the Society. They are selected by committees appointed by the Board and are honoured at the annual conference.

Award of Honour

IPPS Australian Region Award of Honour is the highest award. This year's (2023) awardee Ray Doherty (**Fig. 10**) graduated from The University of Queensland with a Bachelor of Science majoring in Biochemistry & Botany in 1973. In the same year he started working with the CSIRO in legume crop research followed by long periods with Lakkari Native Plant Nursery, Jensen International, Newton Container Trees, Leaf Liaisons, Birkdale Nursery and Azalea Grove Nursery. He has undertaken many different roles throughout his career in horticulture; from propagation to retail sales, tree and shrub production, landscape procurement and supply, contract growing, international import and export, product costing and accounting. This broad experience has been available to members of IPPS as we "seek and share". He has taken on various leadership roles in the industry and has provided support and encouragement wherever possible. Ray is a very accomplished member of the society, holding various roles including 2019 conference convenor and chair of the IPPS Australian awards committee encouraging youth in horticultural careers for which he is passionate.

Figure 10. Ray Doherty with the Award of Honour trophy of IPPS, Australia.

Edward and Mary Bunker Award

The focus of the Edward and Mary Bunker Award is to recognize an outstanding contribution demonstrating the IPPS moto 'To Seek and To Share' for the betterment of the industry at large. The selection criteria include any of the following or related categories:

- Innovation and development in plant propagation, production, or nursery systems,
- Mentoring new propagators, promoting horticultural training, education and extension,
- Plant breeding that has impacted Australian or International markets or resulted in the introduction of new plants or varieties to Australia,

- Facilitation of market access within Australia or internationally,
- Innovation or problem solving in any aspect of business including health and safety, transport, logistics, supply chain management, new technologies, etc.

The award is open to non-members helping raise awareness and recognition of IPPS throughout a wider expanse of horticultural industries.

The recipient of 2023 was Jane Edmanson, a horticulturalist, author, and television and radio personality (**Fig. 11**). As of 2018 Jane is best known across Australia as the presenter for the ABC TV program Gardening Australia. She is a recipient of Australian Honours in 2004 for "service to horticulture, particularly through the promotion of environmentally sustainable gardening practices, and the encouragement and education of young gardeners". She has authored many books: The Australian Garden, Cheap and Easy Propagation, Working Manual for Gardeners, The New Zealand Garden, Jane Edmanson's Favourite Plants to name a few. More details on her career from growing up in Mildura on the banks of the Murray River caring for citrus and other fruit in her parents' orchards to her present-day activities of radio and TV broadcasting can be found in these Proceedings, based on her talk at the presentation ceremony.

Figure 11. Jane Edmanson (left) receiving the Edward and Mary Bunker Award from the President of IPPS Australia Bruce Higgs.

Steve Vallance Pewter Tankard Award

In 1979 the Great Britain & Ireland Region of IPPS, presented the pewter tankard to Australian Region, to be used as an annual award to recognise the contributions of one of our members to the society. It was regularly awarded until 1991. It was then paused until 2010, when at the Freemantle conference, it was awarded to Steve Vallance. Steve really embodied the ideals of the award, contributing without fanfare, but with commitment. In honour of Steve, and the way he went about his 'seeking & sharing', in 2017 the award was renamed the Steve Vallance Tankard.

Tony Vander Staay, Immediate Past President of IPPS Australia was the recipient of the award in 2023 (**Fig. 12**). Looking through past awardees his father was one of the early recipients of this award. Covid presented its difficulties during his Presidency, but he was willing to adapt and adopt new ways to keep the society functioning and he has been of great assistance to the Board and Chair in running the Society in those uncertain years.

Figure 12. Tony Vander Staay (left) at the award ceremony receiving Steve Vallance Pewter Award from the President of IPPS Australia Bruce Higgs.

Rod Taellis Memorial Youth Award

IPPS recognizes outstanding achievements by the younger members of the industry. One way the Society achieves this is through the Rod Taellis Memorial Youth Award which is presented annually to the most commendable achievement by a person under 30 years who is working within or studying horticulture in Australia. The award is named after one of the Society's most respected members, Rod Tallis. Rod was a committed nurseryman with a passion for plant propagating, inspired enthusiasm for the IPPS and had a keen interest in the youth of the industry.

He was the mainstay in many of the early conferences and played a particularly supportive role in running the Sydney Conference in 1981. His commitment and passion were an inspiration to many plant propagators. The winner of Rod Tallis Award will have completed a research project, written an article or series of articles or developed a new process or product. They may be enrolled in an academic institution or maybe a hands-on role in a nursery. Over the years many of the previous winners have proceeded on to establish very successful careers in horticulture and have also become leaders of this society. The winner is invited to the annual conference to present a paper on their research or achievements and receives a one-year membership to IPPS and a commemorative plaque.

The winner of the award in 2023 was Elliott Olumuyiwa Akintola (**Fig. 13**) Agronomist and Category Manager for Plant Health and Protection at Garden City

Plastics for his research on the changes in volatile substances in tomato roots under three levels of phosphorus nutrition. Using an optimised headspace solid-phase microextraction (HP-SPME) technique combined with gas chromatography-mass spectrometry (GC-MS), Elliott demonstrated the manipulation of volatilomic pathways by tomato roots adapting to changes in phosphorus levels. Under limited phosphorus, the plant was making sacrifices to increase the signals for arbuscular mycorrhizae associations. The results were presented by Elliot at the Annual Conference after the award was presented to him and the full paper of this interesting research is published in the current Proceedings.

Figure 13. Elliot Olumuyiwa Akintola (centre) receiving Rod Tellis Award at the 2023 IPPS Conference in Geelong, Victoria.

IPPS Conference Youth Initiative (6 Pack) and South African Exchange

The Support Team and South African Exchange are two programs that are designed to offer two exciting opportunities to new propagators. The Support Team program involves selecting passionate young propagators relatively new to the industry to attend the Annual Conference. Part of the program involves these propagators assisting with the day to day running of the conference. They are affectionately named the 6 pack (**Fig. 14**).

The South African Exchange involves selecting one young propagator to travel to South Africa and stay with local nurseries and to then attend the annual conference in Australia. IPPS Australia hosts one from South Africa in exchange. The 6 pack of 2023 included the two South African exchangees and four young propagators selected by the Committee appointed by the Board of IPPS Australia.

Anita Boucher Award for the best paper of the conference

Each year, a committee consisting of the Editor, Newsletter Editor and a representative from the Hort Journal Australia selects the best paper, announced at the gala dinner. Each presentation is scored on three criteria: a) The quality of presentation, b) The quality of the content, and c) The potential impact of the work in the short or mid-term for the horticultural industry at large or a particular segment of it.

Figure 14. The 6 Pack of 2023 IPPS Conference – from left Joshua Taylor, Sharline Alison, Dolly Prosper, Stephanie Hastie (South African Exchange Awardee from Australia), Laura Coyle and Sizwe Ndabeni (South African Exchange Awardee from South Africa).

The 2023 award for the best paper was awarded to Andrw Laidlaw of the Royal Botanic Gardens Victoria for his paper on the "Importance of Plants in the Landscape". Andrew is a qualified landscape architect and horticulturalist with over thirty years of experience in the industry. Andrew is well known for his in-depth knowledge of plants, design, and his innovative approach to the design process.

His presentation on the importance of plants in the landscape was an informative and well-deserved winner of the IPPS 2023 Best Conference Paper and is published in these Proceedings.

Figure 15. Josh Waterworth (right) receiving the Peter Smith Perpetual Golf Trophy from Tony Vander Staay, the immediate past president of the Society.

Peter Smith Perpetual Golf Trophy

It has been a tradition of the IPPS Australia to have a golf tournament among willing participants before the conference. The 2023 trophy was won by Josh Waterworth, Director of Waterworth's Nursery, Glasshouse Mountains in Queensland (**Fig. 15**).

Honorary Membership Award

IPPS Australia awards honorary memberships to its members who have immensely contributed to horticulture industry. In 2023 two Honorary Membership Awards were announced:

Peter Ollerenshaw was one of the recipients (**Fig. 16**). Peter began work as a horticulture apprentice with Canberra City Parks in 1964 and then worked at the Canberra Botanic Gardens for most of his apprenticeship continuing there for 10 years. After a 2-year stint at Monaro Nursery in charge of plant production he returned to the Australian National Botanic Gardens (ANBG) in Canberra for 11 years, firstly as an overseer in the gardens and then at the nursery. He was supervisor in charge of gardeners both at ANBG Canberra and the Jervis Bay annexe. In 1989 Peter left the Gardens to establish Bywong Nursery concentrating on Australian plants. He is renowned for breeding many new cultivars. Peter attended the first meeting of IPPS in 1973 and has been Australian President and International director of IPPS and has served on industry related bodies. He is a past recipient of the Australian Award of Honour. Peter has always been prepared to share his knowledge with others and has been an advocate for the society for many decades.

Figure 16. Peter Ollerenshaw (centre) receiving honorary membership of IPPS, Australia at its 51st conference in Geelong, Victoria.

Peter Waugh who served in three IPPS Regions was the other recipient of honorary membership. He was a passionate member of the New Zealand Region for many years serving as President and International Director. He then became instrumental in establishing Japan Region and has made outstanding contributions to IPPS for over thirty years.

CONFERENCE PRESENTATIONS

The conference presentations were held on 25th and 26th May 2023 at the Novotel Hotel, Geelong, Victoria. With three sessions for each day and three presentations for each session, there were a total of 18 presentations. We have grouped the presentations according to their theme and give highlights below.

Methods of Propagation

David Hancock is the founder of Natural Area Nursery in Western Australia (WA), one of the largest environmental contractors and consultants in WA with over 100 staff producing over 1.2 million plants from over 350 native species for restoration projects. David's commitment to developing propagation methods for hard to root species has earned him much respect within IPPS (of which he is a long-standing member) and beyond. In his presentation on "Provenance Propagation Methodology of Perth Bushland Species from Seed", David emphasised the importance of keeping records of germination outcomes of seeds from different locations, time of collection, seed processing technique and methodology of seed germination. As a result of this meticulous record keeping, the nursery is now well positioned to supply plants that were hard to propagate a decade ago. David presented selected case studies of valuable bushland species for which propagation methods have been developed in the recent times. Full paper of his presentation is available in these Proceedings.

Amanda Shade, also from WA is the Nursery Curator and Manager of Living Collections for the Botanic Gardens and Parks Authority of Kings Park. Amanda highlighted the unique nature of nurseries of botanic gardens as they have opportunity and responsibility within the industry as the primary producers for organisations that hold documented collections of living plants for the purposes of scientific research, conservation, display, and education. Her presentation covered all these aspects, and the details can be found in these Proceedings.

The presentations on propagation at this conference were absorbing as each of them dealt with nurseries catering to different customers with distinct products. Ian van Zanten's presentation was on ornamentals. After more than 20 years as head nursery manager and head grower for a finished nursery supplying big box retailers in the United States and Canada Ian joined Ball Australia in 2020 and the participants had the opportunity to visit its Victoria facility during the pre-conference tour as already mentioned here. Ian is the Growing Manager of Ball Australia operations, and his presentation covered the vegetative production program at this facility using cuttings. He identified three parts of this operation: mother stock production, sticking cuttings and finishing. Details of these three parts are presented in his full paper in these Proceedings. More about the Ball Australia operations including videos can be found in their website (Ballaustralia, 2024).

Tony Hughes is a lecturer at the local Training and Further Education (Gordon TAFE) college in Geelong and at the same time runs his own propagation nursery, propagating 'difficult' native species, both herbaceous and tree species. His presentation was on grafting native species from drier areas of Australia for better survival in the wet and colder climate of Victoria making these beautiful ornamentals available in the market as grafted plants. His full paper with colour photos of the two species *Eremophila* (Scrophulariaceae) and *Prostanthera* (Lamiaceae) he used as examples is available in these Proceedings. It is interesting to note that for *Eremophila* spp. the best rootstock is *Myoporum insulare* or *M. montanum*, also from Scrophulariaceae.

Peter Lewis, Rod Tallis award winner in 1984 took to consulting work overseas, mainly in Asia and the Middle East after many years of teaching and then managing Birkdale nursery's interests in China. His presentation was based on his consulting work for Landscape Nursery a subsidiary of Red Sea Global, wholly owned by the Public Investment Fund of the Kingdom of Saudi Arabia (KSA). The Landscape Nursery has a target of growing 25 million native plants by 2030 for greening of two world-leading destinations announced by the Crown Prince of KSA - The Red Sea (coastal) and Amaala (inland) (RSG, 2024). The desert environment in KSA whilst spectacular is very challenging for the plant production. With diversification of the KSA economy towards a more environmentally sustainable and healthy place to live, greater emphasis is now being placed on using the native flora from the region to revegetate the desert environment. Some of the key characteristics of the climate and environment in the Red Sea region in which this research work was undertaken includes 360 days of sunshine, hot summer temperatures for extended periods (40 – 50 ºC), warm winter temperatures (> 15 ºC), regular high winds with blowing sand and very low rainfall (< 10mm /yr) with extreme thunderstorms. These conditions result in high daily evaporation rates (up to 15mm a day), soil is generally a fine particle, poorly draining sand with high total dissolved salt concentrations (>2500 ppm), high Ph >7 and high soil sodicity. Peter in his presentation gave details of native species with high ornamental and ethnobotanical potential that were propagated in the nursery and the challenges they encountered in planting them out in the desert and keeping them alive until established.

With over 30 years of experience in plant biotechnology research and research management at several leading nurseries such as Yuruga and Fleming's, and currently serving as the Research Manager at Skybury, Dr Puthiyaparambil Josekutty shared his own experience and lessons learned in the true spirit of IPPS, "Seek and Share". His presentation, also available as a full paper in these Proceedings, contains many aspects of propagation for farms using tissue culture techniques. He discussed important aspects of quality assurance, genetics, production of high-health plants, quarantine during import and export of plant material and the use of tissue culture technologies for crop improvement. He used examples from horticultural crops such as avocado, banana and papaya for which he has a reputation in micropropagation.

Presentations from Award Winners

Rod Teallis Award

Elliott Olumuyiwa Akintola was the winner of the award and his full paper titled "Identification of Changes in the Total Volatilome of Tomato Plant Roots in Response to Phosphorus Availability" is published in these Proceedings.

Edward and Mary Bunker Ward

The winner of the Edward and Mary Bunker award was Jane Edmanson, and at the award ceremony Clive Larkman described

Jane as an epitome of the IPPS motto 'To Seek and To Share'; while practicing horticulture she spent time teaching and in media. She described her involvement in horticulture at the conference and an extended summary of the speech is published in these Proceedings.

Anita Boucher Award for the Best Presentation

This award recognises the best presentation at the conference. It is awarded based on a marking system by a three-judge panel considering the content, presentation, and its impact on the industry in the short and medium term. The winner was Andrew Laidlaw, Landscape Architect at the Royal Botanic Gardens Victoria (RBGV) for his presentation on "The Importance of Plants in the Landscape". He described different projects he undertook over the last 25 years within RBGV as well as for Global Gardens of Peace (an Australian charity that creates gardens to support vulnerable communities around the globe) in Australia and overseas. Full paper covering a number of interesting projects is published in these Proceedings.

South Africa Exchange (Australian award)

Stephanie Hastie from TAFE in Adelaide, South Australia had the opportunity to spend, in her own words, three fascinating weeks in South Africa visiting a range of nurseries, botanic gardens and areas with natural vegetation. Stephanie's visits and her experience in South Africa is now written into a full paper and published in these proceedings. Every year an enthusiastic young Australian gets this opportunity to visit South Africa thanks to the vision of IPPS to encourage youth in horticulture.

South Africa Exchange (South Africa Award)

Thandisizwe Siphenkosi Ndabeni was the recipient of the award from South Africa to visit Australia. His three weeks in Australia took him from Western Australia to Victoria and from there to South Australia - three important and diverse horticultural hubs of the continent. He participated in the pre-conference tours and at the conference he shared his experience visiting nurseries and travelling in Australia. His presentation is published as a full paper in these Proceedings.

Conservation, Restoration and Collection

At the Annual Conference of 2023 diverse themes in conservation from collection of plants, restoration of flora in environments damaged by human activity, conservation in botanic garden settings as well as sustainable management of nursery waste through recycling were presented.

Danielle Saintpierre is the Technical Director at SIRAS Pacifique, a New Caledonian company specialising in environmental conservation and mine site restoration. Nickel ore mining in New Caledonia, a French territory, has led to serious degradation of soil and its unique vegetation leading to considerable damage to the ecosystems and biodiversity of the island. Danielle's presentation emphasised the value of native and endemic species in restoration of these sites and the results of revegetation efforts, particularly using hydroseeding technique. Her full paper with details of this work is illustrated with many photos of mining sites that have been degraded and the results of restoration work undertaken by her team.

Botanic gardens play a vital role in plant conservation as they hold seed banks, herbaria, and ex situ plant collections. Ex-situ conservation has become urgent with climate change threatening plant communities in their natural environments. However, with climate change the botanic gardens need to assess the suitability of plants for coming decades and require a landscape succession strategy. Clare Hart, Manager for Horticulture at RBGV presented the landscape succession strategy (LSS) for RBGV established back in 1846. With 20 plant collections, 6000 trees representing 190 countries of the world it is Melbourne's green space and the challenge is to keep it green under climate change. Hence the need for a LSS as the temperatures are increasing and rainfall is decreasing.

Considering this challenge, on 28th April 2016, RBGV announced LSS that will guide its management through climate change into the next century. Clare emphasised the need to understand the history of the garden and the climatic era in which it was established, which is no more. Although the New Zealand collection for example is at high risk, she emphasised its value for understanding how New Zealand taxa respond to warmer conditions, identifying what plants are more vulnerable and what plants will thrive under climate change. Nevertheless, the LSS will manage the overall succession of the current landscape towards one that by 2090 is dominated by species more likely to be resilient under the projected climate, yet maintaining its landscape character i.e., changing now in order to stay the same. In addition to plants from drier and warmer areas of Australia, some areas of Argentina, USA, Mexico, the Mediterranean and West and South Africa offer plants that are suited for the succession plan. RBGVs LSS initiative was the catalyst for the First Botanic Gardens Climate Change Summit held in Melbourne in December 2018 with participants from 13 botanic gardens that have the climate predicted for Melbourne in several decades. The declaration signed by all the participants was the steppingstone for the Climate Change Alliance of Botanic Gardens currently having 400 members representing over 90 countries (CCABG, 2024). Since then, CCABG has launched the Landscape Succession Toolkit that allows botanic gardens and arboreta a framework by which to adapt to the climate crisis and transition their landscapes and plant collections to ones that will continue to thrive in the future climate. Details of the work carried out at the RBGV on LSS has been described in detail by Symes and Hart (2021).

Alistair Watt from Otway Ridge Arboretum has been on plant collection missions in the Pacific Rim including New Caledonia, Fiji, Chile, New Zealand etc. since 1985 and some of his more than 200 new species introductions are displayed in different collections in RBGV and in Geelong Botanic Gardens as well as Sydney, Adelaide and Hobart botanic gardens. Some specimens are from direct introductions while others are from re-propagated plants. Alistair in his presentation, now published as a full paper in these Proceedings has given lists of all species introduced during his travels and describes where these introductions can be found in Australia. It could be possible that some of these species are in the lists of plants suitable for planting under the LSS of botanic gardens and arboreta.

Matthew Mills is a consultant in operations and strategic management with clients in packaging, manufacturing, agriculture and retail sectors. With 30 years of experience in these sectors (of which 14 at Board level) he has developed a deep understanding of delivery of sustainable profit strategies in a circular economy. His presentation encompassed the initiatives in Australia to manage disposal of plastics used in horticulture in an environmentally responsible way. He described in detail the PP5 initiative, a joint venture between Norwood Industries, Garden City Plastics, and Polymer Processors. As a joint venture, the goal is to build an innovative infrastructure that allows the industry, and the communities that operate within, to sustainably consume and reuse PP5 plastics. The PP5 initiative is now in its advanced stages, and it costs the industry the same to recycle used industry PP5 as it does to source it from outside sources. It is anticipated that by 2025 the industry recycled PP5 will be cheaper than purchasing outside the industry so that the savings can be passed on to growers. The other advantage is of course reduced plastic pollution and reduced landfill fees.

In an enlightening talk, Clive Larkman described how the plant retailing businesses in Australia managed to sustain their businesses through diversification of their products and services. Desire to have edible and ornamental plants in home gardens and inside dwellings made the plant retailing business flourish and in addition, people's desire to socialise around food made the natural transition of garden centres to multifaceted retailers with garden ornamentals, cafes and restaurants. Clive emphasised with examples how growing plants is the key to a happy and healthy society, reminding the participants the theme of 2023 IPPS Australia Conference - plants and their propagation is essential to life.

Acknowledgements

The authors thank Clive Larkman, Pam Berryman and Bruce Higgs for providing support and guidance with the necessary information contained in this Editorial.

LITERATURE CITED

Alowyn Gardens (2023). Nursery. https://alowyngardens.com.au/nursery/ Accessed 25 February 2024.

AV (2023). Agriculture Victoria. Victorian Nurseries and Floriculture Industry, Fast Facts January 2023. https://agriculture.vic.gov.au/__data/assets/pdf_file/0009/921177/Nursery-and-floriculture-fast-facts-Jan-2023.pdf. Accessed 20 February 2024.

Ballaustralia (2024). Ball Australia - Let Us Colour Your World. https://ballaustralia.com/ Accessed 20 March 2024.

CCABG (2024). Climate Change Alliance of Botanic Gardens. https://www.rbg.vic.gov.au/initiatives/climate-change-alliance/ Accessed 24 March 2024.

ChoiceEnergy (2024). Case Study – Kuranga Nursery Benefits from $13,000 in Annual Energy Savings by Making the Switch to Solar. https://www.choiceenergy.com.au/customers/agriculture/kuranga-native-nursery Accessed 24 February 2024.

Diggers (2024). Support a Seed. https://www.diggers.com.au/pages/support-a-seed Accessed 20 March 2024.

GA (2023). Geoscience Australia. Area of Australia – States and Territories. The Australian Government Publication. https://www.ga.gov.au/scientific-topics/national-location-information/dimensions/area-of-australia-states-and-territories Accessed 23 February 2023.

GCP (2024). Custom Moulding. https://www.gardencityplastics.com/custom-moulding Accessed 27 February 2024.

Kum, T., Kendal, D., Symes, P., and Hart, C. (2023). Targeting effective plant conservation using the climate assessment tool. Australasian Plant Conservation: Journal of the Australian Network for Plant Conservation, *31*(4), 3–6. https://search.informit.org/doi/10.3316/informit.138599319435933

Kuranga (2024). Our Story. Available at: https://www.kuranga.com.au/our-story/#Story Accessed 27 February 2024.

LarkmanNurseries (2023). About Us. http://larkmannurseries.com.au/www/content/default.aspx?cid=640&fid=661 Accessed 1 March 2024

Norwood (2024) About. Available at: https://www.norwood.com.au/about/ Accessed 26 February 2024.

Riversofyarrambat (2023). Rivers of Yarrambat, Garden Center. https://riversofyarrambat.com/garden-centre/ Accessed on 24th February 2024.

RSG (2024). Red Sea Global. https://www.redseaglobal.com/en/about-us Accessed 20 March 2024.

Symes, P., Hart, C. (2021). The Climate Change Alliance: botanic garden horticulturists as agents for change. Sibbaldia: The International Journal of Botanic Garden Horticulture *20*: 95–122. https://doi.org/10.24823/Sibbaldia.2021.352

Combined Proceedings IPPS 73: 67-73. 2023.

South Africa Exchange

Stephanie Hastie

505 Fullarton Road, Netherby, SA 5062, Australia

stephanie.hastie@tafesa.edu.au

Keywords: nursery, cultural, Cape Town, Durban, Johannesburg, botanic gardens

Summary

Earlier this year I was fortunate to be given the opportunity by the Australian IPPS to do a 3-week exchange in South Africa. Whilst I was there, I gave a presentation at their IPPS conference during which I attempted to soften their attitudes towards some of our beautiful Eucalypts. An ambitious goal given that several Eucalypts are significant environmental weeds in South Africa. Now I have a different but equally, if not more ambitious, task – to give you a brief overview of my experience as an exchangee in South Africa. While I was there, I got to visit a wide range of different nurseries – from retail to production, exotic to native, specialist to generalist, and traditional to a tissue culture laboratory, and in visiting these places I got to see plants being produced from a range of propagation techniques. I also got to visit a few botanic gardens and areas of natural vegetation. I started the exchange in Johannesburg, then flew to Durban, and finished in Cape Town.

IPPS Vol. 73 – 2023

INTRODUCTION

Johannesburg is South Africa's biggest city and the capital of the Gauteng province. It is located about 1 hour away from Pretoria, which is South Africa's administrative capital. Both Johannesburg and Pretoria have a subtropical highland climate (Cwb), which means they tend to have dry winters, and warm wet summers. Durban is a coastal city on the eastern side of South Africa's Kwa-Zulu-Natal province (KZN, for short). It has a humid subtropical climate (Cfa) (meaning it has no dry season, and hot summers). It was warmer and more humid than Johannesburg. Cape Town is a port city on South Africa's southwest coast. It has a warm-summer mediterranean climate (Csb), characterised by dry summers, and cool winters, similar to Adelaide, which has a hot-summer mediterranean climate (Csa).

TRAVELS

Magaliesberg Mountain Range

One of the first places I visited was the Magaliesberg mountain range. The Magaliesberg is one of the oldest mountain ranges on Earth, formed by volcanic eruptions about 2.3 billion years ago. It has had a long history of human occupation dating back at least 2 million years. In terms of native vegetation, it is a mix of grassland, sub-Saharan savannah, and remnant Afromontane Forest (**Fig. 1a**). Unfortunately, there are also quite a few introduced weedy species, from both South America and Australia, as well as a serious issue with water hyacinth (*Pontederia crassipes*). At the time of my visit, the water hyacinth was covering parts of the crocodile river and the Hartebeespoort dam so thoroughly that from a distance it looked like lush turf covering a large plane.

Sittig's Nursery

The first nursery I visited was Sittig's nursery, which belongs to Hans & Carol-Ann Sittig. They produce a lot of potted colour – both perennials and annuals – and they run both a retail and wholesale nursery (**Fig. 1b**). They described how the nursery had been struggling financially after Covid. Like many nurseries, they were hit very hard from lockdowns. The only thing they were allowed to open for was edibles.

The University of Pretoria

I was able to visit the University of Pretoria and have a guided tour with the curator of their botanical gardens, Jason Sampson, who was pretty chuffed as he just had a *Kalanchoe* named after him. While I was there, I got to see one of the rarest plants in the world - *Encephalartos woodii*, which is extinct in the wild (**Fig. 1c**). As many as 500 specimens exist, but all of them were derived from basal suckers or offsets from the original plant discovered by John Medley Wood in 1895. Unfortunately, the specimen at the university is yet to produce any suckers. Jason told us that this specimen alone is worth more than the rest of the university's cycad collection combined.

A lovely South African native most of us would be familiar with is *Strelitzia*, particularly *S. reginae* and *S. nikolai*. As Jason walked around the garden, he talked of hiding his rare plants "in plain sight" – and pointed out some *Strelitzia juncea* growing at the entrance of a building. *Strelitzia juncea* has tall, rush-like leaf stalks, and is better adapted to tougher, dried conditions.

Figure 1. (a) Magaliesberg mountain range. One of the oldest mountain ranges on Earth, formed by volcanic eruptions about 2.3 billion years ago, (b) Sittig's nursery in Hartbeespoort produce both perennial and annual potted colour, (c) *Encephalartos woodii* at the University of Pretoria Botanic Gardens is worth more than the rest of the Universitty's cycad collection combined according to its curator Jason Sampson, (d) *Strelitzia juncea* has tall, rush-like leaf stalks, and is better adapted to tougher, dried conditions, (e) *Strelitzia juncea* (Narrow-leaved Bird of Paradise) growing at the entrance of a building at the University of Pretoria.

The hotter and dryer the conditions, the less pronounced the leaf 'paddle'. Sadly, this *Strelitzia* is vulnerable in the wild, which is due in part to illegal collecting for the horticultural trade. Plant poaching is a huge issue in South Africa, and a visit to the Kirstenbosch National Botanical Garden nursery will give you heartbreaking insight into the scale of that issue. It is currently housing row upon row, in room upon room, of rare plants that have been poached and then intercepted by law enforcement (**Fig. 2a**).

The University of Pretoria had a beautiful specimen of a Fever Tree (**Fig. 2b**). Fever Trees are an iconic South African tree, which get their common name from early European explorers, who associated these trees with malaria. They didn't know what caused malaria at that point and, of course, the reason the trees were associated with this fever was because they grow close to water sources and areas with more mosquitos. The botanical name is *Vachellia xanthophloea*. The species name can be translated as "yellow bark". It's a bit of a sore spot for South Africans because this is one of the species previously in the genus *Acacia*, which Australia has had the sole rights to for some time now. But, as I felt when I visited South Africa, it is an enduring source of indignation for many South Africans who grew up with their iconic Acacias. And I cannot say I would feel any differently if the roles were reversed.

Figure 2. (a) *Conophytum* is a commonly poached genus, and here you can see one row of benches with trays and trays of specimens being housed at the Kirstenbosch National Botanical Garden nursery, (b) Fever Tree (*Vachellia xanthophloea*) at the University of Pretoria, (c) Walter Sisulu Botanical Garden bird watching area, (d) Harold Porter Botanic Garden.

One thing that particularly stood out about all three botanical gardens I had the chance to visit – Kirstenbosch, Harold Porter and Walter Sisulu – were the mountain ranges that framed all three. They really gave an impressive scale to the gardens – because you can go from examining the minutest flower or insect, to then looking up the cultivated gardens – and then further to the wild and natural areas higher up / further away.

At the Walter Sisulu Botanical Garden they are doing trials to see how much success they can get with aerial layering on *Olinia emarginata*. *O. emarginata* (mountain hard pear) is a small to medium-sized evergreen tree, with attractive flowers and berries that make it a desirable ornamental plant. Although it is common in the wild, they have a lot of difficulty propagating this tree from seeds. You can see all these propagation pods - what they call "turtles" (**Fig. 3b**). Last I spoke with Dr Miranda Deutschlander, she told me they have 10 still alive and growing, although they were not yet ready to remove from the parent tree at that stage.

Figure 3. (a) Kirstenbosch Botanic Garden. (b) Aerial layering on the *Olinia emarginata* (mountain hard pear) at the Walter Sisulu Botanical Garden.

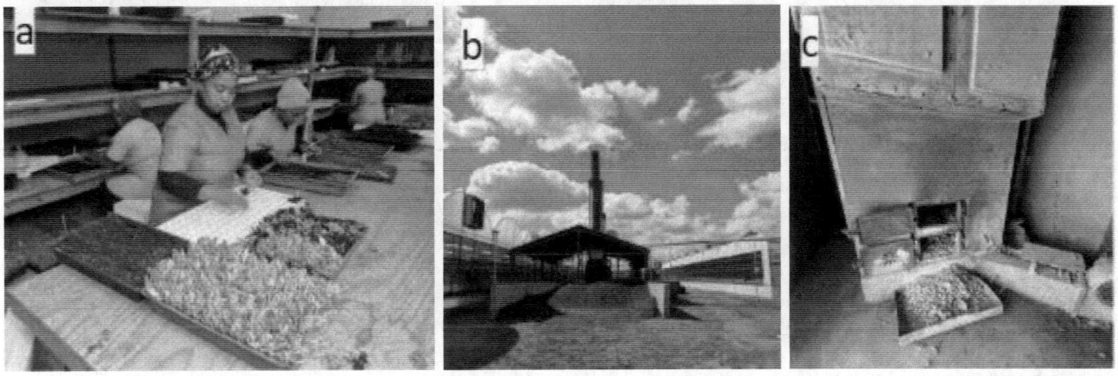

Figure 4. (a) Nursery workers at Samgro Wholesale Nursery, where we had a tour with Siyasanga Yenzela, (b) Coal-fired power at Ball Straathof, (c) Coal-fired power at AT Nursery.

Something that was very distinctly different between South Australian nurseries and South African nurseries, were the number of employees. The socio-economic climate of South Africa – in other words, the high number of people needing employment and the comparatively low cost of wages – means that nurseries that might operate in Australia with about 10-15 staff, might have about 60 to 80 to 120 staff in South Africa, and as a result there is a lot less reliance on automation and a lot more reliance on hand labour.

A serious challenge South Africans have to contend with is load shedding. They are frequently without power for significant periods of time. I take my hat off to them, because running a nursery is difficult enough with reliable power supply, let alone without. But if there is a will there's a way, and many nurseries rely on coal as a backup (**Fig. 4b** and **4c**), some with a view to moving to solar power in the future.

Stephward nursery in Durban is a bit of a specialist collectors' nursery, with some very unusual plants. They told us they stopped counting once they had over 10,000 species. But they are not propagating high numbers of each of those species. At this nursery I got to see a beautiful peacock fern, *Selaginella willdenowii*, with its iridescent blue leaves (**Fig. 5c**). It is not a true fern, but a fern-like lycophyte. This plant is adapted to shaded areas underneath a forest canopy, and it is thought that the blue iridescence helps to reduces the effect of any strong beams of sunlight that might penetrate through the canopy. Stephand had lots and lots of interesting orchids, including a number of *Bulbophyllum*. These are an epiphytic orchid with fantastic inflorescence. *Bulbophyllum* are also known for having many species that produce foul rotting meat scented flowers (**Fig. 5b**). They also had a really healthy specimen of *Mimosa pudica*, known as the 'shy' or 'sensitive' plant, which responds to touch (**Fig. 5a**). It's weedy in Queensland, so it might not be so exciting for some, but for a South Australian it was special to see it. It is believed the trait evolved as a way to avoid predators. If a herbivore comes along and starts nibbling on the plant, bending the leaves inwards and drooping would make it more difficult for the predator to tear off the leaves, and it would also serve to expose the prickles on the stems.

Figure 5. Stephward Nursery (a) *Mimosa pudica*, (b) *Bulbophyllum*,(c) *Selaginella willdenowii*.

Whilst visiting Airely Nursery we got to see their new rolling (side shift) benches (**Fig. 6a**). The staff at Airely rated them highly. They increase the growing area by 20% – 35% and reduce the risk of fungal disease by elevating the plants and allowing under pot airflow.

Whilst in Durban we also got to visit the Dube Tradeport tissue culture lab, where you can sign an agreement with them to give them exclusive rights to reproduce your plants in exchange for them working out the best propagation protocol (**Fig. 6b** and **6c**).

Figure 6. (a) Rolling (side shift) benches at Airely Nursery,(b) Dube Tradeport tissue culture lab display, (c) Dube Tradeport tissue culture lab *Rosa* sp.

CONCLUSIONS

In summary, the 3-week exchange to South Africa gave me the opportunity to experience and learn from a diverse range of nurseries, nursery people, as well as other experts and places in the world of plants and horticulture. Like Australia, South Africa has many different kinds of climates and many different nurseries. I visited specialised nurseries (for example, that grew only natives, or propagated many species of a particular kind of plant (such as Tillandsia) or grew many species of rare exotics).

I also visited more generalist nurseries – nurseries that supplied to the public and offered a little bit of everything (a mix of edibles, potted colour, tried and true perennials, succulents, and natives, etc). Some nurseries were moderately to highly automated, while others relied heavily on manual labour and less so on automation. Like Australia, many nurseries had room for improvement regarding nursery hygiene practices.

There was a good knowledge of best practice, however, like us they must battle with what to prioritise with a limited budget and busy schedule. They must also battle with the pressure of rising energy costs and load shedding. I would highly recommend the exchange to any young propagators.

Acknowledgement
My heartfelt thanks go to the IPPS and all the people who contributed to make my visit to South Africa happen.

Combined Proceedings IPPS 73: 74-83. 2023.

Identification of Changes in Total Volatilome of Tomato Plant Roots in Response to Phosphorous Availability

Olumuyiwa Akintola Elliott
10Ej Court, Off Assembly Drive DanadenONG South, Vic 3175

elliott.akintola@gardencityplastics.com

Keywords: volatile compounds, plant interactions, mycorrhiza, organic compounds, strigolactone

Summary

The volatilome are bioactive volatile organic compounds. They respond to changes in growing conditions and can work as signalling molecules within or between plants. This paper describes changes in volatile organic compounds in plants in response to changes in phosphorus availability.

INTRODUCTION

Volatilome is a term used to describe the total bioactive compounds such as volatile organic compounds (VOCs) produced by plants through its biosynthetic pathways (**Fig. 1**). These wide range of chemical compounds are produced under specific conditions and in response to changing growing conditions such as nutrient deficiencies, environmental stress conditions, pollination, defence strategies, signalling and communication strategies intra / inter plants and organisms. In a nutshell, the plant volatilome is considered as an extended metabolome, reflecting the plant's physiological status (Lee Díaz et al. 2022). This research will use an untargeted approach to identify changes in the whole plant volatilome or VOC profile in response to changes in phosphorus nutrition.

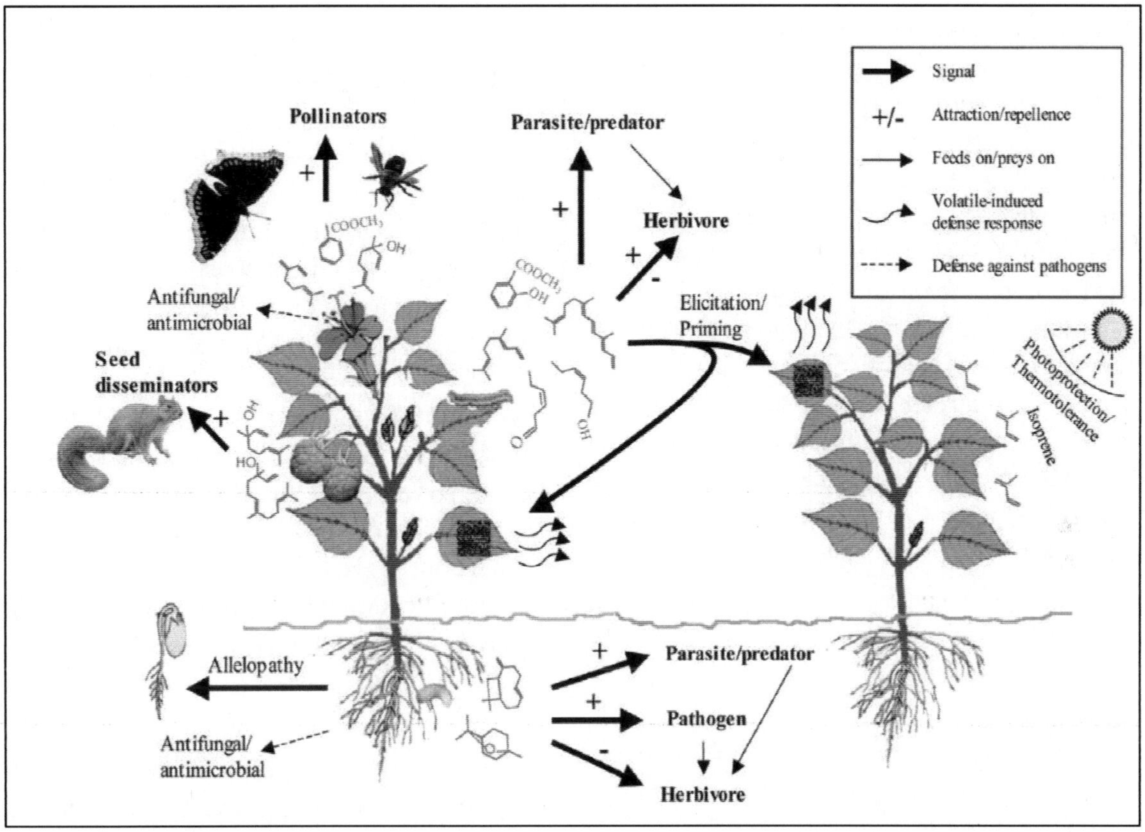

Figure 1. overview of volatile-mediated plant interactions with the surrounding environment (Dudareva et al. 2006).

Phosphorus (P) is an essential microelement required by all plant cells yet, it is both the least mobile and least available of the major plant nutrients (Kovar & Claassen 2005). Owing to its presence in major organic molecules, such as nucleic acids, (RNA, DNA,) ATP and membranes (Péret et al. 2011), plant productivity can be greatly limited by P starvation in the growing environment. In most soils, inorganic P is available at low levels in solution because of its strong adhesion to soil minerals hence, affecting phosphorus chemical mobility and bioavailability (Hinsinger 2001). This makes this essential element almost inaccessible to plants thus, a major limiting factor for plant growth and development (Massalha et al. 2017).

The uptake of P usually in the form of orthophosphate (Pi) from soil solution is made possible through symbiotic associations with arbuscular mycorrhizal (AM) fungi (Chiu & Paszkowski 2019). There is research evidence that suggests that plant roots and AM fungi perceive each other prior to their physical interaction. However, the identities of the diffusible signals are currently unknown, but the plant signal is most abundant in the root exudates of phosphate deprived plants.

Through a combination of growth, developmental and metabolic responses, plants have developed strategies to sense, cope, and respond to P changes in their growing environment. These strategies aim to reduce usage, and increase uptake and recycling (Rouached, Arpat & Poirier 2010). A strategy utilized by plants for acquiring P in the soil is through symbiotic relationships with arbuscular mycorrhizal (AM) fungi. Smith and Waters (2012), reported that the deployment of this strategy in response to P limitation is mediated in part by strigolactone signalling. In general, strigolactones are carotenoid-derived plant hormones involved in the regulation of plant development i.e., aerial shoot branching, and rhizosphere signalling to stimulate root-AM interactions.

Plants exude strigolactones to attract AM in the rhizosphere to increase P uptake through the root. In turn, AM fungi obtain photosynthates from the host plants (Yoneyama et al., 2012). Considering the complexity of the soil microbiome, there is an enhanced competition for limited available nutrients. As a result of this complexity, it is highly unlikely that strigolactones are the only communication mechanism for plants in response to P starvation.

Plants are top emitters of Volatile Organic Compounds (VOCs). An estimated 1,700 chemical compounds out of over 100,000 chemical products known to be produced by plants are volatiles (Spinelli et al., 2011). These compounds are involved in a range of ecological functions including responding to stress conditions (Holopainen & Gershenzon 2010), defence mechanism (Farnier et al., 2012; Penuelas et al., 2014; Vaughan et al., 2013), signalling and communicating with other organisms in the rhizosphere (Wenke, Kai & Piechulla 2010). Therefore, VOCs are viable options for alternative signals that accumulate in response to P deficiency to ensure plant survival. Therefore, the non-targeted approach adopted in this project will attempt to discover the changes in the whole volatilome profile at different P levels which may be key mediators in biotic interactions, signalling and communication belowground.

Tomato (*Solanum lycopersicum*) is an important vegetable crop worldwide and are known to be high emitters of VOCs. Tomato plant responses to stress conditions such as nutrition, drought, salt, insect pests and the influence of AM fungi on its growth and yield has been well documented (Asensio, Rapparini & Peñuelas 2012; Bai et al. 2018; Catola et al. 2018; Chitarra et al. 2016; Rivero et al. 2018b). Also, the characterization of the volatilome under other conditions have been reported but to the best of our findings, there has been no work done to identify the collective volatilome in response to increasing levels of Phosphorous.

It is our expectation that through this project we will be able to identify changes in the total Volatilome of tomato plant roots at increasing concentrations of P compared to controlled nutrient composition over 7 days of application using Headspace, Solid phase Microextraction Gas Chromatography Mass Spectrometry (HS-SPME-GC/MS) technique as described by Rivers et al. (2019).

MATERIALS AND METHODS

We subjected tomato plants to three levels of P nutrition over a times series and adopted an optimized Headspace, Solid phase Microextraction Gas Chromatography Mass Spectrometry (HS-SPME-GC/MS) (Fig 2) to identify and quantify VOCs that provided a competitive advantage in P acquisition.

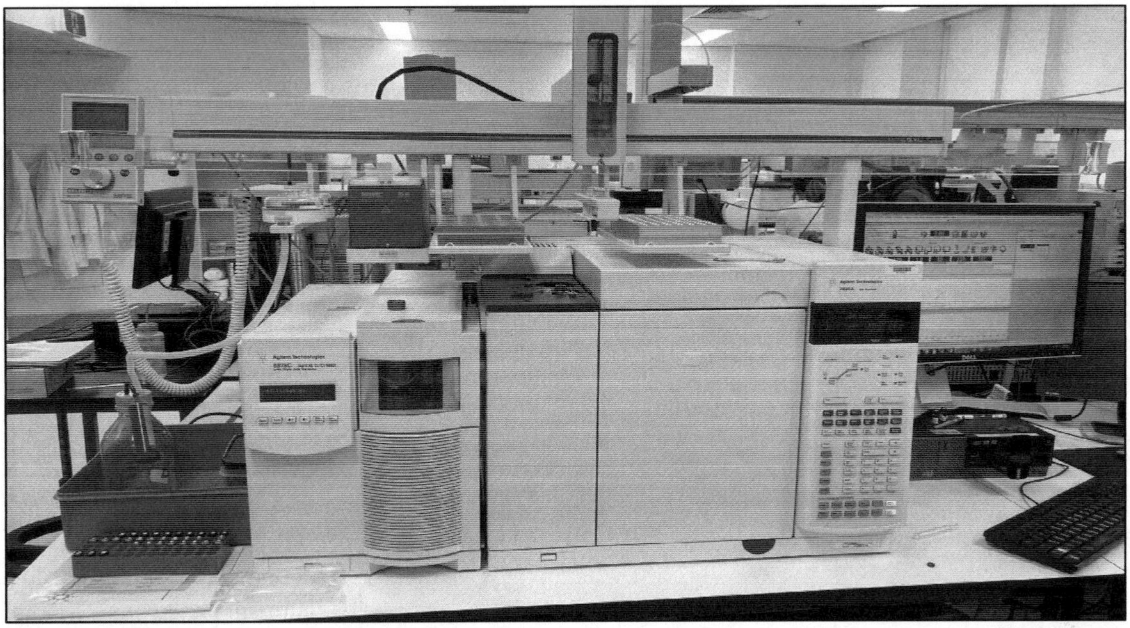

Figure 2. Single quadrupole GC/MSD instrument (Agilent Technologies, Palo Alto, CA, USA) retrofitted with a MPS 2 Gerstel Multipurpose sampler (Gerstel GmbH & Co. KG, Germany).

Briefly the HS-SPME-GC/M is an 8-step process as shown in the schematics (**Fig. 3**). the HS-SPME involves the use of a fibre that is chemically coated with either a solid (sorbent) or a liquid (polymer) adsorption phase to extract both volatile and non-volatile analytes from various liquid or gas phase media. If equilibrium is reached at ambient conditions, the amount of analyte that can be extracted by the fibre will be proportional to analyte concentration in the sample. However, to speed up the extraction time, heat and/or agitation is applied to the sample to induce faster release of the analytes. After extraction, the SPME fibre is transferred to the heated inlet of the Gas Chromatography/Mass Spectrometry for the desorption of the analytes and then the subsequent analysis.

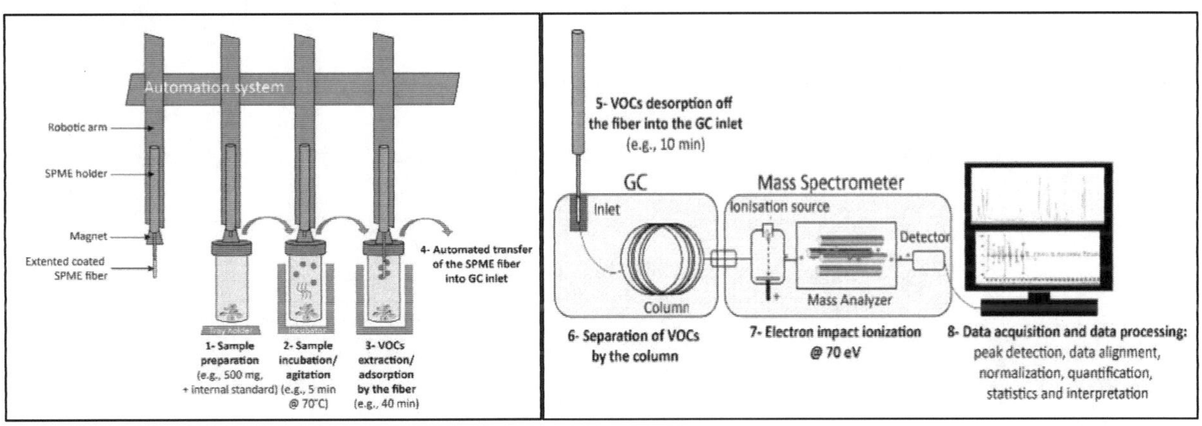

Figure 3. Schematics of HS-SPME – GC/MS technique (Julie Leroux, 2022).

The major advantage of HS-SPME is that the extraction is fast, simple, reproducible, and can often be done without solvents. For certain analytes, the limits of detection (LODs) can reach parts per trillion (ppt or ng/L) trace levels. For this research we adopted an automated HS-SPME sample handling system as shown in B of the image 2 on the right of the screen, the prepared fresh/frozen sample is placed in a sealed HS vial and arranged in the sample tray holder (**Fig. 2B**). The robotic arm of the system grabs the sample vial into the incubator/agitator holder where heat induces the plant to release the volatiles into the headspace of the vial. The SPME fiber is then inserted into the GC inlet for conditioning to release any residual contaminants for 10min at 250°C.

The analytes moving through the GC column, aided by the carrier gas, undergo an optimized temperature program to help separate the analytes according to boiling point and polarity (**Fig. 2C**, step 6). As the analytes enter the ion source of the mass spectrometer, the molecules are captured and undergo the standard GC/MS hard ionization technique known as electron impact ionization (EI). EI involves the production of free electrons from the filament at a constant 70eV (electron volts) to bombard each molecule (molecular ion) to produce characteristic fragmentation ions of low mass-to-charge ratios (m/z). The product ions and molecular ions are then converted and detected as electrical signal by the detector (usually an electron multiplier) (**Fig. 2C**, step 7).

VOC Identification

Untargeted volatilomics aims to identify the whole volatilome (both novel and known VOCs including VAs) in plant samples hence the deconvolution algorithm in the Agilent Masshunter Qulatitative and Quantitative software (**Fig. 4**) was utilized for the separation of overlapping peaks and their respective MS spectrum in the total ion chromatogram (TIC). Each peak MS spectrum is then matched against the NIST mass spectral reference library (**Fig. 5**) and Kovats non-isothermal RI matching (≥70% confidence).

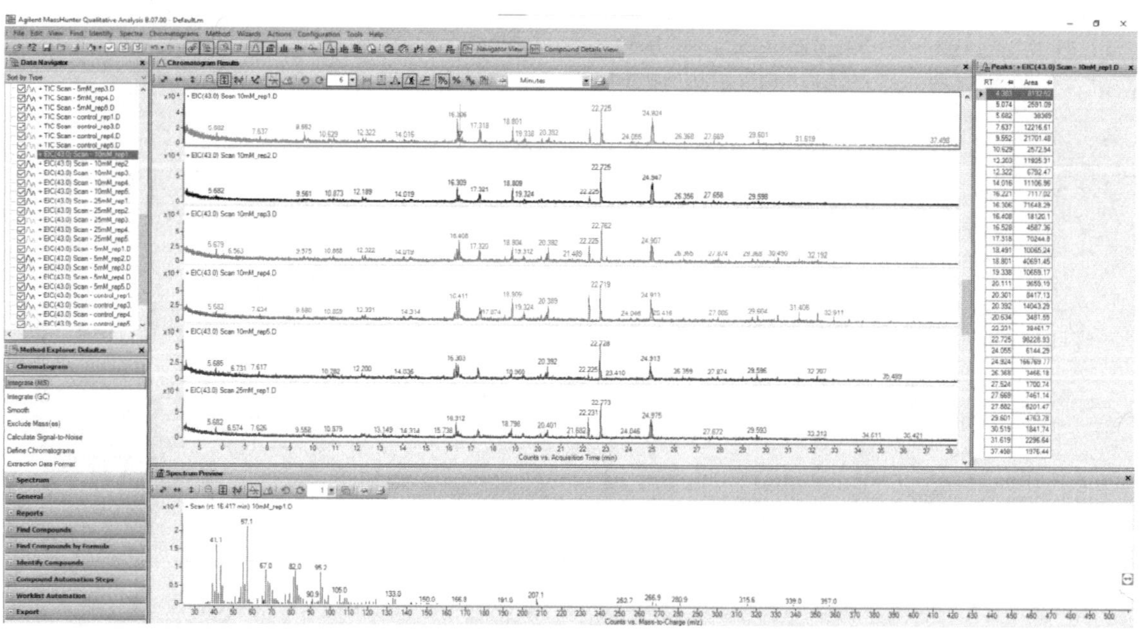

Figure 4. An example of output from Agilent MassHunter Qualitative Analysis Software.

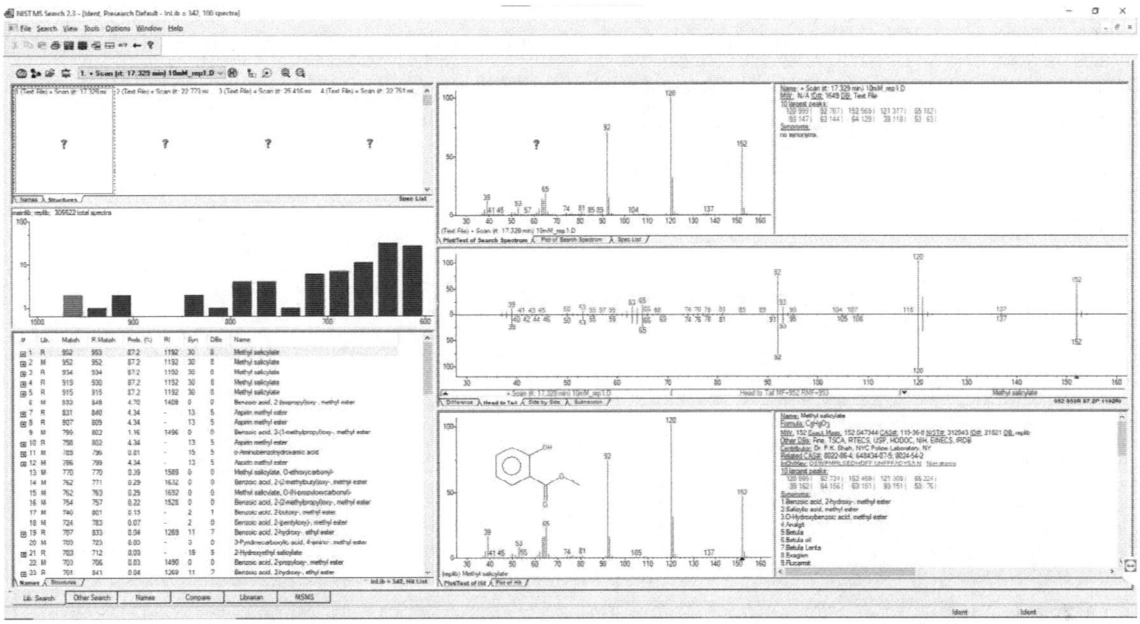

Figure 5. NIST Mass Spectral Library (version 2014).

Quality Control (QC) and Quality Assurance (QA)

QA/QC standards were used as described by Rivers et al. (2019) to maintain the consistency across sequence rounds. The randomisation of samples and standards within statistical 'blocks' were adopted to account for any systemic fluctuations during the SPMS-GC/MS sequence analysis. QC laboratory blanks (empty HS vials exposed to the laboratory environment during sample and standard preparation prior to capping) were also interspersed to check for carry-over contamination between samples, fibre bleed, and to account for other laboratory VOCs.

Statistical Analysis

The relative content of each VOCs obtained directly from GC peak areas and appear as percentage composition (Palá-Paúl et al. 2004). Agilent MSD Chemstation software (version E.02) was used for data acquisition; Agilent MassHunter software (version B.07) was used for data analysis. Statistical analyses were performed using R and Microsoft excel for ANOVA and T-test of significance. For the box plot, the stats were performed using Agricolae package in R where we ran an ANOVA to determine the presence of a significant difference among the treatments and the control followed by an LSD (Least Significant Difference) post hoc test to determine which treatments were significantly different. Identification of VOCs was based on mass spectrometry and the NIST Mass Spectral Library (version 201) was used for mass spectral matching (≥ 70% confidence) and peak annotation. Kovats nonisothermal RIs were calculated for all identified peaks using n-alkanes C9-C22 and compared against scientific literature RIs from the NIST, PubChem and Adams Essential Oils databases was used to determine the chemical composition of the VOCs (Quan & Ding, 2017).

RESULTS AND DISCUSSION

VOCs Changes with P Treatments

We plotted the relative concentrations of each compound eluted from the roots and five distinct regions differentiating the treatments can be seen in the heat map (**Fig. 6**). VOCs such as Eucalyptol, 2-Oxo-1 methyl-3-isopropylapyrazine, Benzene - acetaldehyde, Pyrazine, 2-methoxy-3-(2-methylpropyl), and Butane, 1-chloro-3-methyl has a stronger change or reduction with increasing P levels.

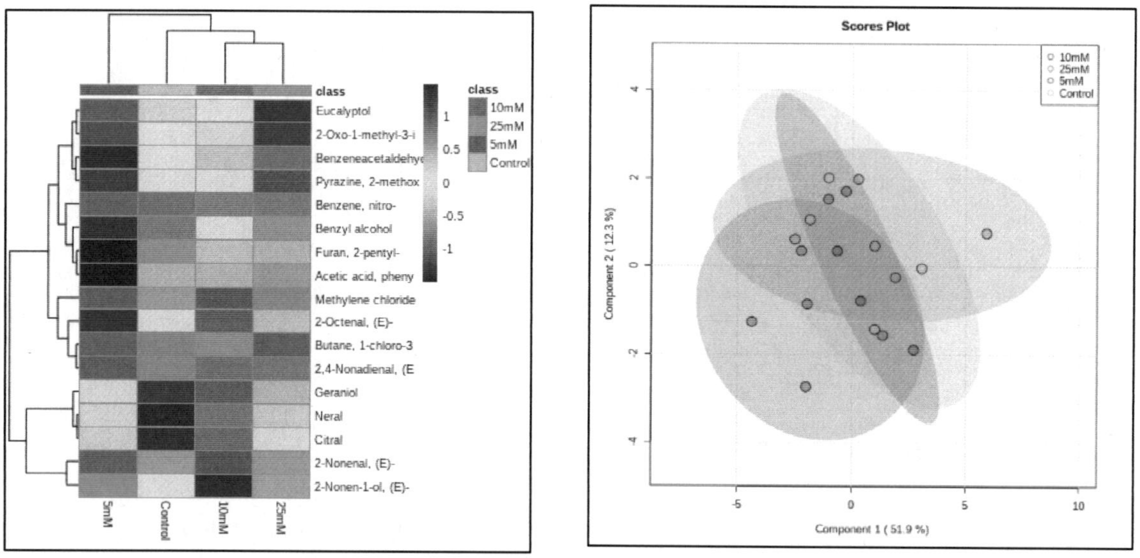

Figure 6. Clustering result shown as heatmap (distance measure using Euclidean, and clustering algorithm using ward.D) (left) and PCA Scores plot between the selected PCs. The explained variances are shown in brackets

It is noteworhty to mention that 2-Nonenal, (E)- and Methylene chloride were eluted in more abundance with increasing P levels. The PCA scores plot were plotted to help identify how the different treatments created a spread in of the VOC compound eluted from the root samples. From our study, we have identified that there are clear differences in the volatilomic profile of the treatments, but when comparing specific VOCs, there were no significant differences to be found for the duration of the treatment. As depicted in the heat map things are very different, the profile is changing however, there is no significant differences between within 4 of the specific volatiles.

Plants have a unique ability to adapt and cope with changing environmental conditions We have potentially looked at the VOCs produced by plants in the soil as a means of determining the efficiency of P fertilization and changes in plant metabolisms to adapt to different levels of availability. The understanding of signalling and communications involved in the intensive exchange of nutrients and metabolites in the rhizosphere is an added layer of information that could be deployed in precision agriculture for site-specific management of production. In this study we have used a broad and untargeted approach to identify those VOCs that have been eluted in P stress conditions and how these would induce a reorganisation of the volatilome to ensure plant survival.

Table 1.Full data for all VOCs identified, including names, Chemical Abstracting Service (CAS) number, molecular formula, molecular weight, observed and NIST Kovats non-isothermal RIs, and NIST forward and reverse matching scores.

#	Compound IUPAC Name	Trivial Name	CAS	PubChem CID	Molecular formula	MW (g/mol)	average RT (min)	Observed RI	NIST RI (Semi-standard non-polar)	NIST Forward Match	NIST Reverse Match	Quantifier ion (m/z)	Biological Class	Subclass	Reference
1	1-chloro-3-methylbutane	Butane, 1-chloro-3-methyl-	107-84-6	7893	$C_5H_{11}Cl$	106	4.63	N/A	693±4 (10)	721	752	70	Amino acid metabolism	Alkane	(Nugroho et al. 2022)
2	Methyl (z)-N-hydroxybenzenecarboximidate	Oxime-methoxy-phenyl-	67160-14-9	9602999	$C_8H_9N_2O$	151	8.40	904	N/A	804	841	133		Acid	Ramani, Krishnaveni and Shalini (2018)
2	Benzaldehyde	Benzaldehyde	100-52-7	240	C_7H_6O	106	10.25	966	962±3 (416)	872	887	106	Benzenoid/ Phenylpropanoid	Aldehyde	(Giridhar, Rajasekaran & Ravishankar 2005)
3	2-pentylfuran	Furan, 2-pentyl-	3777-69-3	19602	$C_9H_{14}O$	138	10.99	991	993±2 (179)	823	873	81	Green Leaf Volatile	Furan	(Tahir et al. 2019)
4	1,3,3-trimethyl-2-oxabicyclo[2.2.2]octane	Eucalyptol	470-82-6	2758	$C_{10}H_{18}O$	154	12.33	1035	1032±2 (580)	727	780	67	Terpenoid	Alcohol	(Sabra et al. 2018)
5	Phenylmethanol	Benzyl alcohol	100-51-6	244	C_7H_8O	108	12.48	1040	1036±4 (174)	907	924	79	Benzenoid/ Phenylpropanoid	Alcohol	(Jaitz et al. 2011)
6	2-phenylacetaldehyde	Benzeneacetaldehyde	122-78-1	998	C_8H_8O	120	12.76	1049	1045±4 (378)	696	844	91	Benzenoid/ Phenylpropanoid	Aldehyde	(Bos et al. 2002)
7	(E)-oct-2-enal	2-Octenal, (E)-	2548-87-0	5283324	$C_8H_{14}O$	126	13.14	1062	1060±3 (124)	869	910	70	Oxylipin/Fatty Acid	Alkane	(Zainol Hilmi, Idris & Mohd Azmil 2019)
8	2-Oxo-1-methyl-3-isopropylpyrazine	2-Oxo-1-methyl-3-isopropylpyrazine	78210-68-1	529437	$C_8H_{12}N_2O$	152	14.02	1090	1225±N/A (1)	808	830	137	Oxylipin/Fatty Acid	Ester	(Gu et al. 2022)
9	(E)-non-2-enal	2-Nonenal, (E)-	18829-56-6	5283335	$C_9H_{16}O$	140	16.21	1164	1162±3 (223)	827	887	41	Green Leaf Volatile	Alkane	(Petropoulos et al. 2014)
10	benzyl acetate	Acetic acid, phenylmethyl ester	140-11-4	8785	$C_9H_{10}O_2$	150	16.31	1167	1164±2 (64)	896	922	108	Benzene Derivatives	Acid	(Piesik et al. 2011)
11	(E)-non-2-en-1-ol	2-Nonen-1-ol, (E)-	31502-14-4	5364941	$C_9H_{18}O$	142	16.41	1170	1176±4 (10)	815	870	57	Green Leaf Volatile	Alcohol	(Knowles & Knowles 2012)
12	2-methoxy-3-(2-Methylpropyl)pyrazine	Pyrazine, 2-methoxy-3-(2-methylpropyl)-	24683-00-9	32594	$C_9H_{14}N_2O$	166	16.64	1178	1183±3 (41)	822	860	124	unclassified		(Ryan et al. 2005)
13	methyl 2-hydroxybenzoate	Methyl salicylate	119-36-8	4133	$C_8H_8O_3$	152	17.38	1203	1192±2 (145)	962	962	120	Benzenoid/ Phenylpropanoid	Acid	(Murungi et al. 2018)
14	(2E,4E)-nona-2,4-dienal	2,4-Nonadienal, (E,E)-	5910-87-2	5283339	$C_9H_{14}O$	138	17.87	1220	1216±4 (99)	705	916	81	Oxylipin/Fatty Acid	Alkane	(Rong et al. 2021)
15	(2E)-3,7-dimethylocta-2,6-dien-1-ol	Neral	106-24-1	637566	$C_{10}H_{18}O$	154	18.41	1239	1255±3 (343)	927	929	69	Terpenes	Alkane	(Lawo et al. 2011)
16	(2Z)-3,7-dimethylocta-2,6-dienal	Geraniol	106-26-3	643779	$C_{10}H_{16}O$	152	18.49	1242	1240±3 (168)	883	892	69	Terpenes	Alcohol	(Vasilev et al. 2014)
17	(2E)-3,7-dimethylocta-2,6-dienal	Citral	5392-40-5	638011	$C_{10}H_{16}O$	152	19.33	1272	1276±N/A (1)	878	921	69	Terpenoid	Alkane	(e Silva et al. 2021)
18	Nitrobenzene	Benzene, nitro-	98-95-3	7416	$C_6H_5NO_2$	123	14.09	1093	1080±15 (14)	934	944	77	Benzenoid/ Phenylpropanoid		(Deb, Roy & Huq 2012)

Several studies have identified the changes in the plant volatilome and its effects on AM – plant interactions under different stress conditions such as salinity and drought (Aroca et al., 2013; Rivero et al. 2018a; Ruiz-Lozano et al., 2016).

Interestingly, the plant – AM association requires a finely regulated molecular dialogue, in which strigolactone (SLs) production – derived from carotenoids are shown to be essential cues for instance SLs production are increased significantly by the in order to maintain the symbiotic association to cope with the stress condition especially in P starvation (López-Ráez et al. 2008). It is important to note that the carotenoid cleavage is a common biosynthetic reaction occurring in the plant biosynthetic pathway, including the production of important plant signalling molecules and our results have identified VOCs that are derivatives from this pathway and changing to P availability.

An interesting observation from our study was the changes we saw in the terpenes Citral, Neral and Geraniol. These compounds were in general showing a decreasing trend with increasing P availability and because there have been scientific debates for these VOCs to be either monoterpenoids or apocarotenoids in their biosynthetic classes, they could very well be apocarotenoids and have been derived from the Carotenoids pathway. This is because we are seeing a similar linear trend as other apocarotenoids such as transgeranylacetone we saw in our earlier research, regulated from the acyclic upstream carotenoids. This then could be an indication that the plants are trying to feed things through the Carotenoids biosynthetic pathways all the way to strigolactones. Its evidence their pathway could be manipulating in a way that increases the production of the strigolactones needed for the AM symbiotic association. In other words, the strigolactones pathway is being regulated in a way that will feed more carotenoids down to beta carotenes which is a precursor to strigolactones to make sure the plants are able to acquire those high levels of phosphorus nearby.

Furthermore, it seems there may be suboptimal phosphorus changes, it may be that there appears to be specific changes between 5mM and 20mM or 5mM and control treatments and once you get higher concentration you lose that change and, it returns. So, it very well could be that be there is a small steady increase of concentration of whatever P treatment we were applying. We believe the P concentration appears to be changing the pathway or manipulating metabolism in a way that once we go above 5mM it then returns and shifts to another extreme. The heat map maybe suggesting there are optimal levels of phosphorus and once you exceed those optimal levels of P, you will see a different change, this needs to be addressed in future experiment, this potential, and this hypothesis needs to be tested in future trials.

From our results, we have clearly seen that the adoption of the optimised HS-SPME GC-MS provides a high-throughput, sensitive means of identifying and quantifying VOCs and is applicable in detecting the changes in the total volatilome of tomato plants. In our study, we observed a change in the entire volatilome of the plants under different treatments although no significant differences were observed within the specific VOCs identified. We believe that this in part, could be due to the age of the tissues in our experiment. McQuinn and Leroux et al (unpublished) have reported an elevated levels of volatiles and a much more dynamic profile in developing buds versus mature or fully developed tissues. Therefore, we believe future studies would be improved by observing the changes in younger tissues and collected at different growth stages.

CONCLUSION

As P is a limited resource the ability to use VOCs as a component to understand the interaction in the rhizosphere whilst providing important information on changes in the biosynthetic pathway that may enhance or decrease the signalling and communication systems. We have seen that the adoption of a fast, sensitive, and high throughput technique such as the optimised HS-SPME GC-MS technique has provided with an insight of the manipulation of volatilomic pathways by plants in adapting to changes in their environment. This is the first time the technique has been adopted for Tomato roots volatilomic studies, but we believe it is applicable to other tissues and species. In summary, we have seen that the plant is making sacrifices to improve the required signals for AM symbiotic associations in limited P conditions.

Acknowledgements

I acknowledge the support of the following individuals and organizations for their invaluable support during my research:
Garden City Plastics, my esteemed employer, for their unwavering support and encouragement throughout the research period.

Dr. Ryan McQuinn and Dr. Mark Williams, my project supervisors from Western Sydney University, for their expert guidance, insightful suggestions, and continuous mentorship.
I am grateful to Dr. Julie Leroux and Prof. Barry J. Pogson's laboratory at the Australian Research Council Centre of Excellence in Plant Energy Biology, Australian National University, Canberra. The mass spectrometry experiments were conducted at the Australian National University RSB/RSC Joint Mass Spectrometry Facility (JMSF). I extend my gratitude to Adam J. Carroll and the entire JMSF team for their valuable assistance and support.
Finally, I am deeply indebted to my family for their unconditional love, unwavering care, and understanding during this research endeavor, especially for accommodating the extra hours dedicated to my work. Their support has been instrumental in the successful completion of this research project, and I am truly grateful for their contributions.

LITERATURE CITED

Dudareva, N., Negre F., Nagegowda D.A., and Orlova, I. (2006). Plant volatiles: recent advances and future perspectives. Critical Reviews in Plant Sciences *25*: 417-40.

Kovar, J.L. and Claassen, N. (2005) Soil-root interactions and phosphorus nutrition of plants, phosphorus. Agriculture and the Environment *46*: 379-414.

Lee Díaz, A.S.,. Rizaludin M.S, Zweers, H., Raaijmakers, J.M. and Garbeva, P. 2022, Exploring the volatiles released from roots of wild and domesticated tomato plants under insect attack. Molecules *27*:1612.

Leroux, J., Truong, T.T., Pogson, B.J. and McQuinn, R.P. (2022), Chapter Ten - Detection and analysis of novel and known plant volatile apocarotenoids. In E.T. Wurtzel (ed.), Methods in Enzymology vol. 670, Academic Press, pp. 311-68.

Massalha, H., Korenblum, E., Tholl, D. and Aharoni, A. (2017). Small molecules below-ground: the role of specialized metabolites in the rhizosphere. Wiley Online Library.

Rivers, J.Y., Truong, T.T., Pogson, B.J. and McQuinn, R.P. (2019). Volatile apocarotenoid discovery and quantification in Arabidopsis thaliana: optimized sensitive analysis via HS-SPME-GC/MS. Metabolomics *15*:1-13.

Stierlin, É., Nicolè, F., Fernandez, X. and Michel, T. (2019). Development of a headspace solid-phase microextraction gas chromatography-mass spectrometry method to study volatile organic compounds (VOCs) emitted by lavender roots. Chemistry and Biodiversity *16*:e1900280.

Wenke, K., Kai, M. and Piechulla, B. (2010). Belowground volatiles facilitate interactions between plant roots and soil organisms. Planta *231*:499-506.
DOI 10.1007/s00425-009-1076-2.

White, RA, Rivas-Ubach, A, Borkum, MI, Köberl, M, Bilbao, A, Colby, SM et al. 2017, 'The state of rhizospheric science in the era of multi-omics: A practical guide to omics technologies', Rhizosphere *3*:212-21. DOI https://doi.org/10.1016/j.rhisph.2017.05.003, <https://www.sciencedirect.com/science/article/pii/S2452219817300666>

Combined Proceedings IPPS 73: 84-88. 2023.

Provenance Propagation Methodology of Perth Bushland Species from Seed

David Hancock

Natural Area Nursery, Natural Area Consulting Management Services57 Boulder Road, Malaga WA 6090, Australia

david@naturalarea.com.au

Keywords: natural areas, native plants, germination, dormancy

Summary

The Western Australian region near Perth is a unique bushland. There are numerous rare plant species in the area that deserve conservation. Seed viability is impacted by collection techniques. Development of germination strategies for these must be developed to complement conservation programs. This paper provides examples of this work over the past 20 years at Natural Area Nursery.

INTRODUCTION

The Natural Area native plant nursery is accredited by the Nursery & Garden Industry Australia (NGIA) with a reputation as the pre-eminent producer of plants for restoration, revegetation and landscape for Perth and WA regions. At our Perth native nursery we tailor plant stock to individual

IPPS Vol. 73 – 2023

customer and project needs and welcome the most challenging of propagation by species, provenance and size. Natural Area's capability to combine seed and propagule collection and plant supply in-house is unsurpassed in the market. Our in-house botanical capability compliments our propagation abilities to produce difficult and recalcitrant species. The nursery has twice been awarded 'Best Small Production Nursery' by the Nursery and Garden Industry Australia and has been a finalist in the Western Australia Water Awards for the water retrieval and recycling systems in use.

How Do We Perceive Provenance?

The customers of Natural Area nursery, including our own internal projects, generally require stock of seed and plants to be specific as to provenance. The customer definition of provenance is not consistent. It varies from descriptions such as "Perth region", or within 50 km of the city and down to individual Local Government Areas and in some cases, bush reserve by bush reserve only - a few KM apart. Therefore, we have adopted a policy of all collections and propagation being recorded by individual locations.

We collect from a very high number of locations and therefore when customers seek stock relative to their provenance requirement, we can very often meet that need or action our collectors to do so. Having a good-sized licensed seed collection team places us well to maintain good stocks and get what is needed when we need it. It also means that we come across the opportunist collections and the unexpected and unusual seed set that enhances our propagation.

Our experience over the last 20 plus years has taught us that viability of seed can vary significantly based upon location. We have developed a very substantial database of seed propagation outcomes based upon collection sites, timing of collection, seed processing techniques and actions to achieve germination. We do not delve into the true science of viability variance as we are focused on the propagation outcome. If we know where to find viable *Gahnia*, *Machaerina*, *Conospermum* & *Lepidosperma* etc. that is all we want to know. It is quite surprising that seed from locations a few kilometers apart have very different outcomes. It is telling also that development and fragmentation of bushland has meant presentation of some species is restricted to a handful of locations. There are species that we have only been able to find two locations that present across 100 km of the Perth plain, and we do a lot of searching. Some are now poorly represented.

The importance of provenance specification to managers and restorers of Perth's unique bushland is important also because of the proven genetic variances that exist. Some of the DNA evidence suggests that similar species have evolved in complete isolation to each other and that variations across genus and species are common for bushland areas that are a short distance apart.

Seed Germination Strategies

The development of germination techniques has been a major focus for our propagators, many of which have been with us for over 10 years. The propagators have been obsessed with testing and recording best seed source, collection timings and germination triggers. They have been encouraged to research the literature, experiment, and use radical treatments if necessary to achieve the desired outcomes. With so many species in demand for quality restoration work in our market, there is huge potential to further develop our intellectual property in propagation. I am happy to provide with more information of our various methodologies upon request, but a few examples;

- The seasons. Some like autumn, some winter, some spring i.e., don't sow everything in one season
- Extensive variations in use of smoked water, i.e., concentrations and duration timings
- Physical smoke, various source material, repetitions and timings
- Temperature control
- Heating and cooling cycles
- Deprivation of light
- Various acid treatments
- Hydrogen peroxide, again various concentrations / timings.
- Weathering and second seasons resowing.

Weathering and second season germination including details on the role of mycorrhiza in propagation and the role of seed germination from 'difficult species' in conservation are outlined in our recent publication (Hancock, 2023). Second season germination recently yielded over 800 plants for us in one species alone. A major success factor can be a combination of various treatments. All this makes for fascinating, challenging and rewarding work and it's a celebration when success is achieved, and we are driven to pursue the other species on our target list.

A Snapshot of Success

To answer the obvious question, "cut the talk and show me the plants'', I have put together a suite of photos of species that not so many years ago, I thought we could never grow, or if so, in any reasonable numbers. There are many others we could cite but the proof is in the demand from our customers, now at 1.2 million and growing, spread over around 350 species.

Myoporum insulare (Scrophulariaceae) has many common names such as Boobialla, Water Bush, Native Mangrove and Blueberry Tree and is distributed along the coastal areas from Shark Bay in Western Australia to north-eastern New South Wales and coastal Tasmania. It is an excellent windbreak in coastal areas growing to 1 – 6 m in height (**Fig. 1a**).

Aotus grassilima (Fabaceae) is native to Western Australia and occurs in swampy areas on the coastal plain from Gingin south to Manjimup. The yellow and red flowers are produced in October and November (**Fig. 1b**).

Alyxia buxifolia (Apocynaceae) commonly known as sea bush or dysentery bush is a spreading, woody native shrub growing to 2 m height with thick, glossy green leaves (**Fig. 1c**). Waxy white flowers appear in Spring, followed by red berries. It is ideal for Coastal positions, also tolerant of shade, growing in Eucalypt forests. It is useful as a screening plant and is mass planted in garden beds. It is widely adapted to different soil types and from dry to wetland areas. Use of the bark by first nation people of Australia for treatment of dysentery gave this plant the name dysentery bush. It is an ideal refuge for many insects, lizards and native birds.

Acanthocarpus preissii (Asparagaceae) is a rhizomatous perennial occurring on coastal dunes in Western Australia (**Fig. 1d**). Commonly known as prickly lily, its white flowers appear between April and May, hosting some rare butterflies including silver-spotted ochre (*Trapezites argenteoornatus*).

Figure 1. Some of the diverse species grown from seeds at the Natural Area Nursery: (a) *Myoporum insulare* (Scrophulariaceae), (b) *Aotus grassilima* (Fabaceae), (c) *Alyxia buxifolia* (Apocynaceae) and (d) *Acanthocarpus preissii* (Asparagaceae).

Figure 2. Some of the diverse species grown from seeds at the Natural Area Nursery: (a) *Adansonia gregorii*, (b) *Brachyloma preissii*, (c) *Calytrix fraserii* and (d) *Conospermum triplinervium.*

Figure 3. Some of the species growing at the Natural Area Nursery, among approximately 350 spp., raised from seeds and ready for planting: (a) *Exocarpus spartius*, (b) *Gahnia trifida*, (c) *Lomandra maritima* and (d) *Hibbertia hypericoides*.

Figure 4. Species from diverse families such as Zamiaceae (*Macrozamia fraserii & M. reidlieii*) (a), Loranthaceae (*Nuytsia floribunda*) (b), Myrtaceae (*Scholtsia involcrata*) (c), Goodeniaceae (*Scaevola crassifolial*)(d) etc. are represented in the diverse spectrum of native plants, particularly adapted to Western Australia, propagated and sold by Natural Area Nursery.

Acknowledgement

Thanks and congratulations to the team at Natural Area Nursery for their dedication and successes.

LITERATURE CITED

Hancock, David (2023) Achieving success with difficult species. The Propagator, International Plant Propagators' Society, Australian Region Newsletter 74:2-4.

Combined Proceedings IPPS 73:89-95. 2023.

Propagating for Farms

Puthiyaparambil Josekutty

Research Manager, Skybury farms, Paddy's Green, QLD 4880, Australia

www.skybury.com.au; josekutty964@gmail.com

Keywords: tissue culture, micropropagation, biosecurity, quarantine, banana, avocado, papaya

Summary

Propagating for farms is to assist with improved crop production. Therefore, several factors ranging from the genetics to quality control have to be carefully considered while propagating for farms. As a veteran tissue culturist and from a decade of tissue culture production and management experiences as a commercial plant propagation Manger in Australia, I shall discuss some interesting experiences and lessons learned at the Yuruga Native Plants Nursery, Walkamin, Queensland (QLD); Fleming's Nursery, Australia and Skybury Farms, Paddy's Green, QLD for the benefit of the wider plant propagator community. I shall discuss some important aspects like the genetics, genetic improvement in vitro, fidelity, disease and pest control etc. in relation to commercial propagation of plants for farms with Banana, Avocado and Papaya as examples. Additionally, the usefulness of in vitro culture systems for rapid crop improvement, plant production, assisting with quarantine compliance etc. are discussed.

IPPS Vol. 73 – 2023

INTRODUCTION

Plants are propagated from seeds and cuttings for Millenia, ever since man ventured into agriculture/ horticulture. Techniques and tools were developed by ancient household cultivators, farmers, horticulturists and scientists over the years. Plant tissue culture is one of the revolutionary tools for propagation based on the totipotency of plant cells described by Haberlandt in 1902. The isolation and characterisation of auxins (Went and Thimann, 1937), gibberellins (Yabuta and Sumiki,1938) and cytokinins (Miller et al. 1955) assisted the rapid progress with plant propagation (Skoog and Miller, 1957). Murashige has comprehensively reviewed plant tissue culture and its applications to plant production (Murashige, 1974).

Discovery of somatic embryogenesis (Steward et al. 1958), the unique ability of plants to regenerate whole plants from a somatic cell, allows rapid, reliable cloning of some plants that are recalcitrant to other cloning methods as well as rapid crop improvement through selection and genetic engineering (Pierroz, 2022). Bergman (1977) reviewed cell cultures and its multiple applications like cell line selection for crop improvement and production of secondary metabolites.

Cocking (1960) reported enzymatic isolation of protoplast and discussed potential use of protoplasts. Takabe et al. (1971), for the first time demonstrated regeneration of whole tobacco plants from isolated protoplasts. Further rapid advancements in protoplast culture by Cocking's laboratory revolutionised this technique as detailed by Cocking et al. (1972). Further advances with protoplast culture and its applications came about in different parts of the world (Cocking, 2000). Protoplast culture technology was reviewed recently (Ranaware et al. 2023), due to its contemporary significance to the novel gene editing technology.

Although, plants can be rapidly cloned in large numbers in vitro using some or all of the above-mentioned techniques, propagating for farms requires additional considerations. Some of these considerations (e. g. genetic fidelity, uniformity, freedom from diseases etc) are highlighted in this article. Propagating for farms is also positively regulated in Australia for good reasons as I point out with specific examples in this paper.

CLONING STRATEGIES FOR FARMS

Of all the plant propagation methods, vegetative propagation ex vitro (using stem cuttings) is the most efficient method when the plant species concerned is amenable to this method of propagation. This is because such cuttings yield true to type plants (clones) demonstrating the characteristics of the mother plant (fidelity) if they are not affected by different growing environments, in other words not subjected to the impacts of G x E interactions. When the cutting production is efficient, it is also the most economical method of large-scale propagation as in the case of propagating ornamentals like *Begonia* sp., *Hibiscus* sp., *Hydrangea* sp. etc.

Banana Micropropagation

Micro-propagation (cloning of plants in vitro from shoot tip/ nodal segments) is a rapid, reliable method of cloning that is widely used in the farming sector because of the efficiency, genetic fidelity, and freedom from diseases ensured by this process when the facility and operating procedures are properly managed. A good example is cloning of bananas (**Fig. 1** and **2**). Most of the banana farms in Australia use tissue cultured banana plants produced according to strict regulations of the quality approved banana nursery program (QBAN) - (https://abgc.org.au/qban/) to ensure quality standards, especially freedom from dreaded virus disease Bunchy top of banana caused by *Banana bunchy top virus* (BBTV) and the fungal disease (Panama wilt) caused by *Fusarium oxysporum* f. sp. *cubense* race 4. Main features of effectively managed QBAN includes NIASA accreditation (https://nurseryproductionfms.com.au/niasa-accreditation/), training, independent, disease indexing of the mother stock by Grow help Australia, as well as best practices in the grow on nursery, regularly monitored by the banana industry - the Australian Banana Growers Council (ABGC).

Figure 1. Stages in banana micropropagation. (a) Initiation from corm, after surface sterilisation and inoculation, (b) 10 days after transfer to proliferation medium, (c) One month after transfer to proliferation medium, (d) Shoot regeneration.

Figure 2. (a) In vitro rooted banana plants and (b) plants ready for acclimation in the greenhouse.

Avocado Micropropagation

Accreditation by the Avocado nursery voluntary accreditation scheme (ANVAS) – (https://avocado.org.au/wp-content/uploads/2019/01/ANVAS-Terms-and-Conditions-V3.pdf), is a scheme developed by Greenlife industry (https://www.greenlifeindustry.com.au/about/industry/greenlife-industry), that regulates avocado nurseries and practices in Australia to ensure that the best possible avocado plants are supplied to the Industry. Although not a common practice like banana, recently developed avocado micropropagation methods (Josekutty, 2019) can revolutionise quality avocado plantlet supply to farms (**Fig. 3**). Additionally, the disease-free avocado plants micropropagated in vitro, could be a choice for exporting planting materials across the quarantine borders.

Cell Culture and Somatic Embryogenesis for rapid genetic improvement of Papaya

Another important aspect of propagating for farms is using appropriate propagation technology to generate genetically improved crops as Skybury farms did in the recent years. One school of thought is that somatic embryos are of single cell origin as demonstrated by (Nagamani et al., 1987). According to the principles of somatic cell genetics, it is possible to grow and screen millions of plant cells (potential plants as they are totipotent) in the tiny space of a 100 ml culture flask in the laboratory (Josekutty, 1998). A high-throughput somatic embryogenic system can then regenerate thousands of variant plants from the micro-colonies originating from individual cells in vitro.

We used this power of plant cell culture to generate new lines of highly productive Skybury, Sweet, Red Papaya lines (Josekutty, 2022; Puthiyaparambil et al. 2023) (**Fig. 4**). Development of such advanced papaya lines enabled Skybury farms to rapidly enhance consumer preference and papaya market capture although papaya is a relatively less preferred fruit in Australia, largely due to the ignorance about its nutritional superiority and additional health benefits compared to other common sugary fruits like banana and grapes.

Figure 3. Avocado micropropagation for farms. (a) Initiation, (b) Shoot regeneration, (c) Shoot proliferation, (d) Rooted plants ready for acclimation.

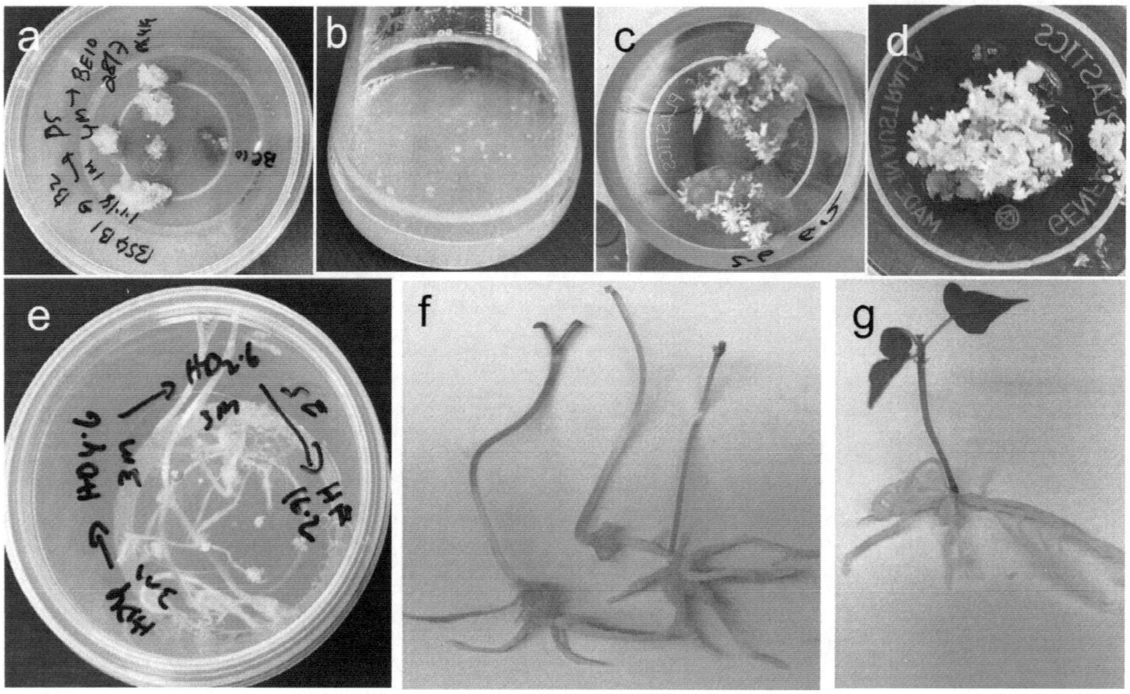

Figure 4. Papaya crop improvement through cell line selection. (a) Initiation of callus, (b) Cell cultures established from callus, treated with viricides, (c) Somatic embryo induction from callus, (d) Advanced somatic embryos ready for plant regeneration, (e-f) Different stages of plant regeneration from somatic embryos.

CONCLUSIONS

There are several methods for commercial propagation of plants for home gardens, conservation activities or commercial farming. Several methods for in vitro cloning of plants have been developed and micropropagation technique that revolutionised the propagation industry remains the preferred mode of propagation for many reasons such as:

(i) the ability to propagate in a mass scale
(ii) ability to produce plants throughout the year with no seasonal effects
(iii) production of high-health, disease free plants
(iv) ability to produce many plants within a limited time and space
(v) minimize requirements for quarantine compliance when sending plants across borders. Propagating for commercial farming requires several careful considerations and it is well regulated in Australia by the respective industries for valid reasons.

LITERATURE CITED

Cocking, E. C. (1960). A method for the isolation of plant protoplasts and vacuoles. Nature *187*: 962-963.

Cocking, E.C. (2000). Turning point article plant protoplasts. In Vitro Cell Dev Biol-Plant *36*, 77–82. https://doi.org/10.1007/s11627-000-0018-2

Bergmann, L. (1977). Plating of plant cells. In: Barz, W., Reinhard, E., Zenk, M.H. (eds) Plant Tissue Culture and Its Bio-technological Application. Proceedings in Life Sciences. Springer, Berlin, Heidelberg. https://doi.org/10.1007/978-3-642-66646-9_18

Haberlandt, G. (1902). Culturversuche mit isolierten Pflanzenzellen. Sitz-Ber. Mat.-Nat.KI. Kais. Akad. Wiss. Wien, *111*(1): 69–92.

Josekutty, P. C. (1998). Selection and characterisation of variant *Solanum xanthocarpum* that overproduce steroids. Phyton, 62 (1/2), 125-130.

Josekutty, P. (2019). Micropropagation of avocado (*Percea americana*). Proceedings of the IPPS – Australia conference 2019, Sunshine coast, Australia.

Josekutty, P. (2022). Accelerated In Vitro Breeding Creates Improved, Designer Papaya, Proceedings of the IPPS-Australia conference 2022, Blue mountains, NSW.

Puthiyaparambil, J. C. Pagie, M., Teressita, S., Jay, P.M., Bongani, N., Paul, F., Candy, M., Mark, M., Marion, M., Ian, M., Sanskruti, L. and Nitin, M. (2023). Rapid papaya crop improvement through accelerated in vitro breeding and molecular diagnostics. Acta Hortic. *1362*: 499-506. DOI 10.17660/ActaHortic.2023.1362.67

Miller, C.O. et al. (1955). Kinetin, a cell division factor from deoxyribonucleic acid. J. Am. Chem. Soc. 77**,** 1329–1334.

Nagata, T and Takabe, I. (1971). Planting of isolated tobacco mesophyll protoplasts on agar medium. Planta, 99: 12-20.

Nagmani et al. (1987). Single-cell origin and development of somatic embryos in *Picea abies* (L.) Karst. (Norway spruce) and *P. glauca* (Moench) Voss (white spruce). Plant Cell Rep., 6(2):157-159. doi: 10.1007/BF00276677.

Murashige, T. (1974). Plant propagation through tissue culture. Annual Review of Plant Physiology, 25:135-166.

Pierroz, G (2022). Making babies: how auxin regulates somatic embryogenesis in Arabidopsis tissue culture. The Plant Journal (2023) 113, 5–6. https://doi.org/10.1111/tpj.16024

Ranaware, A.S. et al. (2023). Protoplast technology and somatic hybridisation in the family Apiaceae. Plants, *12*: 1060. https://doi.org/10.3390/plants12051060

Skoog, F. and Miller, C.O. (1957) Chemical regulation of growth and organ formation in plant tissue cultured in vitro. Symposia of the Society for Experimental Biology, *11*: 118-131.

Takebe, I et al. (1971). Regeneration of whole plants from isolated mesophyll protoplasts of tobacco. Naturwissen. *58*:318-320.

Yabuta, T and Sumiki, T (1938). Communication to the editor. J. Agric. Chem. Soc. Japan. *14*:1526.

Combined Proceedings IPPS 73: 96-105. 2023.

The Importance of Plants in the Landscape

Andrew Laidlaw

Royal Botanic Gardens Victoria, Birdwood Avenue, Melbourne, Victoria 3004, Australia

andrew.laidlaw@rbg.vic.gov.au

Keywords: Royal Botanic Gardens, landscape design, garden design

Summary

Gardens can connect a diversity of people to sense of place. This paper describes several design projects from the Royal Botanic Gardens in Victoria. They include a children's garden, the Guilfoyle's Volcano, the Fern Gully, the Arid garden, White oak area, the Sensory garden and other projects.

IPPS Vol. 73 – 2023

INTRODUCTION

Plants have a connection to health and well-being related to humanitarian work. These include aspects related to:

- United Nations now see built green space as part of the matrix to rebuilding community Goal 11.7: "By 2030, provide universal access to safe, inclusive and accessible, green and public spaces, in particular for women and children, older persons and persons with disabilities" (UN 2020)
- Nature-assisted therapy (gardens to natural areas) have been found to be effective for managing PTSD (Kreski,2016; Poulsen et al, 2016)
- Parks and gardens are renowned for social interaction – with cohesive forces supporting individuals/community resilience (Seltenrich 2015)
- Children with access to safe, stimulating (unstructured) environments have greater capacity for healthy physical and psychosocial development (Moore, 2014)

Design Work

My main work is and has been designing gardens in the Royal Botanic Gardens (RBG) Melbourne and for Global Gardens of Peace over the last 25 years. In both these organisations it is always a team approach and requires the design to be built through a strong design process that starts with a consultative approach with all the main stakeholders and responds to the prevailing environmental conditions. Plants are a key component to all my work, plants bring life, change and seasonality and connection to nature. Every plant life starts with propagation. Propagators are the quiet achievers and often remain invisible to the process. Their ability to grow interesting and healthy plants is the starting point to all my work.

Dermot's passion for his work and his skill in the RBG has made a huge difference to my work and the work of many horticulturalists.

Project and Plant Selection

We take the business of plants and projects very seriously at RBG. We have a series of planning documents that guide our process.

- Landscape Master Plan 2020-2040
- Landscape succession
- Living collections document

Plant selection

- We are transitioning our landscape to be more climate resilient.
- We are actively changing the plant palette
- we use plants that are suitable for the prevailing conditions
- Always looking at new and interesting plants that preform and are non-invasive.
- We propagate over 90% of all our plants.
- We try and collect wild species with known providence
- We try and conserve rare and threaten plants
- Everything is documented

Royal Botanic Gardens Victoria Landscape Projects

TIPCG (Kids Garden RBG) 2004

A garden designed to connect children to nature through their play, a garden where children can find secret places, and are free to run and create.

The garden is made up of a series of plant rooms; Bamboo Forest, Rainforest Garden, Snow Gum Gorge, Grass Maze etc, in all these gardens plants do all the work (**Fig. 1**).

Figure 1. Kids garden at the Royal Botanic Gardens Victoria in Melbourne

Guilfoyle's Volcano 2010

A garden designed to exhibit succulent and arid plants in a dramatic design that reinvents Guilfoyle's original Volcano theme. A series of bold foliage plants that are repeated around the main cone creating a bold statement (**Fig. 2**).

Figure 2. A garden designed to exhibit succulent and arid plants in a dramatic design at Melbourne's Royal Botanic Gardens with a theme of volcano: Before (top left), after (top right), display of colour on the banks of the volcano (bottom).

Fern Gully 2016-2020

A garden designed to provide cooling and solace where the temperature can drop by 5 degrees. Visitors can easily move through the space on an elevated steel boardwalk and discover quiet meditative spaces that allows them to be wrapped by plants and connect to nature (Fig. 3).

Figure 3. Fern gully at the Royal Botanic Gardens Victoria provides visitors with space for solace and meditation in a cooling environment.

Arid Garden 2020

A garden designed as a parterre garden for visitors to wander and explore the amazing arid plants and to learn about the story of the Field collection. This garden uses native wildflowers to bring colour and drama in late winter (**Fig. 4**).

Figure 4. Arid garden in Royal Botanic Gardens Victoria was designed as a parterre garden. Top – general view of the garden, Bottom – stunning display of native Australian arid plants

White Oak 2021

A garden designed to allow a 165-year-old oak tree to continue to give as a meeting place after it fell apart in 2020. With the limbs arranged around the main trunk as they fell, the new garden is providing a new meeting place (**Fig. 5**).

Figure 5. The fate of the White Oak at Royal Botanic Gardens Victoria; this 165-year-old oak tree was given a second lease of life after it collapsed in 2020, as a meeting place for the visitors

Sensory Garden 2021 - A garden designed to provide a series of horticultural sensory experiences (**Fig 6**).

Figure 6. Sensory garden provides visitors an opportunity to stimulate all the five senses: sight, smell, touch, sound and taste. Through this they become more aware of the surroundings, leading to mindfulness

Drylands 2023

A garden that looks at a new palette of Australian plants that have been selected for their climate match and their ability to adapt to their environment.

Calming Garden in Alfred hospital 2020

Two small gardens created in the physic ward at the Alfred hospital, that allows patients access to green spaces.

Gaza Garden in Palestine 2013-2023

A children's garden in the Gaza strip, a place for children and their families to experience some joy.

Ventilation Accommodation Support Service (VASS) garden in Thornbury 2023

A food forest garden for a vulnerable community in the inner-north suburb of Thornbury in Melbourne, where the clients will be able to sit out in the garden with their family.

Figure 7. Species and features used in Ventilation Accommodation Support Service (VASS) garden in Thornbury, a Melbourne suburb.

Acknowledgements

For the Wadawurrung people: Connecting to the deeper history of place is all our responsibility.

LITERATURE CITED

Kreski, B. (2016). Healing and empowering veterans in a botanic garden. Journal of Museum Education, 41(2), pp.110-115.

Moore, R.. (2014). Nature play and learning places. Creating and managing places where children engage with nature. Raleigh, NC: Natural Learning Initiative and Reston, VA: National Wildlife Federation.Poulsen, D.V., Stigsdotter, U.K., Djernis, D., Sidenius, U. (2016) 'Everything just seems much more right in nature': How veterans with post-traumatic stress disorder experience nature-based activities in a forest therapy garden. Health Psychology Open 3(1). doi:10.1177/2055102916637090

Seltenrich, N. (2015). Just what the doctor ordered: Using parks to improve children's health, Environment Health Perspectives, 1 23 (10). https://doi.org/10.1289/ehp.123-A254

United Nations. (2020) United Nations Sustainable Development Goals. Available at: https://sdgs.un.org/goals/goal11 (Accessed 1st August 2023).

Combined Proceedings IPPS 73:106-107. 2023.

Plant Retailing – From Then to Now

Clive Larkman

Larkman Nurseries, 7 Jurat Rd, Lilydale VIC 3140, Australia

clive@larkmannurseries.com.au

Keywords: garden centers, life style centers, on-line purchases

Summary

Major developments occurred to alter the way and reason why plants were grown and sold. Fresh vegetables and herbs became more easily available through development in suburban farming including greenhouse expansion, breeding but most importantly the evolution of large food retailers. The ornamental value became as important as the food value of the variety. Gardening became a serious pastime and gardeners wanted 'new', 'bigger', 'longer flowering' plants that are tolerant of diseases, pests and other biotic stress factors. This paper describes changes in retail garden customers.

INTRODUCTION

Plants enable man to change from nomadic hunter/gatherer to settled community. The original cultivated plants were either food or medicinal as these were the key to the survival of fixed communities. Plants were grown for use in the home and this remained the case for thousands of years.

Retail Customer Over time the average retail customer changed from the middle-aged male gardener to the 55+ female gardener. She chose the plants, and he carried them, planted them and maintained them. The lines were well defined and plant retailing was targeted at elder women. In some countries they also became throw away items and once the flower or perfume ceased, they were dumped.

Garden Centres and Lifestyle Centers

Two trends brought about the 'garden center' or 'lifestyle' center'. In the early years nurseries in countries with extreme winters virtually shut down when the cold weather hit. However, a balanced income was needed, and it was apparent the nursery has to bring in many people who wanted to spend time looking at the plants. They turned into retailers who sell everything to do with life and living. The obvious items on sale were ponds, pets, tools, furniture and even clothing as well as not so obvious items like jewellery, soaps and perfumes.

To keep people there and returning again and again, many nurseries started to add cafes, some of which soon became destination restaurants. This resulted in a huge shift to the way gardens were viewed. Late in the century gardens became a key part of the house and the plant palette as the house and garden became one. People wanted greater variety of food than available in the shops and there developed a desire for 'cleaner/greener' plants. Thus evolved the desire for edimental plants – ornamental plants that are also edible. The palette moved back to edibles and younger people took an interest in their gardens. As a result, the typical customer changed to a 35- to 45-year-old couple. They looked at each purchase as a major issue that had to look good, have an edible use and they wanted the 'story'.

Development History of Food and Flora Retailers

The desire for edible plants sent a whole new generation of people into the nurseries and garden centers. Western culture also developed a desire to go out to eat all meals or just sit and have a coffee or wine. Two trends that gave plant retailing a boost and many restyled their appearance.

Online Purchases of Plants

Online retailing was seen as the devil that would end the existence of retail nurseries. Fortunately, online purchases of plants don't appeal to many gardeners and online plants helped bring even more people into nurseries. Recently the new generation of 'millennials' has seen an influx of quite young people into growing plants. They have a passion for trendy new (or old) indoor plants. They are computer savvy and are well researched on what species and variety they want; how rare it is and how to grow it.

From the first days of settlement the key to a healthier society has been the ability to grow and farm plants. Whether the plants are grown as source of food and medicine, as a way to beautify the home, impress friends or fill a need to collect and own things that others don't, they need to be propagated. This could be through conventional seed sowing, modern asexual forms like cutting, grafting or division or through hi-tech micropropagation.

Growing plants is the key to a happy and healthy society. Whatever the reason it shows that plants and their propagation is essential to life.

Combined Proceedings IPPS 73: 108-112. 2023.

Southern African Region Student Exchange

Thandisizwe Siphenkosi Ndabeni

121 Iligwa Street, Mfuleni, Cape Town, South Africa.

nthandsizwesiphenkosi@gmail.com / sizwe@shadowlands.co.za

Keywords: Australia exchange, nursery, cultural, Western Australia

Summary

The 16th of May 2023 felt like a dream after a very long wait for my trip to attend the IPPS conference in Melbourne as an exchange student. I won the exchange program in March 2020, and I was supposed to leave in May 2020 but due to covid19 I was unable to go. Finally, early this year I was able to go. It was an amazing trip, full of joy and excitement as it was my 1st time leaving the country. I was hoping to gain a lot from this trip, and I sure did. I spent a week in Perth, a week in Melbourne and a week in Adelaide. I did a lot of nursery visits, botanic gardens and other fun activities in each city and the experience I had was amazing. The most remarkable thing during my visit is that I got to witness 1st hand the purpose of IPPS which is to seek and share information. During my stay in Australia, I met a lot of people that were very open to new ideas and were willing to share information with me.

IPPS Vol. 73 – 2023

INTRODUCTION

I arrived in Australia a week before the conference and spent my first week in Perth where I was hosted by Mr David Hancock. He took a lot of places around Western Australia and made sure that I got the best experience.

On the first day we visited Zanthorrea nursery, which is a family-owned garden Centre, located in the hills of Perth, Western Australia. This is one of the nurseries in Western Australia that specializes in Australian natives. I was stunned by how unique and beautiful Australian native plants were. The nursery was very neat, and everything was well set out, with plants grouped in different groups. They sell well established plants, as well as tube stock, seedlings, herbs, and vegetables.

Mr. Ilec Hooper took us for a tour in the nursery where he shared the history of the nursery and explained most of the things that I didn't know about Australian plants. I couldn't stop taking pictures because of how beautiful the nursery was and everything was new to me besides some of the Eucalyptus species that we also have in South Africa.

A. B.

C. D.

Figure 1. A. Nursery layout at Zanthorrea Nursery, **B.** Herb display, **C.** Mr. Ilec Hooper and Mr. David Hancock, **D.** Mr. Ilec Hooper and Sizwe after the tour at Zanthorrea nursery.

On the second day we visited Plantrite Nursery, Muchea Tree Farm, Natural Area Nursery, Bunnings Garden Centre, Kings Park, and Biodiversity Conservation Centre. I found Plantrite nursery to be the most mechanized nursery compared to other nurseries that I visited in Perth. They have less staff members on a big portion of land but because of mechanization they still manage to perform all the daily tasks required to run a successful nursery. From trays/pots feeding, transplanting, watering up to a finished product, everything is done with the use of machines and that eliminates the factor of having too many employees. It also promotes consistency because with manual labour it can be difficult to replicate one thing all the time without making mistakes.

Figure 2. Plantrite Nursery. **A.** Retail area, **B.** Growing under poly tunnel with shade, **C.** Seed smoking device, **D.** Propagation and transplanting area.

Things I found common about these nurseries was that they all grew mostly Australian natives, are members of the Nursery & Garden Industry Victoria with certificates of accreditation.

It was amazing to witness what one can do on a small piece of land when I got to Muchea Tree Farm. They grow some of South African fynbos plants from cuttings and seeds, they also have a small area where they do seed smoking of Proteaceae species. For a moment I felt like I was home.

A. B.

C. D.

Figure 3. Muchea Tree Farm. **A** and **B.** Tray filling, **C.** Transplanting machine, **D.** Sizwe test driving forklift used for moving plants and loading growing media into pot/tray filling.

As hard as it was to adjust to Australian time zones I had to keep up and when we got to Natural area, we met Mr. Andre Nguyen who showed me around the nursery. One thing I loved about the nursery, which was unique and cool, was their irrigation system which allows the user to program irrigation using any smart device such as cellphone/computer without having to go the controller physically.

A. B.

C.

Figure 4. Natural Area Nursery. **A** and **B.** Water filtration system, reservoirs, and irrigation controller at Natural Area Nursery, **C.** Mr. David Hancock & Sizwe after the Nursery tour.

When we got to Bunnings, I realized that their setup was like Builders warehouse in South Africa. The quality of plants on the floor was amazing and the use of labels on plants was on another level. Even plants without flowers looked stunning because of the way they were presented.

Figure 5. Stunning mixed herbs at Bunnings Nursery.

At King's Park we spent some time at the conservation area looking at plants they propagate, methods used and their propagation structures. We also looked at their tissue culture lab which was amazing. It was well organized and clean with personal protective equipment available in restricted areas.

At a later stage we went to the park where I got to enjoy the view of Perth city. It was stunning, full of tourists and people that went there for relaxation on their beautiful lawn.

Figure 6. King's Park. **A.** Tissue culture lab at King's Park, **B.** Propagation of plants such as *Pimelea physodes* through grafting at King's Park, **C.** PPE for going into seed cold storage rooms at King's Park., **D.** Sizwe enjoyed the view of Perth city from King's Park.

On the third day in Perth, we visited Rottnest Island which is a home of Quokkas from Hillarys Boat Harbour. The journey from the harbour to the island was very nice, met new people and chatted all the way. On arrival we hooped on to the bus for a tour, the drive was less than an hour which means the island is not that big. Along the way we saw nice views and got an overview of the vegetation found there. After the bus tour we took a walk and along the way I came across many quokkas.

Figure 7. Rottnest Island. **A.** Rottnest Island Ferry, **B.** Quokkas.

On the following day, which was the 20th of May 2023, Mr. Andre Nguyen, Miss Lara Osborn, and I started our day by visiting Nuts About Natives Nursery. We saw a lot of beautiful Australian native plants including *adenanthos barbiger* which produces a very nice orange flower that they use as their logo, *macrozamia* sp which is loved by Kangaroos and *calytrix fraseri* which has nice pink flowers.

Figure 8. Nuts About Native Nursery. **A.** *Adenanthos barbiger*, **B.** Banksia sp.. **C.** *Lechenaultia formosa.*

Figure 9. Post nursery tour at Nuts About Native Nursery.

Then we went to Mandurah, the aim of going there was for me to experience life outside of work in Australia and try new things. I am not a sea food person but, on that day, I tried oysters and I guess that was the first and the last time. We went on a boat cruise around the seaside home and the experience we had was amazing. It was also nice to know that the same Lara Osborn that made time for me and made sure that I had the best time of my life is the same person that was in the sixpack team with Siyasanga Yenzela in 2018 is one of the people who I look up to in South Africa.

Figure 10. Lara, Andre and Sizwe after the boat cruise at Mandurah.

During the conference activities

On the 21st of May 2023 I flew to Melbourne where the conference was going be held. I met Ms. Pam Berryman, my host, at the airport, then we drove to Larkmans nursery where we did some of the conference preparations. We were part of the group of people that attended the pre-conference tour and during the tour I met new people. That gave me enough time to chat with everyone since we were not that many at that time. We visited a lot of wholesale & retail nurseries, labels & pots manufactures, gardens, and garden centres.

Places visited during the pre-conference tour: Rivers of Yarrambat, Alowyn Gardens, Larkmans Nurseries, Kuranga Nursery, Norwood Industries, Garden City Plastics, Ball Australia, The Diggers Foundation, Van Loons and Boomaroo.

On the conference day, the venue was packed, and I saw a lot of new faces but because I had already met most of the people before the conference, it was easy for me to get to know new people. The guest speakers that presented during the conference were very knowledgeable and I enjoyed listening to their presentations to the point that I even had a chat with most of them after the conference. What I really loved about their presentations was that none of them were pushing sales of his/her products, but they were sharing plant knowledge that benefited everyone.

Figure 11. The 2023 Six Pack team.

I was given the opportunity to present as well and talk about my background in Horticulture and the experiences I had in Perth. I was very nervous but that really helped me a lot because at this stage I am no longer afraid of speaking in front of big crowds and the positive feedback I got really boosted my confidence.

Figure 12. Sizwe during his presentation at the conference in Melbourne.

Places visited during the conference and after are Geelong Botanic Gardens, The Gordon Tafe, Vasili's Garden and Roraima. At the Gordon Tafe we had a grafting workshop which was productive and helpful because everyone got the opportunity to do grafting.

Post-conference activities.

After spending a week in Melbourne, I moved to Adelaide where I was hosted by Mr. Matt Coulter. We visited Mt Lofty Botanic Garden, Urbrrae Campus, TAFE SA, Cleland Wildlife Park, Wittunga Botanic Garden and Adelaide Zoo.

Highlights

Figure 13. Sizwe holding a Koala at Cleland Wildlife Park.

Figure 14. Tour at Garden City Plastics.

Figure 15. Tour at Norwood.

Figure 16. Garden tour at Mount Lofty Botanic Garden.

Figure 17. Mr. Matt Coulter showing me the tuber of the Titan arum.

Figure 18. Gala dinner in Melbourne.

Figure 19. Grafting workshop at The Gordon TAFE.

Figure 20. Stunning designs at Adelaide Zoo.

Figure 22. Networking moments in Melbourne.

Figure 21. Tour at Adelaide Botanic Garden.

Figure 23. Popular Australian pots and trays.

Figure 24. Sizwe in Adelaide.

Figure 26. Tour at the South Australian Seed Conservation Centre.

Figure 25. State herbarium & Library tour in Adelaide botanic Garden.

Figure 27. Tour at Vasili's Graden.

Figure 28. Tour at Roraima Nursery.

Figure 29. Pre-conference tour at Searoadferries on our way to Van Loon's Nursery.

Combined Proceedings IPPS 73: 113-119. 2023.

Challenges and Achievements in Mine Revegetation in New Caledonia

Danielle SaintPierre

c/o SIRAS Pacifique - Lot 19, Allée du Titane – Païta ZIPAD, New Caledonia

danielle.saintpierre@siras.nc

Keywords: soils, hydroseeding, seedling establishment, native species

Summary

New Caledonia, a French territory in the South Pacific, harbors exceptional biodiversity resulting from its geological history and isolation. The presence of ultramafic rocks gave rise to unique soils with high metal toxicity and limited organic matter. While nickel ore mining has enabled economic development, it has also led to significant vegetation and soil degradation, disrupting ecosystems and seriously impacting biodiversity.

Mine revegetation efforts aim to restore damaged areas and preserve biodiversity. The use of local and endemic pioneer species has given promising results, as they adapt well to the challenging conditions of the mining sites. These species require minimal maintenance, integrate well with the environment, reproduce effectively, and contribute to vegetation dynamics.

Revegetation techniques, such as hydroseeding and planting, have been employed on mine sites. Hydroseeding, with endemic or native shrub mixtures, offers cost-effective solutions, while planting provides immediate vegetation presence. Companies like SIRAS Pacifique have pioneered rehabilitation projects, combining ecological restoration and civil engineering works. Additionally, steps have been taken to preserve ecosystems, through seed orchards, and to conserve rare and threatened species, through propagation trials.

IPPS Vol. 73 – 2023

Despite progress, further research and collaboration between government, industry, and research institutions are needed to safeguard endangered species and enhance understanding. New Caledonia should therefore successfully balance economic development with the preservation of its remarkable natural heritage.

INTRODUCTION

New-Caledonia is a French territory located in the South Pacific, about 1 500 km from the Australian East coast. When this archipelago separated from Gondwana, about 80 million years ago, it underwent complex subduction-obduction episodes which resulted in the installation of an ultramafic rock layer, on the New Caledonian basement. Once the land re-emerged, erosion reduced the extent of the ultramafic layer, to about 1/3 of the main island's area.

The ultramafic rocks evolved into two particular soils called ferralic and hypermagesian. These are characterized by a high metallic toxicity (due to high concentrations in Nickel, Chromium, Iron, Manganese and Cobalt), a low organic matter content and a low water retention capacity. All these unfavourable conditions, together with the geographical isolation of the island have resulted in the evolution of uniquely adapted flora and fauna and have led to the development of an extraordinary biodiversity, on this rather small territory. As a consequence, New-Caledonia hosts an extraordinary biodiversity, with many endemic and sometimes archaic species (such as *Amborella trichopoda* – the most primitive angiosperm). Within the plant kingdom, 3 300 vascular plant species have been identified in New Caledonia, 76% of which are endemic.

But on the other hand, because of their nickel ore content, the ultramafic soils have made of the island one of the largest reserves of this valuable metal in the world. The country is therefore faced with the paradoxical challenge of having to manage both:

- preservation of its natural heritage, recognized as one of the most original and precious on the planet.
- economic development based on considerable mining resources.

Mine Revegetation in New-Caledonia and Its Challenges

Mine revegetation is one way of participating in preservation of biodiversity. Unfortunately, many difficulties are encountered.

Open-pit nickel exploitation began in 1873 and represents a very important part in the New Caledonian economy. Up until 1975, extraction was carried out with no environmental precaution and mine waste was spilled down the mountain slopes, leaving many scars in the vegetation. Since the mid 70's, techniques have fortunatly evolved in a positive way, for instance:

- topsoil is reused,
- wastes are transported and stored in dedicated discharges.

Nevertheless, nearly 150 years of mining activities have led to significant destruction of vegetation and soils, disturbance of ecosystems and a loss of biodiversity.

Moreover, the areas left by the mining activity have very different characteristics compared to the original soils on which the vegetation developed:

- their surface is hard and compacted,
- they present severe nutrients shortages,
- high levels in phytotoxic elements,
- no microflora,
- steep slopes,
- extreme rainfall and wind conditions.

Enabling vegetation to resettle in such conditions is therefore particularly difficult.

The first trials aiming to revegetate mining sites were carried out in the early 1970's using exotic and a few native species. The poor results led the researchers to experiment on a mix of local and endemic pioneer species. This work resulted (in 1992) in the establishment of a list of 67 species which can be used in mine revegetation projects. These well adapted species show all the necessary characteristics enabling succes:

- no maintenace is required once they have been planted,
- they perfectly integrate in the environment,
- they are able to reproduce, and colonise the surroundings,
- by growing they create shade, mulch, organic matter and enhance vegetation dynamics.

In the long term, all these steps allow preservation of biodiversity.

They nevertheless present one disadvantage: their growth is extremely slow and results become tangible and visible only many years after revegetation action has been carried out.

Some examples of revegetation projects in New-Caledonia

SIRAS Pacifique is a company specialized in rehabilitation of degraded sites and has been working in this fied for 30 years. With the years, the company developed different skills and revegetation techniques by adding civil engineering works in ecological restoration projects (such as gabions or wooden walls), ecological engineering, outdoor landscaping and erosion control works. It is today recognized as being a major actor of environmental restoration.

The two main techniques used on many ancient or active mine sites are hydroseeding and planting.

Hydroseeding is generally preferred because it is the most cost-effective revegetation method. It also allows to treat steep and limited access surfaces (figure 3). In our operations we only use endemic or native maquis shrub mixtures. All the species are pioneers of the maquis shrub, 70 to 80% of them being Cyperaceae. **Figure 1** gives examples of the results obtained several years after hydroseeding works on 2 mine sites, *i.e.* Opoué mine, (Eramet Group) and the ancient Brisson mine. It shows how the vegetation evolved in a few years and how efficient the technique can be, even on poor lateritic soils.

Figure 1. Old and steep waste dump on Opoué mine, in 2014 (A) before hydroseeding and 2018 (B) showing vegetation now completely covers and heals the opening that had been created by the spills. Remodeled areas, covered with jute nets (in order to control erosion and optimise seed germination), in 2019 before hydrodeesing (C) and in 2021.

Sometimes hydroseeding is not the best technique as vegetation installation can be very long (between 18 to 24 months), especially in the Northern part of New Caledonia, where the climate is much drier.

Planting is therefore recommended in some situations because plants are instantly present. At least 20 different species are generally used, all of them being endemic or native. For instance, on the same ancient Brisson mine, almost 19.000 plants were reintroduced together with reinforcement techniques, such as fascines or windbreakers (**Fig. 2**).

Extra skills and activities

In order to ensure the reliability and efficiency of the company, all the plant material supplies are integrated within SIRAS Pacifique. The Seed Management sector organises the whole pathway from collection of endemic species' seeds (in the natural environment) all the way through to their use (in hydroseeding works or in the nurseries). In the early 2010's, it became obvious that mine revegetation demands would increase drastically and that seed sampling in the natural environment would eventually have a depleting effect the resource.

Figure 2. (A) Planting behind wooden windbreakers on sites where strong prevailing winds occur - (B) Planting behind fascines (or dead wood bundles), a technique used on slopes to reduce erosion.

The need to set up seed orchards thus appeared evident, especially for Cyperaceae. With the help of mining operators, several orchards have been implemented in different locations (**Fig. 3**).

Figure 3. Two of the seed orchards created on mining sites : Kouaoua mine (SMSP group) and Tiébaghi mine (SLN-ERAMET Group).

Endemic plant supply is guaranteed by the Nurseries, which have been set up in three different locations. Altogether, the production capacity is of 350 000 plants and propagation is mastered for about 250 different endemic or native species. 60% of these belong to the maquis shrub, 30% come from the Caledonian dry forest or from the shoreline. The last 10% either grow in the rain forests or are endangered species.

Rare and Threatened Species Propagation

Because of mine exploitation, many rare endemic species are severely threatened in New-Caledonia. The mining operators have the legal obligation to protect them, when they are on their concessions. SIRAS Pacifique is thus sought to help with the protection procedures, to carry out propagation trials and reintroduction operations.

Since 2007, we have been propagating and re-introducing *Araucaria rulei* within a degraded population on Thio mine for SLN-ERAMET Group. 14 000 plants have been re-introduced. More recently, similar programs have allowed us to reintroduce two other Araucariacea species for Ballande Group, on Cap-Bocage and Kaala mines. Propagation of these species is now well mastrered.

Between 2014-2018**,** trials were carried out, for SMSP Group, to develop propagation methods for 18 threatened species, including several orchids, on Pinpin mine. We obtained varying results, depending on the species.

Since 2020, we have been missioned by SMGM company to propagate 26 species which are present on Vulcain mine. Amonst these 26 species, some are quite challenging. For instance:

- As with many Ericaceae, we are confronted with dormancy issues for *Styphelia enervia* and we have not succeeded in producing cuttings.
- *Polyscias scopoliae* is another demanding species, which also has dormant seeds. Although we have some encouraging results with head cuttings, we are here confronted with the limited availibility of plant material.
- An other tough example is *Homalium betulifolium,* because fructification is quite sporadic and our cutting trials have not been very successfull, up until now.

Fortunately, we have some achievements (**Fig. 4**), for instance with:

- *Arillastrum gummiferum* : when we are lucky enough to collect seeds and sow them immediately, the germination rate is satisfactory. Propagation by seedlings transfer has also revealed efficient.
- Our *Dendrophyllanthus conjugatus* var. *conjugatus* assays have also given satifactory results with 74% of cuttings being viable and 85% of seedlings transfered to the nursery surviving.
- *Lepidocupania tontoutensis'* results are slightly below, with cuttings viability of 30% and seedlings transfer giving 80% survival.

Overall, whatever the endangered species considered, the main difficulties are:

- the lack of plant material,
- a lack of knowledge,
- the lack of access to horticultural equipment (especially hormones),
- the lack of information and experiences exchanges with other experts.

Arillastrum gummiferum
(Myrtaceae)

Germination rate = 30 to 80%
Seedlings transfer = 90% survival

Dendrophyllanthus conjugatus
var. conjugatus
(Phyllanthaceae)

Cuttings viability = 74%
Seedlings transfer = 85% survival

Figure 4. Some of the propagation trials on endangered species (*Arillastrum gummiferum,* and *Dendrophyllanthus conjugatus var. conjugatus*) showing encouraging results.

Conclusion

Despite the difficult contexts and thanks to a collaborative approach between government, industry, and research institutions, the revegetation operations conducted in New Caledonia are now giving rather satisfactory results. This approach has resulted in the restoration of degraded areas and the reestablishment of native flora.

However, there is still a long way to go in terms of safeguarding endangered species. Foremost, knowledge about their botanical and ecological characteristics, their geographical localization and of their culture routes must still be improved.

Combined Proceedings IPPS 73: 120-131. 2023.

The Role of the Western Australian Botanic Garden Nursery – Collections, Conservation and Education

Amanda Shade

Kings Park and Botanic Garden, 1 Kattidj Close, Kings Park, WA, 6005, Australia

amanda.shade@dbca.wa.gov.au

Keywords: flora, horticulture, kings park, propagation, recreation, tourism, training, Western Australia

Summary

Nurseries attached to Botanic Gardens are uniquely placed within the industry – as the primary producers for organisations that hold "documented collections of living plants for the purposes of scientific research, conservation, display and education" (reference – BGCI website) they are often in the privileged position to become involved in a range of activities linked to these core functions of a Botanic Garden. Western Australia is home to over 13,000 highly diverse, unique and environmentally specialised taxa. At Kings Park in Perth, home of the Western Australian Botanic Garden, the nursery has developed over time to become a specialised producer of a large range of this world-renowned flora, propagating plants that are variously utilised for collections, ornamental display, important conservation and restoration outcomes and educational and training purposes. All propagation activities, trials, experimentation and production are intertwined and underpinned by these driving influences, and over the past almost 60 years the Kings Park nursery team have refined their propagation techniques to ensure repeatability and reliability in allowing the showcasing and conservation of a diverse range of species, many of which have not been previously introduced into cultivation or are not commercially available. This paper explores the themes of collections, conservation and education with a focus on the responsibilities of the Kings Park nursery in each of these areas.

INTRODUCTION

Nurseries attached to Botanic Gardens are uniquely placed within the industry – as the primary producers for organisations that hold *'documented collections of living plants for the purposes of scientific research, conservation, display and education'* (BCGI, 2023) they are often in the privileged position to become involved in a range of activities linked to these core functions of a Botanic Garden.

Western Australia is home to over 13,000 highly diverse, unique and environmentally specialised taxa. At Kings Park in Perth, home of the Western Australian Botanic Garden (WABG), the nursery has developed over time to become a specialised producer of a large range of this world-renowned flora, propagating plants that are variously utilised for collections, ornamental display, important conservation and restoration outcomes and educational and training purposes.

All propagation activities, trials, experimentation and production are intertwined and underpinned by these driving influences, and over the past almost 60 years the Kings Park nursery team have refined their propagation techniques to ensure repeatability and reliability in allowing the showcasing and conservation of a diverse range of species, many of which have not been previously introduced into cultivation or are not commercially available. This paper explores the themes of collections, conservation and education with a focus on the responsibilities of the Kings Park nursery in each of these areas.

Background

WABG is located within Kings Park, a large urban park that occupies an elevated 406 ha site overlooking the Perth Central Business District. Kings Park was established as a public park in the 1890's and is managed by the Botanic Gardens and Parks Authority (BGPA), a State Government statutory authority. The key functions of BGPA are related to conservation; recreation and tourism; education and interpretation; cultural heritage; horticulture; and scientific research. The park attracts over 5 million visitors annually and is known for its significant cultural and natural history and sweeping views over the Swan and Canning rivers (**Fig. 1**).

About two thirds of the park today is remnant Swan Coastal Plain bushland; approximately 17 ha are dedicated to the Western Australian Botanic Garden, and the remaining area is landscaped parkland that incorporates playgrounds, a range of visitor facilities and numerous memorials including the State War Memorial. The idea of establishing a Botanic Garden in Perth began to gather momentum in the 1950's, and Kings Park was chosen as the future home of the State Botanic Garden in 1959. From the very beginning there was interest in propagating and cultivating Western Australian flora for display, with collections of seed from the wild being undertaken as early as 1963 to be grown on in the Kings Park nursery for research and display purposes. This journey from the field to display continues to this day with most taxa propagated from material collected from the wild under licence, with the nursery providing an essential link in that chain. The WABG was officially opened in 1965. At that time, the founding vision was that the Botanic Garden would be *'designed to interest and educate the public in the conservation and cultivation of the Western Australian flora'*. The celebration of endemic Western Australian genera continues to this day, but the extreme diversity in our flora can often present challenges for those of us attempting to propagate and bring them into cultivation. The WABG is supported by the Western Australian seedbank, the Kings Park Nursery, a comprehensive horticultural database that records all available details of each collection and a herbarium specimen of every collection.

Figure 1. Entrance to the Western Australian Botanic Garden located within Kings Park, a large urban park that occupies an elevated 406 ha site overlooking the Perth Central Business District on one side and Swan and Canning rivers on the other (Image credit Dave Blumer).

The Kings Park nursery was established in 1962, several years prior to the opening of the WABG. Roughly 50-60,000 plants are grown annually for display of the collections to the public within the WABG and wider parkland areas; for restoration and rehabilitation programs in BGPA managed remnant bushland; for tree replacement programs; and for conservation programs managed by other WA state government departments. The nursery is slightly different to commercial and other production nurseries in that it has only one 'client' – Kings Park. Although the nursery is still a primary producer, its programs are very much dictated by the requirements of the state botanic garden and the parkland displays and natural area revegetation/restoration programs – i.e., collections, conservation and education as the driving influences. Because of the nature of its activities, the nursery has a defined main propagation season of spring and summer, when the majority of production occurs. This is to allow for the 'client' requirements of stock size and age for winter planting. As a Botanic Garden nursery, at Kings Park we are uniquely placed to be able to see exactly where the nursery stock is planted – the location of the nursery within a Botanic Garden means the staff is also able to monitor the establishment and performance of the plants once they leave the nursery. This often provides real-time valuable data for the staff to enable constant improvements in propagation programs and methods. It is also a great stock garden for sourcing future material. We are also privileged to potentially have access to the flora from the entire state of WA, which at last count was close to 14,000 taxa.

Rather than focus on a smaller number of species and grow large numbers of these Kings Park nursery does the opposite – it regularly deals with a large number of species (around 3-4000) but tend to produce small quantities of each. While the nursery provides an integral link in the chain from field to garden, this paper covers three key roles– collections, conservation and education.

Collections

A living plant collection is a managed group of plants grown for a defined purpose or demonstrating a particular theme. Botanic Garden collections are underpinned by science - they are formally identified or botanically verified, documented, labelled and actively curated to varying degrees based on their intent. Holding living plant collections are key roles of Botanic Gardens and are what distinguishes a Botanic Garden from a standard garden. They are often valuable scientific and cultural resources and support the typical function of a Botanic Garden – conservation, research, education and training, and public and social engagement (BGANZ, 2023).

Living plant collections may be themed based on aspects such as taxonomy, geography, ecology, conservation, display, culture or education (**Fig. 2**).

Figure 2. A living plant collection is a managed group of plants grown for a defined purpose or demonstrating a particular theme. Southwestern Australia is home to the greatest diversity of *Banksia* spp. (Proteaceae) with 60 species out of a total of 173 recorded, and Australia is home to all but one. This diversity is captured in the *Banksia* collection at Western Australia Botanic Gardens (Image credit Dave Blumer and Amanda Shade, BGPA).

The WABG holds collections of primarily WA flora from all over the state. These are collections that are in-ground and publicly accessible, but we also manage container collections in the nursery, a seedbank, tissue culture collections, cryostorage collections and herbarium (**Fig. 3**). These collections are not publicly accessible but play an important role in supporting the function of the WABG. Our nursery collections at Kings Park are maintained as an insurance policy. We maintain and repropagate collections of plants within our nursery and glasshouses for a number of reasons, in particular as an insurance policy. It is unrealistic to expect that every species from a state as vast and environmentally diverse as WA will grow well in the Perth climate and soils. Also, we have some pathogen challenges in some of our soils, and general security challenges as we are not a fenced park or Botanic Garden and are open and accessible 24/7. We need to secure against theft, damage and vandalism, so hold a range of species in a more secure nursery collection. These may be:

- Threatened flora or those with limited seed material available for future production. These are precious cargo, and we need to ensure we have backups in a secure environment.
- Certain genera that are notoriously difficult to establish or short-lived in-ground in our Perth conditions, or never meet their full potential (eg, *Boronia, Eremophila, Darwinia*)
- From remote or difficult to travel to regions or from plant groups that require significant manipulation of environment that is not possible in-ground. For example, some Kimberley species can

cope with cold and wet to a degree, but not at the same time. We have the benefit of glasshouses in the nursery to modify growing environment over our winters for this group of plants.

- Ephemerals (e.g. *Stylidium*) – often plants in this category can get lost in the wider landscape so it's more practical to secure representation in a nursery container collection.
- For ongoing horticultural research – for example, species previously untried in cultivation or species that require bulking up to generate more material for trials.

Often some species fall into more than one of these categories.

Figure 3. In addition to publicly accessible in-ground collections, the Western Australia Botanic Garden (WABG) manages a seedbank (a), tissue culture collections (b) and a herbarium (c) that are not publicly accessible but play an important role in supporting the functions of the WABG (Image credit Dave Blumer and Amanda Shade, BGPA).

Conservation

Conservation is a major influence on our activities within the botanic gardens, within nursery projects and within our scientific research branch. While it is a key function of a Botanic Garden conservation horticulture is also gaining momentum as an important branch of horticulture. With increasing biosecurity threats, loss of habitat, climate change, and resource limitations, promotion of conservation is at the forefront of modern-day Botanic Garden functions worldwide.

The Botanic Garden nursery is perfectly placed to produce quality plants for ongoing promotion of conservation, and to support conservation projects. Plants produced within the Kings Park nursery:

- Help raise public awareness of threatened flora generally – we propagate and grow plants that are displayed within designated conservation areas in the Botanic Garden; we also pass on knowledge and skills to our volunteer groups who produce this flora for public sales, spreading important conservation messages to the wider community.
- Assist with the promotion of respect for our native flora – people don't tend to care about what they don't know about. Education is power, and the more stories we can tell about the importance of conservation, the better.
- Aid with our bushland restoration and conservation programs – our remnant swan coastal plain bushland is part of a Threatened Ecological Community (*Banksia* woodland), and our nursery contributes to these conservation efforts by producing plants using provenance material.
- Contribute to significant statewide conservation programs of critically endangered flora – through a long running collaboration with other government departments on translocations of critically endangered species, Kings Park nursery grows stock for translocation plantings, but also for projects involving threatened flora such as seed orchards. Many species of threatened flora can be difficult to propagate and require thorough trialling, detailed data recording on processes and success rates, and intensive focus.

We are currently seeing the major influence of climate change and biosecurity concerns on conservation. Kings Park nursery is working on developing and displaying more climate resilient and landscape suitable plants for the future, ensuring we have reliable data on how to propagate and grow these before recommending them for wider use within industry and the community. This emphasises the importance of ongoing trials and monitoring, and the importance of accurate records.

Education

The nursery has an important role in both horticultural training and public education. Some of the ways we contribute include:

* Workshops – we run an autumn and a spring workshop series for the public on propagation, grafting and other horticultural elements of plant production.

* Tours for school, TAFE and university groups.

* Industry events and webinar presentations.

* Horticultural training via schools' work experience

* Professional development placements. Increasingly we are being approached by regional community groups, mining companies and other government departments to take on placements to value-add nursery skills to local indigenous people, students and employees engaged in existing or upcoming nursery-based enterprises. There is a huge appetite at the moment to learn about propagation and cultivation of WA flora, from both a conservation and restoration perspective, but also from a growing business angle, and a lot of it is being driven by regional groups.

Possibly our most important role in the education realm is our long-running horticultural trainee program. This program has been running since the 1960's as a means for developing highly skilled people with a passion for WA flora and horticulture in general. Our trainees are offered unique opportunities to work within a diversity of horticultural teams throughout the park, including three-month stints in the nursery, where they are exposed to everything involved with plant propagation; plant establishment, soils, pest and disease management, general nursery duties and maintenance.

Nurseries are perfect places to inspire students about horticulture. The unbridled joy on their faces when they see germination of seed they've personally sown, or cuttings that they've prepared strike for the first time– it unlocks something and builds a foundation for connection to plants and horticulture. Propagation and grass-roots horticultural activities involved in the production of plants is the perfect place to start if looking to inspire the next generation about the value of plants, and about important environmental and conservation messages **(Fig.4)**.

Figure 4. Botanic Gardens and Park Authority trainees in the nursery (Image credit Amanda Shade)

These three responsibilities can't and don't exist in isolation – they are intertwined and underpin everything we do. In all of our roles we propagate and cultivate a range of species each year - including annuals, herbaceous perennials, shrubs and trees from just about every corner of Western Australia – for primarily collection, conservation or education purposes (and often a combination of all three). Alongside this production role, we regularly conduct propagation and growing trials of rare species, difficult species, or species with desirable display potential that have not previously been cultivated. No one year will be the same in terms of species propagated, number propagated, or methods employed.

It's our role with these trials to think outside the box, and perhaps most importantly, document our findings such that others can replicate what we may find is the best practice for the propagation and subsequent cultivation of often rare, temperamental, or new to cultivation WA species.

Following are some brief case studies that highlight our collections-conservation-education focus and also our significant propagation experimentation activities.

Case studies

1. Calytrix brownii

In winter 2022, a team of six Kings Park staff travelled to the Kimberley region of WA on a field collecting expedition to collect primarily material of Myrtaceous species as a preparedness action against the potential of Myrtle Rust (*Austropuccinia psidii*) existing in other eastern states spreading into WA. We focussed on species we didn't already have seed or plants of in our collection – so were aiming to boost our collections and add to our conservation activities. We had a long target list of species that we were looking to collect, including the species *Calytrix brownii.* We collected both seed and vegetative material, with the latter being flown back to Perth for processing (**Fig. 5**).

Figure 5. Collection of material at Bell Gorge (Image credit Emma Dalziell)

There were some immediate challenges faced by the nursery team when the material arrived. Firstly, we had never propagated this species before, although we had experience with other Southwest *Calytrix* species. Secondly, the material was delivered mid-winter, having come from a much warmer and drier environment in the north, so some significant environmental manipulation was required. To address these challenges, we needed to do a bit of pre-planning and research beforehand. We looked at the natural habitat and environment this species grows in and modified the propagation media and glasshouse environment and performed a range of treatments based on what we thought would be most suitable, carefully recording every detail of all that was done, from media, containers, growth regulators, pre-treatments, glasshouse environment, frequency and method of irrigation, size of cuttings and temperatures (the latter being monitored via temperature data loggers). Recording of the best combination of variables means we will be able to repeat the same in the future.

This propagation resulted in production of a large number of plants from three different collection sites. Some have recently been planted for the first time in the WABG, adding to the diversity and increasing the value of our collection of Kimberley species, and contributing to public education by expanding the species on display. From an education perspective it also taught our trainees a valuable lesson in the importance of understanding the complexities in propagating species out of season and out of their natural range, and the research needed to make these undertakings a success. In terms of conservation, we will use this species as one of our sentinels in monitoring for possible Myrtle Rust outbreaks in the Botanic Garden – a task that is even more important now, given this pathogen has recently been detected in the north of WA and has the potential to move south.

2. ***Eucalyptus* x *impensa* (Eneabba Mallee)**

This taxon is a critically endangered, naturally occurring hybrid from the northern sandplains of WA. Any seed stored has high conservation value and therefore has limited availability for use – and results in significant variation due to its hybrid status. Rooting of cuttings for Eucalypts can be quite challenging so we decided to try grafting. Not knowing the exact parentage, we had to make some considered decisions about potential rootstock from a compatibility perspective. We consulted with Eucalypts experts and looked carefully at species with similar characteristics that grew naturally within a similar range, and other species within that classification series. Using these tools, we decided on two different species for initial trials – *Eucalyptus macrocarpa* and *E. burracoppinensis*. We employed the mummy grafting technique for this species – it's our go-to method for a range of other *Eucalyptus*, *Corymbia* and genera from drier inland regions such as *Eremophila* and *Verticordia*, as we tend to focus on this technique primarily over our hot summer months. Mummy grafting uses essentially the same techniques as standard cleft or wedge grafting, with the added step of removing all foliage from the scion and wrapping the entire scion in grafting film. We place our mummy grafts in a very hot glasshouse (which can reach up to 50-60 degrees on very hot Perth summer days) and often see bud burst within a week. Since we've started using this technique for selected genera, we've seen significant increases in success rates. *Eucalyptus* and

Corymbia in particular produce very encouraging results consistently (**Fig. 6**).

Figure 6. An example of standard grafting of *Pimelea physodes* onto *Pimelea ferrugi-nea* rootstock (Image credit Dave Blumer).

What this development has meant for our three themes is that it has enabled a greater range of species to be successfully produced in greater quantities, more diversity within our collections, more conservation messages getting out to the public, and has helped build on the horticultural knowledge that we can share with others. Production of more threatened species of plants vegetatively in this manner also means we are not impacting the natural populations.

We now have established this taxon in the collection in-ground. These specimens are continually monitored for their performance, and we've gathered some useful data so far about how the two different rootstocks have differed in terms of longevity and long-term compatibility, not to mention overall condition. Feedback on longevity helps us fine-tune rootstock selections for long-term survival in-ground.

3. ***Physopsis chrysophylla* (Golden Lambstails)**

This species is a perfect resilient plant of the future – it grows naturally in a very restricted distribution in the Shark Bay region of WA, has very low water requirements, grows well in sandy soils and is highly ornamental. Traditional vegetative propagation methods proved largely unsuccessful, so we needed to start experimenting with the variables.

Several furry dry/arid region species we've propagated have shown encouraging success with the use of rockwool as a growing medium. *Physopsis* has a very tactile and furry stem and foliage so we naturally assumed this would perform. It didn't. But that's OK – as we teach our trainees, a zero result is still a result, and can help us understand what we may need to change in order to improve. We tried altering media by the addition of different products to aid drainage; different hormone treatments; anti-transpirants as a pre-treatment; tried different timing; different glasshouse environments; and finally landed on the right combination after several years of fine-tuning and tweaking the details.

A team member had a lightbulb moment – ordinarily we would remove the very soft small tips of species such as this one as they are very prone to wilting. Her suggestion was - what if we kept these, placed them in very shallow propagation trays to enable adequate heat transfer, use lower concentration hormone, and place them in an area of slightly higher humidity with no overhead watering to combat any drying effects. Sometimes it's a small adjustment to your process, sometimes it's a combination – in this case it was a combination. By modifying the type of cutting, the depth of the container used, the glasshouse environment and the timing we can now successfully strike this species with consistently good success rates. We also found the best results were in a somewhat restricted timeframe over summer. Timing is a really important factor that sometimes gets overlooked, and this is a species that we have found has quite a specific 'window of opportunity' for successful propagation.

CONCLUSION

Working within the nursery industry is such a rewarding experience, but to be able to also work for a Botanic Garden nursery is an absolute privilege. Having the opportunity to make a significant contribution towards conservation and education outcomes is both rewarding and humbling.

A colleague at another state Botanic Garden recently described nurseries as the engine room of a Botanic Garden. If that is true, then I think the staff of these nurseries are the fuel for that engine. I'd like to acknowledge my very dedicated and enthusiastic team at Kings Park nursery – not only are they skilled in nursery activities but they are also called upon at various times to be field operators, seed collectors, public relations specialists, tour guides, social media contributors, teachers, and scientists. The support of this dedicated nursery team significantly and consistently allows our Botanic Garden to achieve displays of quality collections, promotion of conservation, and a vitally important contribution to ongoing research and education.

LITERATURE CITED

BGCI (2022). Botanic Gardens Conservation International. https://www.bgci.org Accessed 12 August 2023 BGANZ (2023). Botanic Gardens Australia and New Zealand. https://www.bganz.org.au/ Accessed

Combined Proceedings IPPS 73: 132-140. 2023.

Modern-day Plant Hunting

Alistair Watt

Otway Ridge Arboretum, Lavers Hill, Victoria, Australia.

alistairwatt@bigpond.com

Keywords: plant collecting, plant introduction, Chile, Fiji, New Caledonia, New Zealand

Summary

From 1985 to 2000 I made a number of plant-collecting trips to Chile, Fiji, New Caledonia and New Zealand. This paper lists the principal introductions into Australia and in a few cases, a number of species received from other collectors overseas. Several botanic gardens hold a good range of the original introductions, as well as a large number of newly re-propagated plants. This includes the Geelong Botanic Gardens at its Pacific Rim 'Southern Hemisphere' section. In addition, in a dedicated New Caledonian section adjacent to their New Zealand bed, the Royal Botanic Gardens Melbourne displays an interesting range of species mainly collected by the author. The botanic gardens of Sydney, Adelaide and Hobart also hold some valuable collections propagated from the original plant introductions.

INTRODUCTION

The great plant hunter Robert Fortune was principally sponsored in his travels in the mid-1800s by the Horticultural Society of London (later the Royal Horticultural Society) and also by the mighty East India Company. Those who received his plant novelties were the rich and famous of British society who desired them for their new gardens. Some 150 years later, the situation was different, in Australia in particular. Very few new plants were being introduced, even into our botanic gardens. As an enthusiastic collector of conifers, I had only one possible avenue for obtaining new species and that was to go and get them!

I was lucky. By building a good case, I was able to obtain support from the IUCN specialist conifer group of which I am a member, help from the CSIRO Forestry Research section, assistance (with quarantine etc.) and direct sponsorship from various universities and botanic gardens. The Maud Gibson Trust in Melbourne and private individuals such as the late John Silba from New York, provided a degree of funding for some expeditions.

Not only will the publication of the plant lists below allow those growing the material to have some idea of provenance, but it will also establish the date of introduction of a number of new species such as the now widely grown *Lobelia tupa* and *Astelia chathamica* (**Fig. 1**) into Australia (and some introductions sent elsewhere, e.g., the Royal Botanic Garden Edinburgh).

Figure 1. a) *Metrosideros vitiensis* introduced from Fiji, b) *Acmopyle sahniana* introduced from Fiji, c) This *Astelia chathemica* introduced from New Zealand in 1991 was planted in 1995 as a single small plant is now over three metres across, d) *Geranium traversii* introduced from New Zealand.

The tables below refer essentially to my own plant collecting activities in Fiji (**Table 1**), New Caledonia (**Table 2**) and New Zealand (**Table 3**) etc., but additionally includes some of the material collected in Chile in 1985 while on expedition with staff from the RBG Sydney (**Fig. 2, Table 4**). The species which are listed on CITES (the Convention on International Trade in Endangered Species of Wild Fauna and Flora) were formally cleared through Sydney's Royal Botanic Garden and the relevant Chilean authorities. The list of collections given below only represents the significant new plants and does not include repeat collections of the same species.

Table 1. Species collected in two expeditions to Fiji by Alistair and Julie Watt (1986 - AW 860001-860198 collection numbers) and by Alistair Watt and Bob Cherry (1988 - collection numbers AW 880100 to 880160)

1986	1988
*Acacia richii**	*Acmopyle sahniana*
Agapetes (as *Paphia) vitiensis**	*Balaka longirostris**
*Agathis macrophylla**	*Balaka macrocarpa**
*Bulbophyllum vitiense**	*Collospermum montanum*
*Cordia subcordata**	*Cyathea alta**
Cycas seemannii	*Cyathea lunulata**
Dacrycarpus imbricatus var. *patulus*	*Dacrydium nausoriensis Dacrydium nidulum*
*Dendrobium mohlianum**	*Davallia fejeensis**
*Dendrobium prasinum**	*Dendrobium platygastrium**
Gardenia hutchinsoniana	*Homalium nitens**
Pittosporum rhytidocarpum	*Huperzia squarrosa**
Podocarpus neriifolius	*Metrosideros collina*
	Podocarpus affinis
	Podocarpus degeneri
	Podocarpus hybrid
	Retrophyllum vitiense
	Saurauia rubicunda
	*Scaevola taccada**
	*Spathoglottis pacifica**
	*Turrillia ferruginea**

*These species I know came into limited cultivation here in Australia, but I presently can find no evidence of whether they still survive.

Table 2. List of species collected in expeditions to New Caledonia in 1987 by Alistair Watt et al. (AW 870500 to 870581 collection numbers) and in 1993 (collection numbers AW 930100 to 930150), 1995 (collection numbers AW 950003 to 950204) and in 1996 (collection numbers AW 960001 to 960064) by Alistair and Julie Watt.

1987		1993
Acmopyle pancheri	*Dubouzetia confusa*	*Alphitonia neocaledonica**
Agathis lanceolata	*Libocedrus yateensis*	*Araucaria humboldtensis*
Agathis montana	*Myodocarpus fraxinifolius*	*Astelia neocaledonica*
Agathis ovata	*Neocallitropsis pancheri*	*Cordyline neocaledonica*
Araucaria bernieri	*Neoveitchia storckii**	*Cunonia bullata**
Araucaria biramulata	*Nepenthes vieillardii*	*Cunonia macrophylla**
Araucaria goroensis	*Nothofagus aequilateralis**	*Dacrycarpus veillardii**
Araucaria laubenfelsii	*Podocarpus decumbens**	*Dacrydium balansae*
Araucaria luxurians	*Podocarpus longifoliolatus*	*Dacrydium lycopodioides**
Araucaria muelleri	*Podocarpus sylvestris*	*Dianella ensifolia**
Araucaria nemorosa	*Podocarpus gnidioides*	*Dicksonia thyrsopteroides**
Araucaria scopulorum	*Retrophyllum minor*	*Dodonaea viscosa**
Araucaria subulata	*Xeronema moorei*	*Dracophyllum* sp.*
Codiaeum peltatum		*Falcatifolium taxoides*
Dacrydium araucarioides		*Gymnostoma deplancheanum*
Dacrydium guillauminii		*Joinvillea gaudichaudiana**
		Libocedrus chevalieri
		Melaleuca quinquenervia
		Metrosideros operculata
		Podocarpus lucienii
		Podocarpus sylvestris
		Schefflera candelabrum
		Xanthostemon aurantiacus
1995		**1996**
Carpolepis laurifolia (dwarf form, summit Mt Humboldt)		*Storckiella pancheri*
*Dracophyllum humboldensis**		*Xanthostemon macrophyllus**
*Gardenia aubryi**		*Xanthostemon laurinus**
Grevillea exul		
Grevillea gillivrayi		
Metrosideros tetrasticha		
Nothofagus codonandra		
Podocarpus afin. sylvestris (now P. colliculatus)		
Stenocarpus milnei		
Stenocarpus umbelliferus		
Syzygium tripetalum		
Xanthostemon longipes		

*These species I know came into limited cultivation here in Australia, but I presently can find no evidence of whether they still survive.

Table 3. List of species collected in 1991 (collection numbers AW 910001 to 910047) and 1993 (collection numbers AW 930100 to 930092) by Alistair and Julie Watt in New Zealand.

1991	1993
Astelia chathamica	*Aristotelia serrata*
Caldcluvia rosifolia	*Ascarina lucida**
Cordyline indivisa	*Astelia solandri*
*Cordyline kaspar**	*Beilschmiedia tarairi*
Cordyline pumilio	*Beilschmiedia tawa*
Dianella nigra	*Collospermum hastatum*
Elaeocarpus dentatus	*Coriaria pteridoides*
Elingamita johnsonii	*Corynocarpus laevigatus*
Fuchsia excorticata	*Dracophyllum latifolium*
Gaultheria antipoda	*Dysoxylum spectabile**
Geranium traversii	*Fuchsia procumbens*
Griselinia lucida	*Gunnera prorepens*
Halocarpus biformis	*Hibiscus trionum*
Halocarpus kirkii	*Laurelia novae-zelandiae*
Knightia excelsa	*Melicope ternata*
Laurelia novae-zelandiae	*Pennantia baylisiana**
Lepidothamnus intermedius	*Phyllocladus alpinus*
Leptospermum scoparium	*Podocarpus hallii*
Libertia pulchella	*Pomaderris elliptica*
Libocedrus bidwillii	
Libocedrus plumosa	Kingdon-Ward wild collected rhododendrons ex. Pukeiti gardens:
Lilium mackliniae	*Rhododendron protistum* 'Pukeiti' KW 21498
Macropiper excelsum	*Rhododendron ciliicalyx* KW20280
Manoao colensoi	*Rhododendron crassum* KW 20939
Nothofagus solandri	*Rhododendron johnstoneanum* KW 20305
*Olearia ilicifolia**	
*Ourisia macrophylla**	
Peperomia urvilleana	
Planchonella costata	
Pomaderris kumeraho	
*Rhopalostylis cheesemanii**	
Schefflera digitata	
Taxus brevifolia	
Weinmannia silvicola	
Xeronema callistemon	

*These species I know came into limited cultivation here in Australia, but I presently can find no evidence of whether they still survive.

Table 4. The significant species collected during an extended plant collecting expedition to Chile by Dr Ben Wallace and John Forlonge, both of the Royal Botanic Garden Sydney, and the author in 1985. These have BJW 850001 to 850354 collection numbers.

Acacia caven	*Laurelia serrata*
Aextoxicon punctatum	*Laureliopsis philippiana*
Araucaria araucana	*Lepidothamnus fonkii**
Austrocedrus chilensis	*Libertia sessiliflora*
Caldcluvia paniculata	*Lithraea caustica*
Citronella mucronata	*Lobelia tupa*
Cryptocarya alba	*Nothofagus alessandri*
Drimys andina	*Nothofagus betuloides*
*Empetrum rubrum**	*Nothofagus glauca*
Escallonia alpina	*Nothofagus leonii*
Escallonia pulverulenta	*Nothofagus pumilo*
Escallonia revoluta	*Ovidia andina**
Fascicularia bicolor	*Persea lingue*
Fitzroya cupressoides	*Peumus boldus*
Gaultheria insana (as G. furiens)	*Pilgerodendron uviferum*
Gaultheria leucocarpa	*Proustia pyrifolia*
Gaultheria littoralis	*Pseudopanax laetevirens*
Gevuina avellana	*Puya berteroniana*
*Gomortega keule**	*Puya chilensis*
Greigia sphacelata	*Quillaja saponaria*
Griselinia jodinifolia	*Ribes magellanica*
Griselinia scandens	*Senecio candidans*
Gunnera magellanica	*Tepualia stipularis*
Latua pubiflora	*Viola rubella*

* These species I know came into limited cultivation here in Australia, but I presently can find no evidence of whether they still survive.

Figure 2. The misery of it all! Collecting cushion plant seeds in the snow on Tierra del Fuego, Chile in a blizzard.

In addition to the above-mentioned species, *Papuacedrus papuana*, *Phyllocladus hypophyllus* and *Podocarpus rubens* were obtained in 1986 from Dr Nancy Bowers under import permit directly from Mt Hagen, Papua New Guinea. In 1988, *Pseudotaxus chienii, Dacrydium beccarii* and *Amentotaxus formosana* were sent under import permit directly from Pinetum Blijdenstein, Hilversum, the Netherlands and quarantined aby the Adelaide Botanic Gardens. In the year 2000 we sent seeds of *Alphitonia zizyphoides, Fitchia speciosa* and *Fagraea berteroana* to the Brisbane Botanic Gardens (collection numbers AW 20001 to 20018).

Those last two decades of the 20th century were perhaps a golden age for plant hunting and new plant introductions. Not only were plant handling facilities well developed, but ethylene-absorbent plastic bags were available, many quarantine facilities were in existence here in Australia, and air carriage was quick and efficient. In those days it was relatively straightforward for collectors to obtain an import permit for live plant propagation material, and it was also quite legal to import a wide range of species as seeds. The recent, expensive and extremely clumsy, Nagoya Protocol to the Convention on Biological Diversity (CBD, 2011) is aimed at 'controlling' the movement of plants between countries and will probably now make new plant introductions in this way all but impossible.

Places To See Living Plants

At present several botanic gardens hold a good range of our original introductions, as well as a large number of newly propagated plants. Including the Geelong BG with its Pacific Rim 'Southern Hemisphere' section. In addition, in a dedicated New Caledonian section adjacent to their New Zealand bed, the Royal Botanic Gardens Melbourne displays an interesting range of species mainly collected by the author. The botanic gardens of Sydney, Adelaide and Hobart also hold some valuable collections propagated from our original plant introductions.

Although I am growing most of the new conifer species here in our arboretum at Lavers Hill, the cool Otways climate of southern Victoria has proved to be too cold for many of the tropical low altitude plants such as *Storckiella* and the *Xanthostemon* species, for example. The latter genus, with many beautiful species, although easy from seed appears to be particularly difficult to keep growing.

Certain species – *Astelia chathamica* (**Fig. 1c**), *Caldcluvia rosifolia, Metrosideros collina, Xeronema moorei* and *Metrosideros laurifolia* among them – are now being offered by some 'rare plant' nurseries. I myself believe that some of the other plants still also have much to offer. The Chilean lobelia, *L. bridgesii* (**Fig. 3a**), I consider to be better than *L. tupa. Sophora macrocarpa* is the most robust and best flowering of the genus. My dwarf form of *Metrosideros laurifolia* (previously *Carpolepis*), the natural bonsai-like *Metrosideros tetrasticha* and *Syzygium tripetalum* from the 1600-metre-high summit of Mt Humboldt on the central ridge of New Caledonia, are fantastic and reasonably hardy plants.

Figure 3. a) The Chilean lobelia *Lobelia bridgesii,* b) *Puya chilensis* from Chile, c) *Lomatia ferruginea* introduced from Chile, d) *Dacrycarpus imbricatus* var. *patulus* from Fiji, e) *Latua pubiflora* and (f) *Proustia pyrifolia* from Chile.

Figure 4. a) *Codia pancheri*, (b) *Xanthostemon aurianticus*, (c) *Parasitaxus usta* flower and (d) fruit and (e) *Dubouzetia confusa* all introduced from New Caledonia.

A range of very many other not so well-known species manage to do very well here in the Otways, including *Persea lingue, Ribes magellanica, Lilium mackliniae, Latua pubiflora, Drimys andina, Nothofagus codonandra, Nothofagus glaucus, Metrosideros pophryrea, Amentotaxus formosana, Podocarpus gnidioides,* the climber *Proustia pyrifolia, Senecio candidans* etc., and are definitely well worth a place in the garden. On the other hand, warmer climate shrubs such as *Dubouzetia confusa, Geissois hirsuta, Grevillea exul, Callistemon pancheri, Dacrydium nausoriensis* and *Stenocarpus milneii* are flourishing and flowering in other parts of southern Victoria.

Expedition And Plant Details

When collecting plant material in the wild, either for propagation or herbarium purposes, it is vital to provide details of the provenance of the collection locality of any specimen taken, such as the latitude/longitude, altitude and environment. As an initial step, the collector assigns his or her own unique 'collection number' to each and every specimen. Generally, this number includes an indication of the year of collection. It is the collector's numbers which tie the individual specimen to the collection records maintained for the particular plant specimen, perhaps to determine cold tolerance for example or soil requirements.

All the material listed herein was collected with the required permissions from the authorities of the countries or native landowners involved (for example, in Fiji) and processed by the Australian Quarantine Service. All of the material imported was initially handled by one or another of our major botanic gardens (identifications, quarantine etc.). Species which are listed on CITES (the Convention on International Trade in Endangered Species of Wild Fauna and Flora) were formally cleared through relevant authorities.

So, What Was The Point of it All?

It should be stressed that our own ventures were never intended to be for profit. As a result of our plant collecting expeditions, for example, the Royal Botanic Garden Sydney got many new species for its 'Araucaria lawn', Melbourne Botanic Garden was able to create its New Caledonian bed, and I in return eventually obtained those rare conifer species that had never been cultivated anywhere. It was only on one early expedition that a large seed collection was made for a commercial nursery and for various reasons this was never repeated.

It is quite likely that only a few of the plants introduced between 1985 and 2000 may ever have a major commercial value but as known-provenance collections, the plants have definite value for *ex-situ* conservation (conservation outside the original habitat). The species marked with an asterisk in the tables (*) are those which I know came into limited cultivation here in Australia, but I presently can find no evidence of whether they still survive. The author would welcome any feedback from those who may have information as to whether these are still in existence in Australia. These novelties have changed my own garden and also those of several botanical institutions. And, despite the views of those who argue for the status quo – an all – 'Australian natives' garden, for example. How dull would our gardens actually be if the plants in them had never changed over the centuries! It is perhaps rather sad that these days the selection of plants for a garden is largely dictated by the nexus between TV lifestyle shows, a couple of landscape writers/designers and the ubiquitous huge garden centres.

LITERATURE CITED

CBD (2011) Nagoya Protocol on Access to Genetic Resources and the Fair and Equitable Sharing of Benefits Arising from Their Utilization to the Convention on Biological Diversity. Secretariat of the Convention on Biological Diversity, United Nations Environmental Program, Quebec, Canada. https://www.cbd.int/abs/doc/protocol/nagoya-protocol-en.pdf Downloaded on 12 September 2023.

Combined Proceedings IPPS 73:141-145. 2023.

Mass Propagation in Plugs

Ian van Zanten

Ball Australia, 735 Western Port Hwy, Skye, VIC 3977, Australia

ianz@ballaustralia.com

Keywords: vegetative propagation, cuttings, unrooted cuttings, tissue culture

Summary

Bedding and perennial plant propagation includes programs for unrooted cuttings (URC), seed raised plugs, and tissue culture (TC) plants. The focus of this paper is our vegetative production program using cuttings.

INTRODUCTION

This paper will move through the vegetative propagation program at Ball Australia in its Melbourne facility from planning to dispatch. Our process is broken into three parts.

Part 1: Mother stock production

The mother stock production process starts with sales planning. This process starts 1-2 years ahead of the first plug being shipped. From there we will look at the Nucleus production, quarantine requirements, hygiene, and timelines. After that we will review the main mother stock production compartments. This will review climate, harvest planning, cutting spec, cutting collection, crop maintenance, staffing, cutting transport.

Part 2: Sticking cuttings

Sticking process starts with planning the benches for production flow. We will also look at cutting storage, staffing, sticking lines, watering tunnel, and tray media.

Part 3: Finishing production

Finish the production process includes reviewing our four climate zones, vapor pressure deficit settings and theory, movement days, watering by weight, plant growth regulator (PGR), trimming, preparing for dispatch, and shipping.

Mother Stock Production

I am the Growing Manager at Ball Australia responsible for all the growing in both the mother stock and the young plants nursery. Ball Australia is part of a global 4th generation family-owned business with its head office in Chicago, Illinois.

Ball Australia grows a wide product offering: Vegetative propagation, unrooted cuttings (URC), Seed raised plugs, and Tissue Culture (TC). The focus of this paper is our vegetative production program where we grow over 600 different varieties.

The whole process starts with a sales plan. We look at the sales curve for each variety and plant mother plants around demand. Availability is entered 18 months ahead of the current sales week. All of this is handled by our vegetative planning and product development departments. The same variety could be planted at three separate times in separate compartments to match up with the sales curve. All this planning involves long term thinking. An example of a planning/production schedule is presented in **Table 1**.

Table1. An example of a production planning schedule at Ball Australia.

Action	Timing	Notes
Liner Sales	WK 22, 2023	4 weeks to root a cutting into a liner.
Take Cuttings	WK 18, 2023	12 weeks minimum from planting to 1st harvest 4 -6 weeks rooting 8 weeks build up (pinching)
Plant Mother Plants	WK 6, 2023	4 weeks to root a cutting into a liner.
Start Mother Liners	WK 2, 2023	13 weeks to root out Elite stock and build up MS plug numbers needed.
Plant Elite Pots	WK 41, 2022	4 weeks to root Elite TC
Root Elite TC	WK 37, 2022	TC does not all arrive on time. TC started to arrive week 10, 2022

TC is ordered 6 months to one year ahead of arrival in week 5. The planning for new products in 2024 started in early 2022.

The Elite (Nucleus) house is the heart of the mother stock facility. We have strict hygiene procedures for entry into this growing area. The procedures are listed below:

- Change from street clothes into clean scrubs
- Change shoes into area restricted shoes
- Wash hands
- Put on gloves
- Bring a disinfection bottle of chlorine 0.6%
- Walk through air shower
- Walk through foot bath

Only a limited number of certified staff can work in this area. We carry a minimum of 4 Elite plants in over 600 varieties. Each plant is labelled and tracked via crop code through our production system.

Due to Australia's strict bio security requirements we need to operate our own Mother stock facility. Other countries can import cuttings from large cutting farms. This does not work in Australia. If we cannot source TC of a specific variety, it needs to go through 12 weeks in one of our three quarantine facilities. Once cleared these are moved in sealed containers to the Nucleus house, or into a production growing space.

Hand washing, disposable gloves and disinfection spray is mandatory in the whole facility. In the Nucleus house and the backout house no outside clothes are allowed. Staff must change into scrubs before entering. In the main production houses, lab coats are required for entry. All items are changed and washed daily. Within the mother stock facility, we have five production houses. These houses contain larger blocks of each variety based on yield, number and product demand. Four compartments are rotated to supply year-round production. One is a black out facility for short day crops. Only certified staff can work on these compartments. Each compartment is emptied and cleaned once a year for hygiene reasons. Thrip nets on all vents and concrete floors to keep pest pressure at a minimum.

Climate is monitored daily for discrepancies. This ensures the same quality cutting is available year-round. Some rows have day length extension lighting giving a longer supply season for some crops or, the ability to supply outside natural growing seasons. All areas use fog to cool and maintain humidly during warmer months.

Total harvest and maintenance team is 18 people. These staff are trained over a 6-month period. Each variety has a cutting specification. Harvesting teams use this to compare each week. This ensures consistency every week and helps with training new staff. Product is counted into bags of 105 and stored in cool room until sticking. The bags are collected in the greenhouse and moved to a fridge at 6 degrees in each compartment. Within an hour they are collected to a temperature-controlled transport cart and moved to our main cool room. There they are stored at 9 degrees for up to two weeks before sticking.

Sticking Cuttings

This section reviews the sticking procedure. Carts are prepared before sticking. Each shelf of a cart is one bench of product. Benches are created based on rooting timelines, movement days, and growing requirements. This is done by two staff at the end of each week to prepare for the start of the next week. We use two types of media, a loose filled that we fill on our sowing line and a Proforma glue plug. Specific crops are grown in each media. This depends on the final growing requitements of the specific variety.

We use a progressive sticking method. There are 5 staff members on the sticking team. Each staff member is responsible for a portion of the tray. The tray moves down the belt. The last staff member finishes and checks the tray before it goes through the watering tunnel.

Finishing Production

This section will finish the production process. During production we use a few different processes to ensure that our plugs are consistent every week. The first tool is vapour pressure deficit (VPD) cameras. Each fog zone has a VPD camera. These cameras measures plant leaf temperature, relative humidity, and humidity deficit. VPD is calculated each minute, and a running sum is tracked **(Fig. 1**).

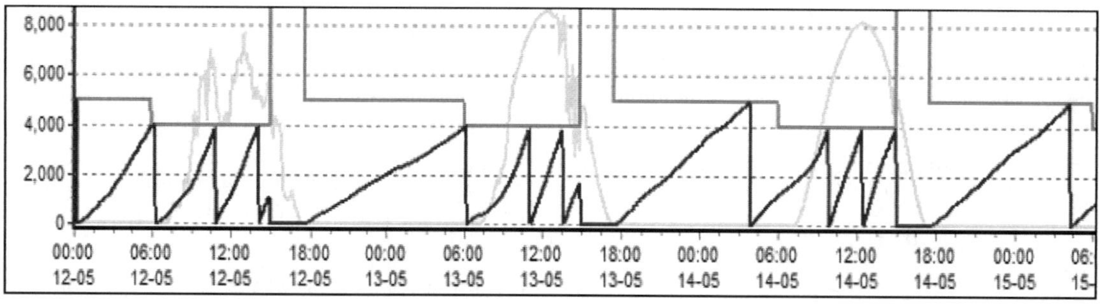

Figure 1. An example of an output from a Vapour Pressure Deficit (VPD) camera in a fog zone in the production facility of Ball Australia (Yellow line - radiation, red line - VPD sum target, blue line -the measured VPD sum).

Each time the VPD sum meets the target the watering boom is started to rehydrate the plants. This results in more passes during sunny days and less during overcast days. Also mean less thinking for the grower and a more consistent product with Melbourne's ever-changing weather. By day 10 we have gradually increased the VPD, reduced the mist and moved the product out of our fog compartments. **Figure 2** illustrates the VPD set points for different types of crops.

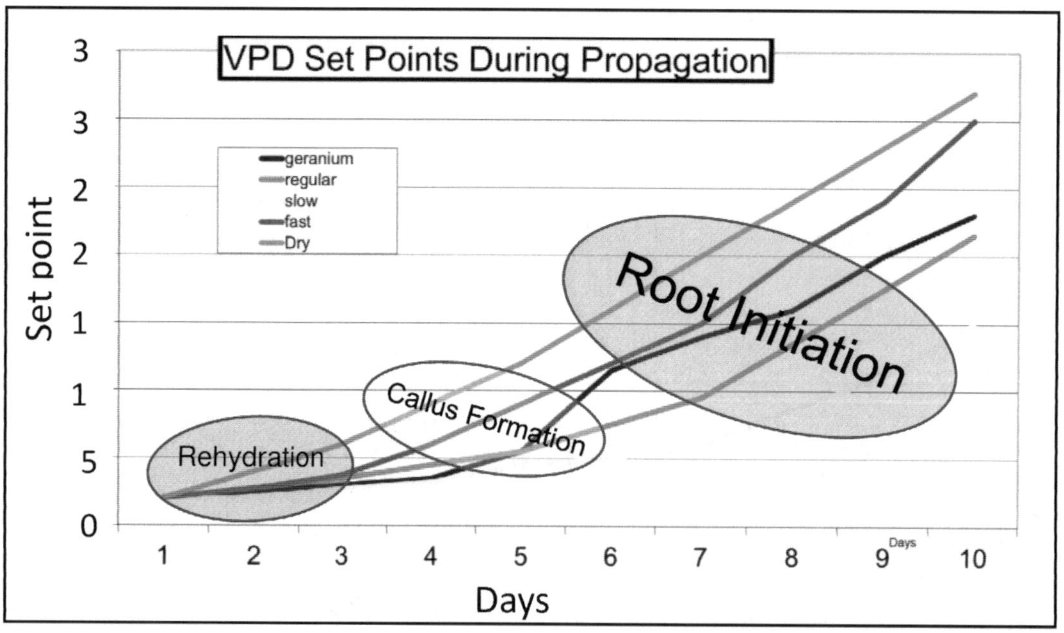

Figure 2. Vapour Pressure Deficit set points for crops with different requirements during propagation with geranium as a control (mid-range of requirements.

To water the crops, we use a water by weight system. All this data is tracked by an app that we designed in-house. We weigh indicator crops twice a day to measure dry back and calculate water up events. This teaches growers to keep the trays within the required range. There is no "too dry", or "too wet", only the weight.

For example, the discussion could go like this: "The tray weighs 900 grams this morning, it is losing 200 grams in 24 hours. The target weight to water up at is 700 grams. We will water it tomorrow". This also aides in diagnosing crop issues. We can tell if a crop has been over-watered or dried back too hard.

The next area of discussion is the different climate zone for each stage of propagation. The first stage of propagation is Fog 1: average temperature 24 degrees, average humidity 90%. Plants spend between 5 and 7 days in this zone.

The next area is Fog 2: average temperature 21 degrees, average humidity 86%. Plants spend between 2 and 7 days "Zoned On". Then they spend 3 to 5 days "Zoned Off". This means that the watering is controlled by the VPD (zoned on) or controlled by the grower (zoned off). This is where water by weight starts. Once first roots start, they are moved to the next climate zone.

The next stage is Climate Zone 3: average temperature 21 degrees, average humidity 70%. Plants spend between 7 to 14 days in this zone. Temperature sensitive varieties never leave this zone. Plants finish rooting in this climate zone. Once roots are established, they are moved to the next climate zone and trimmed if necessary.

The last zone is the Dispatch Zone: average temperature 16 degrees, average humidity 70%. Plants spend between 7 to 14 days in this zone. This is also where we harden off plants. This is also where they are tagged ready for dispatch to our customers. This is a cabrio style open roof building. It is maintained very close to the outside temperature. If the plants need to be trimmed this is completed two weeks before dispatch. Each variety has a set trim height specification. Some varieties will get up to 3 trims before dispatch.

The plants leaving this week were part of the sales planning process over two years ago bring us back to the beginning again.

Combined Proceedings IPPS 73: 146-152. 2023.

Plastic Propagation Additives and Recycling

Matthew Mills

MMGN Consulting Pty Ltd, Thomastown VIC 3074

mmgnconsulting@gmail.com

Keywords: environmental waste, containers, polystyrene, polypropylene

Summary

While plant propagation plastics are commonly used in Australia, there are initiatives in place to manage their disposal in an environmentally responsible way. These initiatives include the PP5 stewardship, a recycling program participated in by many nurseries and major horticultural suppliers. Unpacking plastic use as an input to propagation and its recycling for contained reuse in Australia, is a complex and engaging story of an industry's dedication to succeed.

INTRODUCTION

Plant propagation plastics are commonly used in Australia for the production of seedlings and cuttings in horticultural and agricultural industries. These plastics include various propagation trays, pots, and tubes, stakes and labels made of materials such as polystyrene (PS6) and polypropylene (PP5).

These plastics have proven effective for plant propagation and improved outcomes however, they can contribute to environmental pollution if not disposed of properly. In Australia, the recycling of plant propagation plastics is a fast-developing area. There are initiatives in place to manage horticultural plastics away from disposal streams toward closed loop recycling in an environmentally responsible way. These initiatives have been driven by unity of desire between component manufacturers and green life propagators combined within the Horticulture sector Australia-wide and are unique to fit Australian conditions.

PLASTIC RECYCLING INITIATIVE

Through the challenges of the combination of material requirements, landmass, population and cost-efficient circularity, a leading initiative involving many nurseries and 3 major horticultural suppliers in Australia, has started to offer recycling programs for their plant propagation plastics. This program collects used plastic pots and trays from customers and recycles them through specialist recycling processes which have a guaranteed end customer.

The combination of complicated steps to work through, has been achieved through dedication from Horticulture Sector businesses right across the country co-operating together to achieve a meaningful outcome for consumption and our responsibility to manage landfill. **Figure 1** symbolically presents the initiatives and support base to manage the proper disposal of plant propagation plastics used commonly throughout Australia.

Figure 1. A symbolic presentation of the required initiatives to manage the proper disposal of plant propagation plastics used commonly throughout Australia. The commonly used plastics (hand symbol) require proper disposal methods (circle). PP5 initiative by three major horticultural suppliers has support from many leading nurseries and the community.

What is Polypropylene Plastic and why is it problematic?

Polypropylene plastic (PP5) is an endlessly recyclable plastic used prevalently across a wide range of industries, and the horticulture industry (Industry) is no different. PP5 reaches every garden centre, retail and commercial production nursery in Australia in the form of plant pots, punnets, trays, labels, and stakes. The Industry sells over 12 million kilograms of PP5 as finished goods per year. Currently only 10% of this PP5 is recycled — almost all by us!

The Industry has historically used PP5 recycled from other post industrial polypropylene and often colours its PP5 'carbon black' to standardise the colour of the recycled material and enhance its UV stability for use in full sun. However, this colouring makes the PP5 unrecognisable on Materials Recovery Facility (MRF) sortation lines; thus, creating millions of kilograms of recyclable landfill even when it is placed in kerbside recycling systems.

Our Solution:

The PP5 Initiative (Initiative) is a joint venture between three major Australian businesses supplying horticulture: Norwood Industries, Garden City Plastics, and Polymer Processors. As a joint venture, our goal is to build an innovative infrastructure that allows our industry, and the communities that operate within, to sustainably consume and reuse our own PP5.

To that end we have developed a sustainable circular model (**Fig. 2**) whereby the PP5 sold by us is returned to us and converted into 'polymer granules' that are then remoulded into a form that can be resold. This video explains how we do it! We started from collecting 4,000 kilograms of PP5 in 2021 to collecting 100,000 kilograms per month in the second half of 2022! That is 10% of the PP5 the Industry sells!

Working as a closed loop system within Horticulture, manufactured recycled, carbon black PP5 from Polymer Processors is passed onto Norwood Industries and GCP to produce plant pots, containers, and labels for production nurseries throughout Australia. Nurseries then collect and return end of use PP5 or on sell growing plants, passing the products to landscape projects or retailers for sale to home gardeners. Landscapers, retailers, and home gardeners are then all included in this recycling initiative by being either able to provide a PP5 collection point or return material to the nearest one.

Figure 2. Sustainable circular model of PP5 Initiative.

Since the Initiative's launch in 2020, we have collected over a million kilograms of PP5 to recycle that would have otherwise been sent to landfill. We have fostered a community of over 100 partners, not to mention the countless members of the public. Our partners have installed over 700 collection sites that deliver otherwise end-of-use PP5 to one of our shipping and processing facilities.

How did we get here?

First, we had to develop the infrastructure. We have seven major shipping and processing facilities all around Australia to convert end-of-use PP5 products to polymer granules; including a specialised facility in Queensland that sanitises collected PP5 of an invasive fire ant species (*Solenopsis invicta)* harmful to Queensland's ecosystem (WPSQ, 2022) before the PP5 is transported to other states. We then expanded our existing delivery and transportation network so that PP5 products can be collected and delivered to Melbourne for manufacturing without any cost to the community!

Next, we had to engage our industry community. To amplify the collective action, we ran education and awareness campaigns regarding how our community can collaborate with the Initiative. Underlying these campaigns is our collaboration with various organisations with the platforms to extend our reach such as 'Recycle Mate'; a partnership between the Australian Council of Recycling, the federal government and Greenlife Industry Australia (RM, 2023), Australia's national peak body for production and retail nursery businesses.

Finally, we needed commercial partners to host our collection sites. As our goals resonated with the values of our community, we had no shortage of interest from prospective partners. However, hosting collection sites incur some level of expense. These expenses made some prospective partners reluctant to sign onto our Initiative at first. But as we built the Initiative's brand and reputation, the more members of the public have come to expect our collection sites as their preferred retailers, the stronger the reputational benefits of hosting our collection sites have become.

Our commercial partners include Bunnings and Mitre 10 among some 100 others. The organic demand for more collection sites is now so overwhelming that we anticipate having over 1000 collection sites with over 200 collection partners by 2025! The key to our success has been developing an infrastructure that adds benefit to all contributing participants. Our Initiative not only supports, promotes, and strengthens sustainable green life production in our communities, it also makes the process as economically viable and convenient as possible. Our three pillars of success are presented in **Figure 3**.

We are more than economically viable. We have completely overcome our cost of transport. Once PP5 is returned to our collection sites, our three-step-process — identify, tap, and stack — efficiently transfers that PP5 to our shipping and processing facilities free from impurities, contaminants and wasted space on transports. This process is now so refined that it costs us the same to recycle used Industry PP5 as it does to purchase other recycled PP5 from outside industries or import PP5 into Australia. We anticipate that by 2025 it will cost us less to recycle our industry's own used PP5 than it would be to source elsewhere or import PP5 into Australia, savings which we will pass onto growers via holding down prices on recycled PP5 products. Other economic benefits the growers enjoy include reduced landfill fees and fewer costs managing scrap plastics. We have created an engaging financial incentive to responsibly consume and reproduce our PP5!

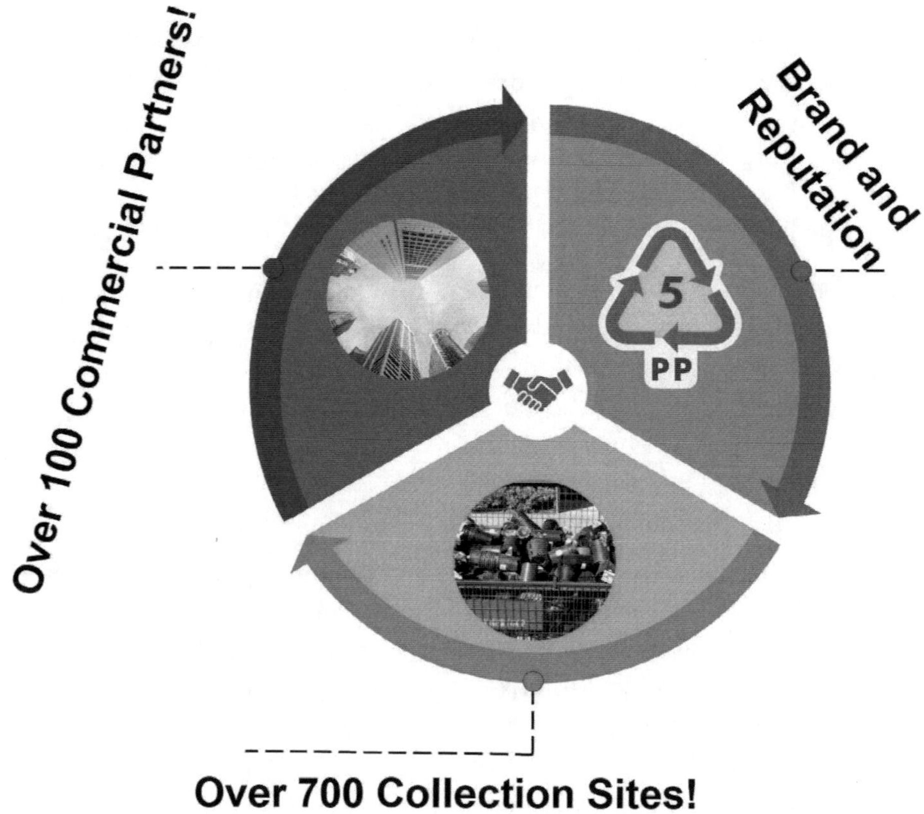

Figure 3. Three pillars of success of horticultural plastic recycling in Australia: increasing number of commercial partners and their collection sites with the development of an incredible reputation of our brand.

To amplify the collective action, we run education and awareness campaigns regarding how our community can collaborate with the Initiative. Our community of sustainably-minded members of the public are core to our Initiative. We market ourselves online through several channels (**Fig. 4**):

- The PP5.com.au website (PP5, 2023)
- PP5 Recycling For Schools App (PP5, 2023)
- Through social media on Instagram and Facebook.

We also engage our community directly via:

- Various commercial industry events
- Conferences with industry peak bodies and the public
- Talkback radio interviews
- Displays at the Melbourne International Flower and Garden Show
- Horticultural Society events
- Delegations to State Parliaments

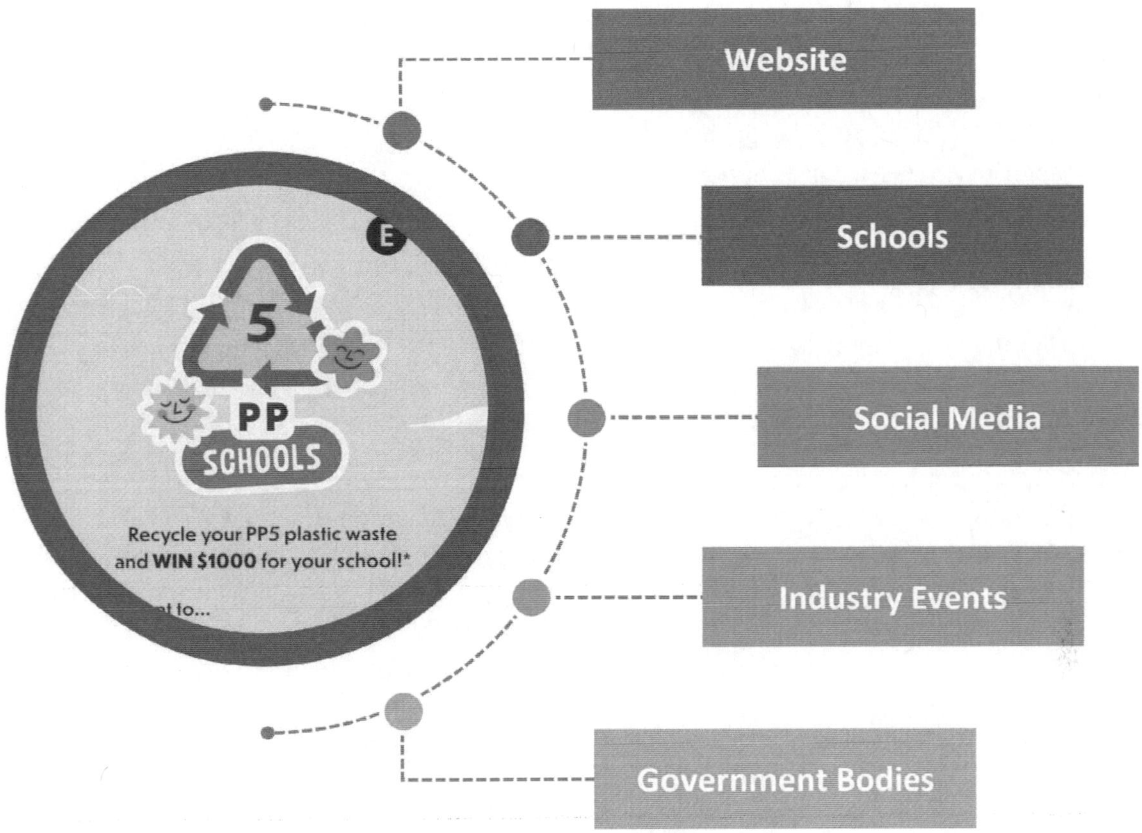

Figure 4. Members of the public are core to recycling effort. PP5 Initiative has a multifaceted approach to educate and collaborate with the members of the public.

Future Perspectives

Given Australia's increasing focus on ethical and sustainable use of plastics, our list of prospective partners is limitless. Our focus for 2023 is three-fold:

- Continue expanding our lists of commercial partners to install even more strategically located collection sites
- Collaborating with local councils to require transfer stations that can serve as both centralised collection sites, and a governmental platform to promote the Initiative
- Collaborating with the Nursery & Gardening Industry Victoria (NGIV) to run several programs in schools across Victoria to educate the leaders of tomorrow, as well as engaging NGIV's interstate counterparts to expand our school programs across Australia!

The versatility of our process extends beyond horticultural PP5 products. We are expanding the capacity of our collection sites to pick up PP5 products outside of the Industry. Our partnership with HexaCover® enables their floating cover system for controlling and deterrent of unwanted waterfowls to be a part of our sustainable circular model. Our goal is to eventually be a one stop shop for all 'carbon black' PP5 recycling!

Last, but not least, is the scope for our community to be used as a platform to promote other sustainability efforts! This is why we hope to partner with the Banksia Foundation (BF, 2021). This foundation has the cross-sector reach that we need to leverage our growing community of like-minded individuals and organisations to promote other sustainability initiatives.

LITERATUE CITED

BF (2021) Banksia Foundation. https://banksiafdn.com/ downloaded on 18 August 2023

PP5 (2023) Polypropylene # 5 recycling in Australia. https://www.pp5.com.au/ downloaded 18 August 2023

RM (2023) Recycle Mate. https://recyclemate.com.au/ downloaded 18 August 2023

WPSQ (2022) Wildlife Preservation Society of Queensland. https://wildlife.org.au/fire-ants-threaten-wildlife-and-ecosystems-in-australia/ downloaded 18 August 2023.

Combined Proceedings IPPS 73:153-154. 2023.

Propagation in the Community

Jane Edmanson

Prahran, Vic 3181, Australia

edmanson@bigpond.net.au

Keywords: restoration, society, schools, education

Summary

I was born in Mildura, in North-West Victoria, growing up with horticulturist parents who farmed oranges, grapefruit, avocados and lemons, all on the banks of the Murray River. It was a family affair, from the growing and harvesting the fruit, delivering to the markets, plus a venture showing tourists how a citrus farm operates. My profession started as a secondary teacher in rural Victoria, and after a few years on, an impulse got a job at the State Schools Nursery. Quickly my love of plants became an obsession, with propagation of Australian natives in tubes and perennial plants to the fore. I loved giving garden talks on all sorts of topics, and still give many to garden clubs and schools. I moved onto becoming a radio and television personality, and a voice for the gardening community. The long running ABCs Gardening Australia has been a joy for all its 34 years, and I continue to enjoy presenting it.

I have had so much pleasure in the many different aspects that a love of spreading the word about gardening takes me. I co-owned a retail nursery in Melbourne, showing how to deal with common problems that arise in suburban gardens (how many lemon tree questions arise?). I have been involved in establishing a variety of organisations to bring all sorts of people to the joy of gardening, often involving propagation as a domestic activity. For instance, setting up Men of the Trees branch to replant a large number of trees, the Rail Trail Association, the Australian Garden History Association, the Ornamental Plants Collections and Horticultural Therapy Association. The Schools Gardens Awards is a wonderful garden-based program for all schools and students of all ages, and I have been part of it for its 40 years plus in existence.

Gardening in the community is very much alive and well, and I would like to promote young people to venture into our wonderful horticultural world.

INTRODUCTION

I grew up in Mildura, a rural town in northwest Victoria, and was surrounded by orchards of apple, pear, oranges, plums etc. growing in the rich alluvial soils of the mighty Murray River. Growing up in a citrus farm, I saw the operations that needed to be accomplished to get the fruit into the store and from there to the market. Although I started as a rural teacher, I quickly moved to run my own nursery and worked for State School Nursery. My passion is not only gardening and planting but getting others to grow plants. With this passion, I moved on to become a radio and television personality and a voice for gardening in Victoria for the past 30 years through radio and long running ABC (Australian Broadcasting Corporation) program Gardening Australia. Here is my view of gardening in the community.

Plant Propagation for Home Gardeners

Where does the mystery and magic of growing your own plants start? Invariably when you are starting out, maybe as a wee kid, around six years old. Probably from a family member, grandmothers especially, are full of enthusiasm. Many early attempts at propagation, start at either seed in early days, bulbs and then moving on to cuttings, sometimes with lots of failures. However, by asking many questions, enthusiastic help from family who were horticulturalists, my success rate improved, and it became a lot of fun.

The process of plant propagation is intriguing, most people being really interested in the final product—- seeds for lettuce, acorns for future shadedown the track, blueberries for their fruit. Once hooked, it is a discovery of how challenging and exciting the hobby can be. The "laws" that govern propagation will dispel a lot of the mystique, it is not rocket science for a beginner, but can have more elaborate setups; hot beds, small greenhouses, a misting system as one improves.

Whether you decide to start off with only a few plants as a hobby or growing for a "cause" for a local conservation group, or for the local school or church— it doesn't matter. As William Blake, the English poet wrote, by planting a seed "you hold infinity in the palm of your hand".

Combined Proceedings IPPS 73:155-162. 2023.

Grafting Australian Native Species

Tony Hughes

Gordon TAFE, Corner of Latrobe Terrace and Gordon Avenue, Geelong, Victoria, Australia 3220

thughes@gordontafe.edu.au

Keywords: *Eremophila, Myoporum, Prostanthera, Westringia*, propagation

Summary

Eremophila is a diverse genus mostly from the drier regions of inland Australia. This has presented several difficulties for propagators in the southern states with colder and wetter climate. Constructing microclimates is one way to overcome the issues of cultivation, but more success can be achieved if the plant is grafted onto one of two different *Myoporum* species. Similarly, the genus *Prostanthera* also has several species originating in drier, more arid climates and cultivation of these species in colder and wetter climates can also be achieved with grafting onto a more suitable rootstock. The wider range of *Prostanthera* spp. are only just starting to become more widely known and the scope for cultivation of grafted plants is becoming larger as the less common species are being propagated. This paper introduces *Eremophila* and *Prostanthera* species as well as the rootstock genera of *Myoporum* and *Westringia.* The grafting techniques used, and best matches of rootstock and scion spp. that have been refined over the years, and maintaining grafted plants and their post-grafting best practices for handling are discussed.

IPPS Vol. 73 – 2023

INTRODUCTION

The genus *Eremophila* (Scrophulariaceae) is fascinating due to the sheer variety and diversity of species in the genus (**Figs. 1, 2, 3**). They are a very useful genus, with a shape, size and flower colour to please nearly everyone. *E. gibbifolia* (**Fig. 2**) is in my top ten of favourite plants. It is a great plant to capture the interest of new students.

Figure 1. The contrasting flower structure among species used at TAFE, Geelong to teach students the basic principles of identifying plant families, genera and species. In this example, *Eremophila* has five calyx parts and a notch between top petals (a and b), Genus *Prostanthera* has two calyx parts and a gap between two top petals that separates only half-way up (c) and *Westringia* with five calyx parts, but the five petals are fused at the base (d-f).

Prostanthera (Lamiaceae) is another fascinating genus with many Australian natives (**Fig. 1**). They grow in almost every situation and have a big range of flower colours and leaf textures. Not to mention the amazing range of aromas. *Prostanthera* spp. have that in abundance. *Prostanthera magnifica* has purple flowers that can attract attention in any flower show.

Figure 2. *Eremophila gibbifolia* is among my ten favourite plants. It can attract immediate attention of students in the classroom.

The genus *Westringia* gets a mention here mainly because of its use as rootstocks, but there are a number of species and cultivars that are great additions to any garden. I like using these three to help teach students the need for good observation techniques to tell different species apart (**Fig. 1**).

Grafting For Survival

The main reason we need to graft certain species of both *Eremophila* and *Prostanthera*, is to help them survive the wetter conditions of the winters here in Geelong in particular and southern Victoria and South Australia all the way to Southern New South Wales in general.

All of these plants dislike 'wet feet', and the Geelong winters can not only be wet but have constant coastal drizzle, which keeps the foliage damp and eventually leads to rot. While a lot of the species can survive if kept in pots, cultivation on the ground, even in raised beds, needs a bit of assistance.Some of the popular *Eremophila* spp are shown in **Figures 2 and 3**.

Myoporum is the genus to use for grafting *Eremophila*, with *M. insulare* and *M. montanum* being the two most reliable (**Fig. 4**). My personal preference is to use *M. montanum* as the end sizes of both the rootstock and the scion are roughly the same.

Figure 3. Some of the popular *Eremophila* spp. *E. foliosissima* (a), *E. fasciata* (b), *E. mulleriana* (c), *E. macdonellii* (d) and *E. splendens* (e)

Figure 4. *Myoporum* is used as rootstock for grafting *Eremophila* spp. (a, c) *Myoporum insulare* (b, d) *M. montanum,* the more compatible species in terms of stem size.

For *Prostanthera,* (popular species shown in **Fig. 5**) I have stopped using *Westringia* 'Wynyabbie Gem' as a roost-stock as it also dies under wet conditions and have swapped to *Prostanthera scuttelaroides*, which seems to have a greater tolerance for damp conditions (**Fig. 6**).

Incompatibilities are mostly based around the end sizes of both the rootstock and the scion. **Figure 7** shows the unequal growth, but it hasn't necessarily meant the end of the plant. These examples have still lived for a number of years, so if you can stand the slightly ugly look, you still get the benefits of a stronger plant.

Figure 5. Some of the popular *Prostantheara* spp. *P. striatifolia* (a), *P. teretifolia* (b), *P. ovalifolia* (c) and *P. calycina* (d). Photo credit Miriam Ford.

Figure 6. The two species used as rootstocks for grafting *Prostanthera* spp. (a) *Prostanthera scutellaroides* (b) *Westringia 'Wynyabbie Gem'.* The former is more tolerant of 'wet feet' and hence preferred for grafting.

Figure 7. Size of scion and rootstock can result in incompatibilities, but some grafted plants survive for long (Photo credit Amanda Shade, Kings Park Botanic Gardens, WA).

Rootstock Preparation, Cleft Graft, Parafilm and Tying the Graft Union and Planting

In the rootstock plant to be used, create a normal sized cutting, but leave 2-3cm above the top leaf for the graft site (**Fig. 8 a-c**). Scions should have a maximum of 3 or 4 nodes, and the leaves should be cut in half to help reduce moisture loss. Cut the bottom of the scion to a wedge to expose the vascular tissues on each side. (**Fig. 8 d-e**). It must be a wedge, not a point. If you cut to a point, you will have removed all the vascular tissues. Thereafter, cut a slit in the middle of the rootstock to 3mm longer than the wedge of the chosen scion. Then gently slide the scion into the wedge until all the exposed tissues are inside the slit.

Parafilm™ is a laboratory tape that has a waxy coating and is used like cling wrap to cover beakers etc. and can be sourced from laboratory suppliers or online marketplaces. Cut a 1cm wide strip of Parafilm from the roll. This one strip will stretch enough for 3 or 4 grafts. Peel the backing off the strip and hold it between your fingers to warm it up a bit. Then stretch out the first third of the strip to about 3cm. Start by rolling the Parafilm around the very bottom of the slit, go around once and let it stick to itself. Continue up the graft in 5mm increments, keeping a slight tension on the Parafilm.

Working upwards closes the wedge on the scion material, making for good contact of the tissues inside the graft union. Go to at least 5mm above the top of the graft to seal off the wound, and then go back down about halfway before breaking or cutting the Parafilm. Roll the end until it sticks to itself.

The aim is to have the wound nice and tight. This keeps the two lots of vascular systems in contact with each other, hopefully making the healing process quicker. It is also necessary to have the union waterproof, as the presence of water may encourage the wrong types of tissue to form. The stages of establishment of a cleft graft are shown in **Fig. 9**.

Figure 8. Preparation of rootstock (a – c) and scion (d-e) for grafting. Wedge shaped cut in the scion shown in (e).

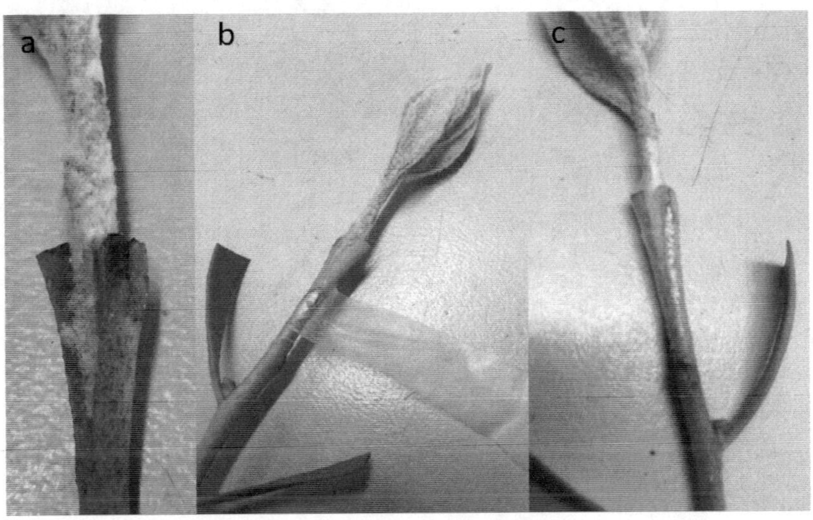

Figure 9. The stages of making the cleft graft. a) Sliding the wedge of the scion to the slit made in the rootstock, b) Starting the wrapping process with Parafilm from the bottom of the wound and c) complete coverage of the wound with Parafilm, ensuring water repellence and good contact of vascular tissues of the scion and the root stock.

It is important to keep the stems moist while keeping the tops dry. Dip the cutting in hormone and plant in cutting pots. To induce rooting IBA 4000 powder can be used and for planting Perlite and Peat moss mix with a ratio of 4 Perlite to 1 Peat Moss is recommended. Capillary pots can keep the tops of grafts dry, as they are watered from the bottom. Pots used for capillary watering have extra holes on the base of the pot, this enables the media inside the pot to have good contact with the mat underneath, allowing water to move from the mat into the pot via capillary action. The capillary mat is made from a hydrophilic material that holds water and allows it to move up into the pot that is placed on it. The material is usually a synthetic material that will last for a number of years.

Water the grafts making sure not to wet the tops. It is best if the foliage stays dry, so use a small watering top to aim between the cuttings. Small bottle tops are ideal or drill out the lid of a soft drink bottle with 1mm drill bit. Providing increased humidity is important, but it is also necessary to keep the foliage dry. Various types of domes are available for this purpose.

Managing the Grafted Plants

It is common practice to have small plastic bags over the grafts of Grevilleas, but the furry foliage of *Eremophila* makes this problematic. This type of foliage has the potential to rot when kept wet by the contact with the plastic. Any growth from the rootstock should be removed, but the original leaves from your cutting should be left behind. These leaves continue to supply energy to the scion and should not be removed until the top growth is the size of a tennis ball. Watch out for shoots off the side of the slit used to form your graft, these must be removed straight away and can be cut flush with the stem. Perlite attached to the roots can be left at the time of potting. Potting the grafts is similar to a normal cutting; one must be careful not to handle the top of the graft so as not to stress the new union. Leave the Parafilm on, it will fall off itself later. Use 60mm tubes or 75mm. An *Eremophila* graft ready for transplanting is shown in **Fig. 10**.

Figure 10. Grafted *Eremophila* ready for planting out.

CONCLUSION

Grafting *Eremophila* and *Prostanthera* onto rootstock that are more adapted to the wetter conditions of Geelong and similar environments allows higher survival and better growth of the plants on soil.

Combined Proceedings IPPS 73:163-168. 2023.

A New Plant Variety Rights Law

Chris Barnaby

Plant Variety Rights Office, Intellectual Property Office of New Zealand, Ministry of Business, Innovation and Employment Private Bag 4714, Christchurch

chris.barnaby@pvr.govt.nz

Keywords: Plant Variety Rights Act, Treaty of Waitangi, principles of The Treaty, UPOV Convention, indigenous plant species, scope of protection, essential derivation, harvested material, examination and testing, costs of protection, infringements, parallel laws

Summary

Plant Variety Rights (PVR) are an intellectual property Right specifically developed for plant breeders, providing a tool for the commercialisation of cultivars and the opportunity to make a return on their investment in developing new plant varieties.

The new law meets obligations under the Treaty of Waitangi, the 1991 UPOV Convention and the Comprehensive and Progressive Agreement for Trans-Pacific Partnership (CPTPP). The Waitangi Tribunal report for Wai262 has formed the basis of change in the management of applications for taonga species and the 1991 UPOV Convention has provided guidance and recommendations on what is included in the new law including the greater scope of Rights, the addition of Essential Derivation and limited Rights over harvested material.

The new law provides a more comprehensive coverage of administrative elements including objection process, use of Hearings, rules of evidence, the Right to be heard and appeals. The examination and testing process in practical terms is largely unchanged with some new provisions regarding the payment of fees, the supply of photos and access to plant material for variety testing. Infringement provisions are clearly set out and cover what is authorisation of the breeder, what constitutes an infringement, when an action can be taken and types of relief.

For the next twenty-five or more years there will be two parallel PVR laws in operation. All varieties granted or applied for under the PVR Act 1987 will continue unchanged because there are no retrospective provisions in the PVR Act 2022. The benefits of the new law will only apply to applications and grants under the PVR Act 2022.

INTRODUCTION

Plant Variety Rights (PVR) are an intellectual property Right specifically developed for plant breeders, providing a tool for the commercialisation of cultivars and the opportunity to make a return on their investment in developing new plant varieties. The PVR Act 1987 provided for the grant of a fixed term of intellectual property to breeders or owners over their new plant varieties, with an exclusive grant of Rights only applying to the production for sale and selling of propagating material of new cultivars. This now superseded law had become dated, had not kept up with new breeding and plant technologies and has had minimal amendment over the last thirty-five years. A new law is now finally a reality, bringing with it new obligations, requirements, and opportunities.

THE NEW LAW

Initial discussions on a new law began in the mid 1990's and progress was slow and challenging over the intervening decades. Momentum picked up in 2017 with the beginning of a series of consultations and huis spread over several years, a Bill was drafted in 2020, the first reading of the new PVR Bill occurred in Parliament in May 2021 and passed the third reading on 16 November 2022. The final step was the PVR Act 2022 (the Act) and the associated PVR Regulations 2022 coming into force on 24 January 2023.

The purpose of the new law is to provide an effective and relevant plant variety rights system that revises and consolidates the law on plant variety rights, consistent with New Zealand's obligations under the Comprehensive and Progressive Agreement for Trans-Pacific Partnership (CPTPP) in relation to the 1991 UPOV Convention, to protect kaitiaki relationships with taonga species and mātauranga Māori in the plant variety rights system and to promote innovation and economic growth in New Zealand by providing incentives for the development and use of new plant varieties while maintaining an appropriate balance between the interests of plant breeders, growers, and others so there is a net benefit to society as a whole

The Treaty of Waitangi

Ko Aotearoa Tenei, the Wai 262 report, provided the basis for the way in which PVR and taonga species were handled and central to meeting the Crown's obligations under the Treaty. The report is an extensive document and covered intellectual property and taonga works, genetic and biological resources of taonga species and the environment and mātauranga Māori. Although a relatively small component in the context of the full report, PVR is specifically addressed and the four recommendations in relation to PVR and taonga species were incorporated into the new law.

Part 5 of the Act recognises and respects The Treaty of Waitangi and provides for the additional procedures which will apply to applications for varieties belonging to taonga species. A key provision is the requirement to establish the Māori Plant Varieties Committee (MPVC). The functions of the MPVC will be

i. To administer additional procedures for applications for taonga species

ii. To recognise, protect and determine kaitiaki relationships

iii. To advise the Commissioner on mātauranga Māori in the plant variety rights scheme

iv. To refuse a grant of Rights for a variety that has adverse effect on kaitiaki relationships

The MPVC is currently in the process of being established and is anticipated to be in place late in 2023. One of its first tasks will be to set out how it will function and to draft guidance regarding how plant breeder's and Māori will engage with the MPVC. Engagement with Māori during the PVR review period highlighted the importance of kaitiaki being involved with breeders of taonga species at an early stage, before any PVR application is made. The MPVC will play a central role in fostering partnerships between native plant breeders and local Māori.

Taonga plant species

The MPVC has responsibility for all applications for varieties belonging to taonga species. The Act has no formal definition of taonga species and refers to indigenous plant species and non-indigenous plant species of significance. Indigenous plant species include all native plant genera and species and non - indigenous plant species include species brought to New Zealand in waka by early Māori. The list of non - indigenous plant species of significance is limited and included in the Regulations.

Varieties belonging to indigenous plant species or non - indigenous plant species of significance which are bred outside of New Zealand will not be required to be considered by the MPVC.

Impact on native plant breeding

The new law will require breeders using these species to have a level of engagement with Māori as a part of breeding activities and be aware that the PVR application process will include submission of the variety to the MPVC. The MPVC is now in the process of being established and at the present time there is a level of uncertainty regarding how the MPVC will carry out its work and how this important change will practically impact PVR for taonga species. The Act requires that kaitiaki relationships be acknowledged and addressed for taonga species and the MPVC and native plant breeders should aim to develop a collaborative way of working together for mutual benefit.

Using application data from recent years it is estimated that around 7% of applications are belonging to taonga species, in the order of 7-11 varieties per year will be required to be submitted to the MPVC. All other varieties belonging to non-indigenous species, over 90% of applications, will not be submitted to the MPVC and there will be no Treaty of Waitangi provisions applied to applications for those varieties.

Convention of the International Union for the Protection of New Varieties of Plants (UPOV)

Plant Variety Protection legislation in most countries is based on either the 1978 or 1991 Convention. The Convention consists of a series of Articles which list the requirements for national law compliance. The majority of UPOV member states are aligned with the 1991 Convention and the PVR Act 2022 upgrades New Zealand law to this standard.

The 1991 Convention provides for stronger Rights and introduces the novel concept of essential derivation.

SCOPE OF PROTECTION

The scope of protection has been expanded from a focus on commercial propagation and the sale of propagating material including whole plants to a much broader objective of commercialisation or exploitation of the whole variety. The existing Rights over commercial propagation, reproduction and multiplication are retained and continue to encompass offering for sale, selling, and marketing of plants of the variety. The scope has been extended to include conditioning for propagation, exporting, importing, and stocking for any of these activities. All these activities will now require the permission of the breeder.

An example of how the new law could make management of a Right easier is the situation where a breeder becomes aware of a nursery stocking one of the breeder's varieties, which the breeder did not know about. Under the new law, the presence or stocking of plants of the variety alone may be sufficient to initiate a conversation between the breeder and the nursery. There would be no need to first establish that the nursery was commercially propagating and selling the variety in order to potentially take infringement action.

A second example is the unauthorised export of plant material of a protected variety. The new law will remove the need to establish commercial propagation activity because the export itself is an infringement and who carried out the unauthorised propagation and how sale of the material occurred becomes a secondary matter.

The greater scope of protection has required an understanding of what is propagating material and what is harvested material. Due to modern propagation techniques identifying the two types is not so straightforward. The Act provides a definition of propagation material but does not provide a definition of harvested material. Harvested material is perhaps a more legal concept, not a technical one, and takes into account the intent of the user and what is usual or standard practice for the species concerned.

Essential Derivation

This provision is an entirely new concept nationally and there is no current equivalent in any other intellectual property. The concept of an essentially derived variety (EDV) has its origins in genetic engineering and the concern that a commercially successful variety could be genetically engineered by another breeder to create a different variety but remain genetically very similar, with essentially the same characteristics as the initial variety. One variety being genetically similar to another is not confined to genetic modification and could include in bred lines, repeated back crossing and mutations. In recent years, the development of new breeding technologies such as gene editing has brought more attention on essential derivation. Essential derivation provides the owner of a protected initial variety the possibility to share in the commercialisation of any other variety predominantly derived from that original variety. The derived variety must be determined distinct from the initial and all other varieties and can be protected.

Essential derivation is something of a balance between the important provision that protected varieties are freely available for further breeding and that of the second breeder acknowledging the contribution of the first variety to the second variety. The greatest challenge to Essential Derivation is the definition of a derived variety and how that determination is made. The Convention Article sets the framework for defining an EDV, which is mirrored in PVR Act 2022, however some national laws have a narrower and restricted interpretation where other national laws interpret the same Article in a broader sense. At the present time there is no consensus and discussion and debate regarding EDVs continues.

Harvested Material

The Act makes specific provision for assertion of Rights over harvested material because the scope of protection now covers broader commercial exploitation of propagating material and not just commercial sales as in the old law. Harvested material could include fruits, vegetables, cut flowers or grain. The 1991 Convention provides for the owner of a protected variety to have the possibility of asserting their Rights over harvested material, including entire or parts of plants, where there has been unauthorised use of propagating material. This can only be applied where the owner has been unable to assert their Rights at the propagation stage. This provision does not provide a choice for a breeder on when to assert Rights because the assertion of Rights over harvested material is not acceptable if this could have been achieved at the propagation stage.

An example may be where the owner of a pineapple variety protects the variety in New Zealand and then uses that Right to manage the importation of fruit of that variety from a Pacific Island nation. The owner may assert their Rights in New Zealand on the imported fruit because the Pacific Island may not have a PVR scheme, and the owner was unable to do this at the time of propagation.

Examination and Variety Testing

The existing examination and variety testing system will largely continue as it is with no substantive operational changes. The updated law provides the Plant Variety Rights Office with improved administrative options and sets out more clearly the processes involved. The new law also makes it clear that a growing trial is necessary for all applications and that DUS testing is an essential legal requirement but retains the existing flexibility available with respect to testing arrangements.

The new legal provisions will be noticeable to applicants during application with set time limits for the payment of trial or examination fees, changes to the requesting of plant material and photo requirements for applications for vegetable and potato varieties.

Cost of protection

In parallel with the development and drafting of the new law, PVR fees have also been under scrutiny. It was obvious early on that the cost of running a viable PVR scheme was not being met by fees paid by applicants and an increase was necessary. In addition, applicants expressed some confusion regarding the types of fees and when they are paid.

The PVR Regulations 2022 set out the new fees with the aim of providing a clearer fee set. There is a single application fee for all plant species and a single examination fee charged towards the end of examination, prior to a decision. Trial fees will be charged when testing begins and are based on the plant species and testing arrangement.

Hearings

The new law provides for a more transparent and clearer process for a breeder or a third party to object to the granting of a Right. During the review there was some criticism that the previous process for objections was unclear. The new provisions include the possibility of using the Intellectual Property Office of New Zealand (IPONZ) hearings systems. The hearings system is entirely separate from PVRO and potentially provides greater objectivity and neutrality.

The Hearings provisions in the new law are just one part of a more comprehensive coverage of related administrative process and the right of objectors to Rights to be heard. The provisions include timeframes for actions to occur, rules of evidence, proceedings process and appeals.

Infringements

The new Act provides guidance for infringement actions as seen in other intellectual property legislation. Part 3 specifically sets out details regarding what is a Right's holders authorisation and defines what may constitute an infringement. In addition, the types of relief are stated and when infringement proceedings can begin. Under the PVR Act 1987 it is possible to take infringement action at any time from the date of application. This includes provisional protection, during examination and testing, and after the Right's has been issued. The PVR Act 2022 limits when infringement proceedings can be taken to after the Right's decision only. This can be retrospective to include infringements during provisional protection but action during provisional protection itself is not permitted. The onus remains on the variety owner to assert their Right and use civil action when they think their rights have been infringed, in common with other intellectual property regimes.

Parallel laws

It is important to recognise that the PVR Act 2022 has no retrospective elements and any grant of Rights made under PVR Act 1987 will continue under that law for the life of that Right. This includes existing applications made before the new law came into force and grants for these varieties made in the future will be under the PVR Act 1987.

Taking into account the length of testing and if made, the term of grant, there will likely still be varieties protected under PVR Act 1987 into the 2040's.

Further Information

More information on Plant Variety Rights in New Zealand may be found at the following link:

Plant Variety Rights (IPONZ website): https://www.iponz.govt.nz/about-ip/pvr/

Gap in pagination due to formatting issues.

Pages 169-171

Combined Proceedings IPPS 73:172-182. 2023.

Mealybugs Demonstrate Feeding Preference Differences Between Different Grapevine Varieties

Ross Bicknell[1], Nigel Joyce[1], Manoharie Sandanayaka[2], Vicky Davis[2], Catherine Sansom[3], John van Klink[3], Michelle Thompson[1], Philippa Barrell[1], Lisa Watkins[1] and Adam Friend[4]

[1]Plant & Food Research, 74 Gerald St. Lincoln; [2]Plant & Food Research, 120 Mt Albert Rd, Auckland; [3]Plant & Food Research, Science Building 2, University of Otago, Dunedin; [4]Plant & Food Research, 55 Old Mill Rd, Motueka

Ross.bicknell@plantandfood.co.nz

Keywords: phylloxera, transmitted viruses, insect feeding assays, grape rootstocks

Summary

Rootstocks are almost universally used for grapevine because they provide protection against the root pest phylloxera. We are interested in other roles that rootstocks might provide, in particular the possibility of deterring mealybug feeding. Mealybugs feed by drawing sap from the plant's phloem tissue. While phloem feeding can weaken the plant, it is of particular concern because it efficiently enables the transmission of viruses. Mealybug feeding/survival was initially tested on eight grape varieties. Five varieties were then selected with varying responses and these were used to test mealybug feeding under a range of environmental and experimental conditions. Although the absolute numbers of insects surviving varied between treatments the ranking of the varieties remained constant. This information is now informing the development of a bioassay for mealybug feeding on grape. It is also being used to study the possible chemical basis of feeding deterrence in this material.

IPPS Vol. 73 – 2023

INTRODUCTION

European grapevine (*Vitis vinifera* L.) propagates readily by dormant hardwood cuttings. Given the large number of known varieties, their widespread distribution, and the mention of varieties in early writings, the striking of cuttings clearly dates back at least to the time of the Roman Empire and probably much further still. However, in the mid-1800s a series of pest and disease incursions struck European vineyards with devasting impact. The causal organisms were of American origin, introduced unintentionally by plant collectors who were gathering species from around the globe and bringing them to Europe for display and use. The full impact of these 'American Plagues' is unknown, but it is recorded that 40% of French vineyards died within a 10-year period, between 1863 and 1873 (Campbell, 2006).

Two foliar diseases: downy mildew (caused by *Plasmopara viticola*) and powdery mildew (caused by *Erysiphe necator*) and an insect pest called phylloxera (*Daktulosphaira vitifoliae* Fitch) proved to be the most difficult to control. Foliar sprays of sulfur and copper were subsequently found to control the mildew diseases, but the root pest phylloxera was largely immune to any applied substance. The solution to phylloxera control came with the discovery that some American grape species could be used as rootstocks, providing enough resistance to the pest that no other measure of control was required. The American species found to be resistant to phylloxera were *V. riparia*, *V. rupestris* and *V. berlandieri* (Munson, 1909). Although each species can be used as a rootstock alone, each also has limitations. *Vitis riparia*, for instance was easily grafted but not suitable for use on alkaline soils, which are common in Europe. Conversely, *V. berlandieri* grows well on alkaline soils but is hard to graft (Jackson, 2008).

In response, hundreds of crosses were conducted throughout the late 1800s, testing a wide range of species in different combinations to develop interspecific rootstocks that provided protection against phylloxera, as well as ease of propagation and good survival in a range of soil types. The rootstocks produced at that time are the same ones used today in most of the world. Phylloxera was first identified in New Zealand vineyards in 1902 by the notable viticulturist Romeo Bragato (Bragato, 1906). He immediately introduced the new European rootstocks. Subsequently, grafting became a standard practice for the establishment of commercial vineyards. Even today the main rootstocks used in New Zealand are Couderc 3309 (*V. riparia* x *V. rupestris*) bred in 1881, Schwarzmann (*V. riparia* x V. *rupestris*) bred in 1891, Millardet et De Grasset 101-14 (*V. riparia* x *V. rupestris*) bred in 1882, and *V. riparia* variety Gloire (https://www.vivc.de/index.php).

Our interest in grape rootstocks began with the observations that mealybugs appear to prefer some grape varieties more than others, and also that plants on different rootstocks vary in their rate of virus infection. Two mealybug species are common in New Zealand vineyards: the long-tailed mealybug (*Pseudococcus longispinus*) and the citrophyllus mealybug (*P. calceolariae*). Both feed by drawing sap from plant phloem tissue. While feeding can result in damage when infestation levels are high, a far greater concern is the transmission of viruses, mediated by this feeding behaviour (Petersen and Charles, 1997; Tsai et al.,

2010; Sandanayaka et al., 2013). In New Zealand, the virus of greatest concern is grapevine leafroll-associated virus 3 (GLRaV3). Infection by GLRaV3 leads to lost vigour, reduced yield and changes juice chemistry (Bell et al., 2021). GLRaV3 infection is incurable, so control options focus on the early detection of the disease, the replacement of diseased vines and insect vector control. The aims of the current study were to quantify differences in insect feeding on different varieties of grape, then to explore the potential metabolic basis for this phenomenon, with the long-term aim of developing mealybug-resistant rootstocks.

MATERIALS AND METHODS

Eight grape cultivars were used in the study: Cabernet Franc (the susceptible control), Malbec, *V. riparia* Riparia Gloire, Couderc 3309, Millardet et De Grasset 101-14, Schwarzmann, *V. labrusca x V. vinifera* Isabella, and Siebel 5437. Potted plants from each variety were propagated from dormant canes and maintained under glasshouse conditions until the vines contained a minimum of six leaves. One mealybug species was used, *P. calceolariae*, with insects sourced from a laboratory colony maintained on sprouting potatoes. *Pseudococcus calceolariae* was chosen as it is known to colonise both the root and shoot tissues of grape, while *P. longispinus* typically colonises grape shoots (Charles et al., 2006).

Mealybug feeding and survival

A longitudinal study was conducted to determine survival from neonate settlement through to sexual maturity. It also provided an estimate of the length of time required for mealybugs to complete their life cycle under the assay conditions, and to explore possible differences between a diverse set of grape varieties. The cultivars used for this experiment were: Cabernet Franc, Malbec, *V. labrusca x V. vinifera* Isabella, Siebel 4986 and Siebel 5437. An excised-leaf assay was used to facilitate the frequent counting of live insects. Individual leaves, removed from potted plants, were held in Perspex chambers with the leaf base immersed in a vial of water. Mealybug neonates were released onto greenhouse-grown leaves and the numbers of surviving insects recorded every 3–5 days. The detached leaves needed to be replaced every 3–4 weeks throughout the study period with the insects moved onto the fresh leaves on each occasion.

An attached-leaf, no-choice greenhouse study was then conducted to confirm the preference results observed in the detached leaf assay. The grape cultivars tested were: the rootstocks Millardet et De Grasset 101-14, Schwarzmann and Riparia Gloire; and the scion cultivars Cabernet Franc and Malbec. Ten healthy plants from each variety were chosen. Six leaves from each plant were then selected and labelled prior to the inoculation of mealybugs. The greenhouse was maintained at a minimum of 10°C and a maximum of 28°C, under natural lighting. An Eppendorf tube containing approximately 30, newly emerged first instar mealybugs was attached to the abaxial side of each individual leaf close to the base of the midrib with a small amount of Blu-Tack. Once attached to the leaf, the lid of the tube was opened to allow the mealybugs to find their food source and the leaf was then covered with a zip-lock fine mesh net bag (13 x 16 cm) (**Fig. 1**). Two weeks after the inoculation, the first infested leaf from the bottom of each grape plant was removed from the stem and brought into the laboratory to examine the settlement and development of mealybugs. Accordingly, a mealybug infested leaf from the bottom of each grape plant was removed once a week to examine for mealybug development; 2nd leaf after 3 weeks from inoculation and 3rd, 4th, 5th and 6th leaves were removed after 4, 5, 6, and 7 weeks, respectively.

Figure 1. Handling of mealybugs in the greenhouse. A) Release of mealybugs onto a leaf. B) mesh bags used to contain the insects during the study period.

A further (no-choice) study of survival was conducted using field-grown plants to test whether the greenhouse study was also representative of field insect survival. Field grown plants of: Cabernet Franc (the susceptible control), Malbec, Millardet et De Grasset 101-14, and Schwarzmann were used. Plants of Riparia Gloire were unavailable for this study. Neonates were transferred to leaves within mesh bags on the trial plants and insect survival was assessed from the second week, for 7 weeks after inoculation.

To determine whether the insects would demonstrate a preference for different varieties when given a choice, plants of Millardet et De Grasset 101-14, Schwarzmann, Riparia Gloire, Cabernet Franc and Malbec were grown together within a mesh cage and neonate insects were introduced onto a paper platform that touched each plant. Insect colonisation of each variety was then assessed 4 and 8 weeks after inoculation.

A final study was performed using root tissues. Neonate insects were bound within a mesh bag, tied around a fleshy root on each replicate plant. The varieties used were Millardet et De Grasset 101-14, Schwarzmann, Riparia Gloire, Cabernet Franc and Malbec. Mealybug survival on the roots was assessed at a single timepoint, 12 weeks after inoculation.

Phytochemistry

To analyse whole leaf chemistry, leaves were collected from two sites (Lincoln field and Auckland glasshouse) and freeze dried. In addition, leaf surface waxes were extracted from the Lincoln vines. Duplicate extractions were analysed throughout the study. For the whole leaf samples approximately 10 mg Dry Weight of the sample was mixed per 1mL solvent and for the wax analysis 1 mg wax was mixed with 1mL solvent. The phytochemistry was then analysed following ethanol extraction, followed by liquid chromatography–mass spectrometry (LCMS). A 2 µL aliquot of each prepared extract was separated with a mobile phase consisting of 0.1 % formic acid in type 1 water (A) and 0.1 % formic acid in acetonitrile (B) by reverse phase chromatography, maintained at 40°C with a flow rate of 400 µL/min. A gradient was applied: as 0-1 min/5%B, 7-10 min/95%B, 11-14min/5%B. The eluent was scanned from 1-11 minutes by API-MS (Orbitrap) with heated electrospray ionisation (HESI) in the negative and positive mode. Data were acquired for precursor masses from *m/z* 110–1200 amu at 70K resolution with data dependent ms/ms for product ions generated by normalised collision energy (NCE:30) at 17.5K resolution. Data were processed with the aid of Xcalibur®4.1 and Compound Discoverer 2.1. Each dataset was compared using Principal Components Analysis (PCA).

RESULTS

Insect Feeding Assays

The results of the longitudinal study of mealybug survival on different varieties are summarised in **Figure 2**, panel A. Mealybug neonates suffered significant losses immediately after transfer to the leaves of all the varieties tested but after the first week numbers stabilised. The highest survival rate was for the *V. labrusca* hybrid Isabel, indicating that this species is an unlikely source of mealybug resistance. The lowest survival rate was for the *V. vinifera* scion variety Malbec. On susceptible plants, at the time of settlement, the insects typically grouped together over a leaf vein (**Fig. 2, panel B**), while this was not seen on the more resistant varieties (**Figure 2, panel C**). At the end of the study (8 weeks after neonate transfer), male cocoons and fourth instar females were seen, indicating that 7–8 weeks of observations are required to follow development through the complete life cycle of these insects under the experimental conditions.

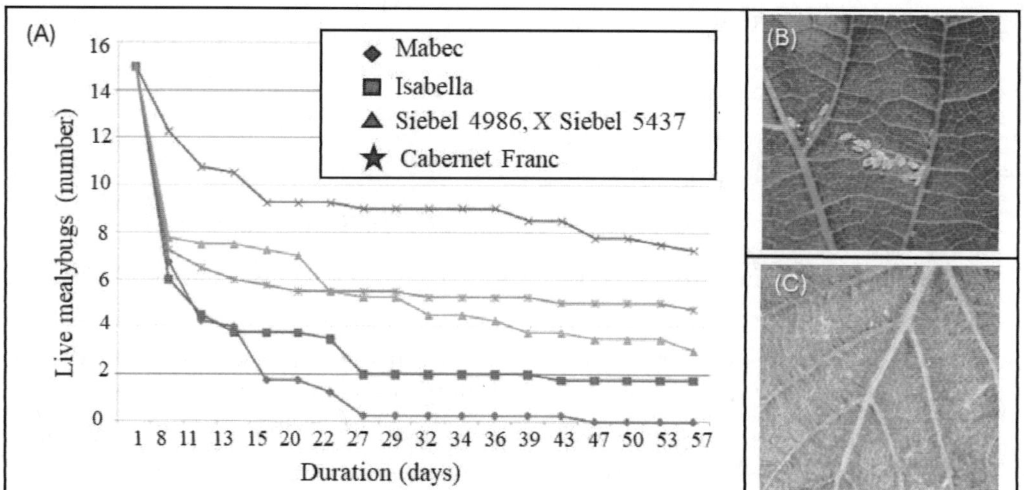

Figure 2. Mealybug survival on detached leaves. A) Longitudinal study of survival over 8 weeks. B) Mealybugs on susceptible variety Cabernet Franc after 4 weeks. C: Mealybugs on resistant variety Malbec after 4 weeks.

The results of the no-choice attached-leaf assay are summarised in **Figure 3**. At the 2-week time-point, no significant differences in mealybug settlement were apparent for the five varieties studied. However, after 7 weeks of observation, significantly more insects were surviving on the Cabernet Franc and Riparia Gloire plants than on Millardet et De Grasset 101-14, Schwarzmann, or Malbec.

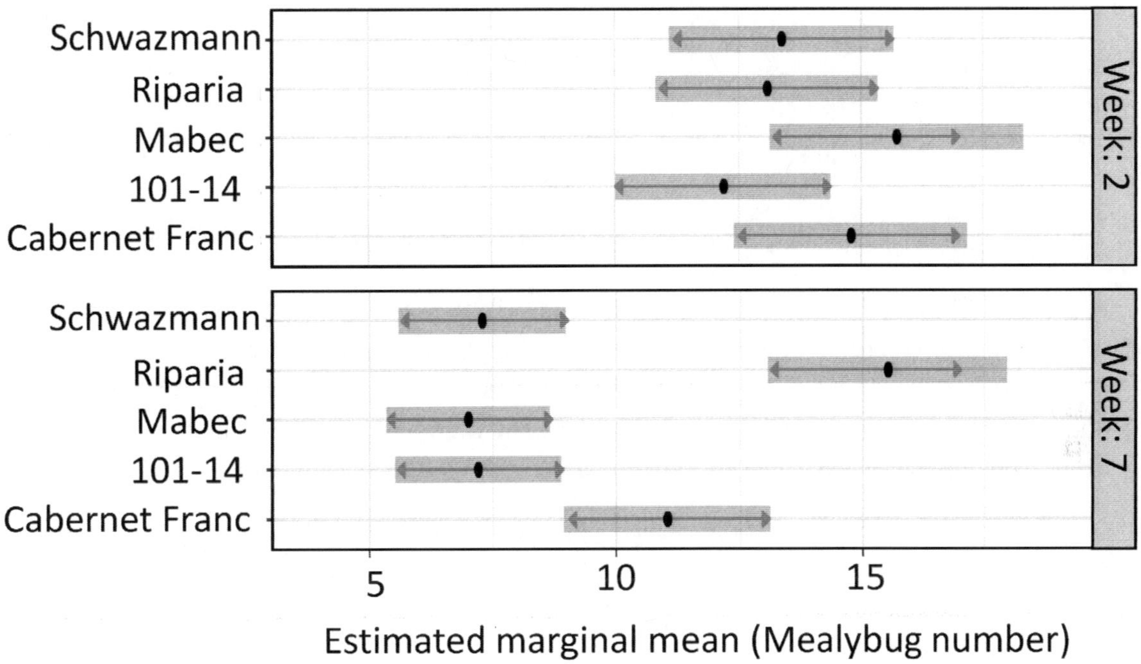

Figure 3. Attached leaf no-choice assay. Estimated means of mealybug numbers on leaves of five different grapevine rootstock varieties after 2 and 7 weeks from inoculation under no-choice conditions. The purple bars show confidence intervals for the estimated marginal means, and the red arrows are used for comparisons among them. If an arrow from one mean overlaps an arrow from another group, the difference is not significant.

Similar results were observed for the field-based no-choice leaf test **(Fig.4)**. Overall, mealybug survival on all varieties was lower in the field than in the glasshouse and the development rate of mealybugs in the field was slower. However, the rankings of the varieties were similar with insects having the highest rate of survival on Cabernet Franc and the lowest survival on Schwarzmann and Millardet et De Grasset 101-14.

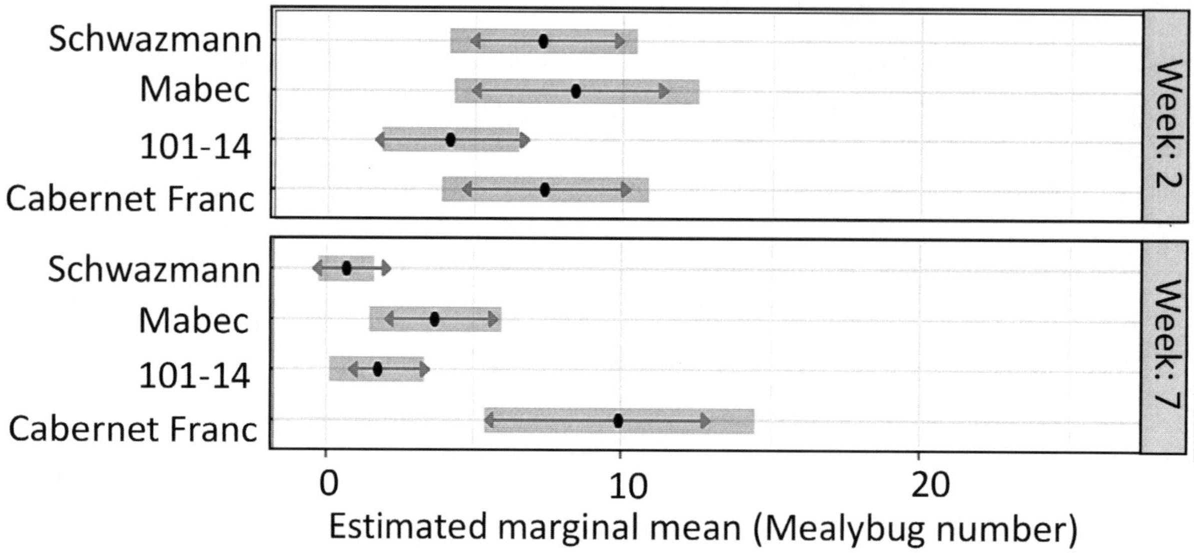

Figure 4. Field-based leaf no-choice assay. Estimated means of mealybug numbers on leaves of five different grapevine rootstock varieties after 2 and 7 weeks from inoculation under no-choice conditions. The purple bars show confidence intervals for the estimated marginal means, and the red arrows are used for comparisons among them. If an arrow from one mean overlaps an arrow from another group, the difference is not significant.

The results for the glasshouse choice test are summarised **in Figure 5**. At the 2-week observation point, insect preference for Riparia Gloire and Cabernet Franc is apparent as seen in the no-choice test, but this effect was no longer seen at the 8-week observation point.

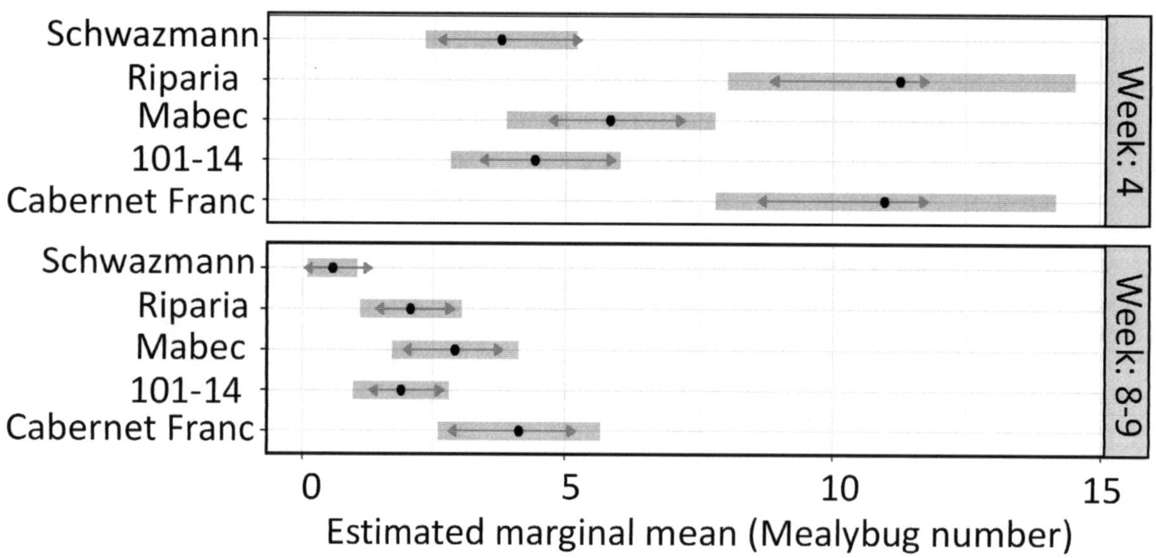

Figure 5. Glasshouse leaf no-choice assay. Estimated means of mealybug numbers on leaves of five different grapevine rootstock varieties after 4 weeks and then 8–9 weeks from inoculation under no-choice conditions. The purple bars show confidence intervals for the estimated marginal means, and the red arrows are used for comparisons among them. If an arrow from one mean overlaps an arrow from another group, the difference is not significant.

Finally, the results of the root, no-choice survival test are presented in **Figure 6**. A preference for the roots of Cabernet Franc is clearly demonstrated.

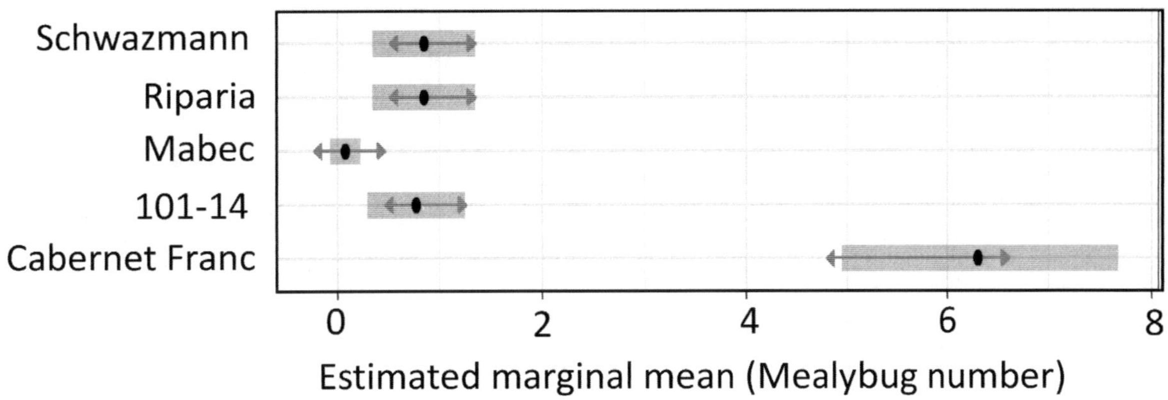

Figure 6. Root no-choice assay. Estimated means of mealybug numbers on roots of five different grapevine rootstock varieties after 3 months from inoculation under no-choice conditions. The purple bars are confidence intervals for the estimated marginal means. If a red arrow from one group overlaps an arrow from another group, the difference is not significant.

Phytochemistry

The results of the mealybug feeding assays were used to rank varieties by insect preference (**Table 1**) to determine whether any classes of metabolic chemical markers correlated with the observed feeding preference scores.

Table 1. Vine and grape relationship to mealybug feeding preference scores

Vine ID	Grape variety	Mealybug feeding preference
VID1159	Cabernet Franc	High
VID305	Malbec	Low
VID848	Millardet et De Grasset 101-14	Low
VID888	*Vitis riparia* gloire	High
VID890	Schwartzman	Low
VID858	Couderc 3306	Unknown

Chromatographic data from negative (Cn) ion mass features were used in a principal component analysis (Figure 7), that showed the Cn data representing polar small molecules, better explained (74%) the data variation compared to Cp (46%), across the first two principal components. The selected mass features from these reverse phase LC-MS analyses were then isolated and manually interpreted to identify their structure. This process identified hydrolysable tannins and some tentatively identified phenolics (see below) as compounds whose levels negatively correlated with mealybug feeding preference. The hydrolysable tannins of greatest significance were confirmed to be ellagitannins which are derived from β-

1,2,3,4,6- pentagalloyl-D-glucose. The key components of the extracted wax associated with low feed preference were tentatively identified as hydroxy tyrosol (HT) derived straight chain lipid esters of palmitic (C16), stearic (C18) and arachidic (C20) acids.

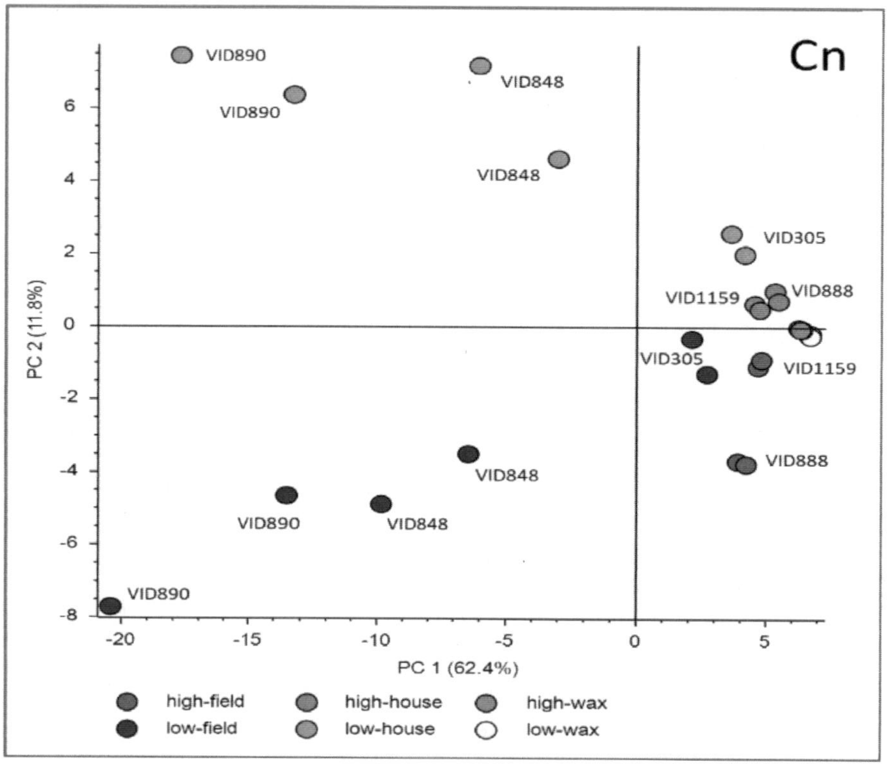

Figure 7. Principal Components analysis of negative (Cn) ion mode liquid chromatography-mass spectrometry.X-axis=Principal Component 1 and Y-axis = Principal Component 2, of leaf extracts from field (Lincoln vineyard) or house (glasshouse at PFR Auckland) and wax (Lincoln field leaf surface extracts) with high and low mealybug feeding preference; VID =vine identification (Cn, only).

DISCUSSION

Five feeding assays were designed to test mealybug preference, survival and developmental progress under a range of environmental conditions, including laboratory (detached leaf assay), greenhouse and field-grown plants. Also, both root and shoot tissues were inoculated, and both choice and no-choice tests were conducted to determine whether these factors influenced insect survival. An initial longitudinal study indicated that mealybugs complete a single generation in 8 weeks, under laboratory conditions. Subsequent tests were, therefore, typically analysed for mealybug survival at week 7. The exception was the root test, which for practical reasons was analysed between weeks 8 and 9.

Absolute mealybug survival rates varied significantly between the different feeding assays. The highest survival rates were observed in the greenhouse leaf assay, whilst the lowest were in the root and field assays. However, although the absolute values differed, the relative rankings of insect survival on the different cultivars tested remained similar throughout. In order of preference (from highest to lowest), the varieties ordered: Cabernet Franc = Riparia Gloire > Malbec > (Schwarzmann = Millardet et De Grasset 101-14). Cultivars Isabella, Couderc 3309, and Siebel 5437 were only used in a limited number of the assays so are not listed. The consistency of preference/survival rankings suggests a standardised test could be used as a 'bioassay' for mealybug feeding in grape. We recommend the leaf, no-choice, greenhouse test for this purpose as it had the highest differences in survival rates between varieties after 7 weeks (Figure 3) and it was among the easiest test to perform.

The consistency of the rankings between environments also suggests that genetic and/or metabolic factors are important in determining mealybug feeding preference and survival on grape hosts. A preliminary metabolomic study was, therefore, conducted to identify constituent chemistries that correlated with the mealybug feeding preference. From this study soluble tannins were implicated as potential feeding deterrents of importance. The roles of tannins in plant herbivore defence can be grouped into three functional mechanisms: 1/ protein precipitation capacity (PPC); 2/ reducing nitrogen (N) digestibility and 3/ oxidative activity at high pH (Marsh et al 2020). Mapping our data onto this classification system, it appears that polar ellagitannins and hydrolysable tannins (Karl et al 1983) play a role in grape herbivore defence. The mode(s) of functional defence would most likely be oxidative activity (OA) from the polar HHDPs with some protein precipitation PPC from the less polar HHDPs and their derivatives.

Leaf surface chemistry may also be important in grape plant defence. Higher levels of hydroxytyrosol fatty acids were seen to correlate with mealybug feeding deterrence in the current study. Structurally similar phenolic lipids were implicated as antifeedants against caterpillars using a leaf disc choice assay post, sprayed with isolated natural compounds (Sharma et al 2007). The esters of hydroxy tyrosol have also been shown to reduce nitrous oxide production in biological assays (Plastina et al 2019) which could further support the biological defence hypothesis. It is important to note that our chemical analyses were preliminary and exploratory in nature. Only limited replication was used and only a limited number of analyses were undertaken. Consequently, it is inappropriate to claim any discovery at this stage and future research is planned to validate the findings.

In conclusion, herbivore defence was observed in different grape varieties and the results suggest that this may be due to the presence of hydrolysable tannins and phenolic lipid wax components. Further research work is planned to explore the potential causative links between the marker metabolites and mealybug feeding preference. If causative links are demonstrated, the chemical markers will be used to guide a grape rootstock breeding effort, aimed at blocking the transfer of GLRaV3 and other insect-vectored viruses into the grape plant by discouraging insect feeding.

LITERATURE CITED

Bell, V.A., Lester, P.J., Pietersen, G., and Hall, A. (2021). The Management and Financial Implications of Variable Responses to Grapevine Leafroll Disease. Journal of Plant Pathology *103*: 5-15.

Bragato, R. (1906). Viticulture in New Zealand. Government Printer, Wellington, NZ

Campbell, C. (2006). The botanist and the vintner: how wine was saved for the world. Algonquin Books.

Charles, J., Cohen, D., Walker, J., Forgie, S., Bell, V., and Breen, K. (2006). A review of the ecology of grapevine leafroll associated virus type 3 (GLRaV3). New Zealand Plant Protection 59: 330-337

Jackson, R.S. (2008). Wine science: principles and applications. Academic press

Karl, C., Muller, G., and Pedersen, A. (1983). Ellagitannins from the leaves of *Vitis vinifera*. Zeitschrift für Naturforschung C *138*:13-16.

Marsh, K.J., Wallis, I.R., Kulheim, C., Clark, R., Nicolle, D., Foley, W.J., and Salminen, J.P. (2020). New approaches to tannin analysis of leaves can be used to explain in vitro biological activities associated with herbivore defence. New Phytologist *225*:488-498.

Munson, T.V. (1909). Foundations of American grape culture. Orange Judd Company.

Petersen, C., and Charles, J. (1997). Transmission of grapevine leafroll-associated closteroviruses by *Pseudococcus longispinus* and *P. calceolariae*. Plant Pathology *46*:509-515.

Plastina, P., Benincasa, C., Perri, E., Fazio, A., Augimeri, G., Poland, M., Witkamp, R., and Meijerink, J. (2019). Identification of hydroxytyrosyl oleate, a derivative of hydroxytyrosol with anti-inflammatory properties, in olive oil by-products. Food Chemistry *279*:105-113.

Sandanayaka, W., Blouin, A., Prado, E., and Cohen, D. (2013). Stylet penetration behaviour of *Pseudococcus longispinus* in relation to acquisition of grapevine leafroll virus 3. Arthropod-Plant Interactions *7*: 137-146.

Sharma, R., Negi, N., Gibbons, S., Otsuka, H., and Negi, D.S. (2007). Chemical constituents and antifeedant potential of *Boenninghausenia albiflora*. Natural Products: An Indian Journal: *3*.

Tsai, C-W., Rowhani, A., Golino, D.A., Daane, K.M., and Almeida, R.P. (2010). Mealybug transmission of grapevine leafroll viruses: an analysis of virus–vector specificity. Phytopathology *100*: 830-834.

Combined Proceedings IPPS 73:183-187. 2023.

Fifty Years of New Zealand Rose Society Trials

Hayden Foulds

40 Gordon St, Woodville 4920, New Zealand

haydenfoulds@gmail.com

Keywords: *Rosa*, rose trials, New Zealand, plant evaluation

Summary

The first Southern Hemisphere rose cultivar evaluations originated in New Zealand beginning in 1969. They have continued for over 50 years. This paper discusses rose evaluations in New Zealand including breeding trends and future prospects.

INTRODUCTION

The New Zealand Rose Society trials in Palmerston North, New Zealand, are the oldest international rose trials in the Southern Hemisphere with the first trial roses planted in 1969. They are also not the first rose trials in New Zealand with early efforts in trialling roses made in Parnell in Auckland and Morrinsville in the Waikato. There were also rose trials conducted at Massey University near Palmerston North in the 1940's and 50's. After these trials ceased, there were no independent trials available to test new roses under New Zealand conditions. Rather, rosarians had to rely on reports on varieties from overseas trials that had different growing conditions to those found in New Zealand.

In the late 1960's, the New Zealand Rose Society Council discussed setting up rose trials and the Dugald Mackenzie Rose Garden in Palmerston North was chosen as the location of the trials. Palmerston North's central geographical location, good soils and moderate climate makes for an area that grows good roses.

The strong support of the Palmerston North City Council and the local Manawatu Rose Society was also a key factor in choosing Palmerston North to host the trials and this support is still strong over 50 years later.

Figure 1. The New Zealand Rose Society Council inspecting the first trial September 1969. Photo credit June Hocking/Rose Trial Archives.

The main aim of the trials is to test new roses under standard conditions for their suitability for New Zealand's growing conditions. It also allows New Zealand rose breeders to compare their roses against those from overseas breeders.

In the early days of the trials, entries came from overseas breeders via their New Zealand agents while others were solicited directly by the trial grounds committee, imported and quarantined before being planted in the trials. Some also came from New Zealand amateur breeders.

Figure 2: The quarantine area for new entries to the trials from overseas November 1970. Photo credit June Hocking/Rose Trial Archives.

Having world renowned rose breeder Sam McGredy move to New Zealand in the early 1970's was also of immense benefit to the trials as many of his new roses were entered and we were often the first ones in the world to see his latest varieties coming through.

By the 1980's direct imports had ceased but overseas bred roses were still entered via New Zealand agents and representatives. There were also more amateur bred roses from New Zealand and overseas breeders being entered with some going on to receive awards.

In the early 2000's things changed again. Firstly, the retirement of Sam McGredy meant that the great line of his creations came to an end. Fortunately, Sam had encouraged other New Zealand rose breeders and soon more of their roses were being entered in the trials. Several long-established rose nurseries also closed their doors for a variety of reasons and so there was a changing of the guard to other growers who were now entering roses. The increased cost and difficulty of importing new varieties also led to fewer overseas bred varieties being entered.

This trend has continued since then to the current day with importing new varieties getting even more difficult and it is entirely feasible that one day, the border will close to new material. On the flipside, there are some great roses being bred by New Zealand breeders like Rob Somerfield, Bob Matthews, John Ford, Mike Athy and Doug Grant. Many of these New Zealand bred roses are now winning many of the awards and getting great publicity.

In 2020, the trials celebrated 50 years since the first awards were presented in 1970. Unfortunately, Covid 19 meant that many of the planned activities were scaled back or postponed until 2021.

OPERATION

The trials are run as a partnership between the New Zealand Rose Society and the Palmerston North City Council under a set of by-laws which set out how they will operate. The trial grounds committee, a subcommittee of the New Zealand Rose Society carries out the administration of the trials. The local and national rose societies, the rose industry and the city council are all represented on the committee. The maintenance of the trials is done by staff from the Palmerston North City Council.

Each trial is conducted over a two year period, although climbing entries are grown for an extra year prior to being judged to allow them to put on extra growth. Six plants of each type are required for trial except for climbers which require two plants. There are 40-50 varieties entered into each trial which are not available on the market in New Zealand, and many do not have names at the time of entry. It is free to enter roses apart from the cost of freighting plants to Palmerston North.

Figure 3. The New Zealand Rose Society Trials 2020.

During the trial period, the roses receive an average level of care aimed at replication of what an average gardener gives roses growing in their garden. They are pruned, fertilised and watered as required. Spraying with pesticides and fungicides has been reduced considerably over recent years.

After each trial is finished, the winners are replanted into beds in the main part of the rose garden while remaining plants are returned to the entrants or destroyed.

Each trial is judged by a panel of 20 judges which range from rose society members to gardeners with a few roses in their gardens. Entries are scored on a 1 – 10 system.

In the first year of the trial, each entry is judged four times over the growing season under the categories of Plant Quality, Freedom of Flowering and Flower Quality which contribute one third of the final score. Fragrance is also assessed separately by the judges.

The second year's judging consists of five assessments over the growing season from November to May. Categories are habit and growth, plant health, freedom of flowering, flower form and flower quality. This contributes two thirds of the final score. A separate panel of judges for novelty.

AWARDS

The winning roses from each trial are announced at a function held in Palmerston North in late November/early December each year. The Mayor of Palmerston North is often on hand to present the awards and successive Mayors have taken a keen interest in the rose trials.

An entry must reach an average of 70% across all judging categories and must be released commercially in New Zealand to receive an award.

Certificates of Merit are presented to those entries that gain 70% or more. The June Hocking Award (Fragrance) and Nola Simpson Award (Novelty) are presented to any winner that scores highly in these areas. Both these awards were renamed after two long serving members of the trial ground committee who each gave over 30 years' service.

To recognise the efforts of New Zealand amateur breeders, the Silver Star of the City of Palmerston North is awarded to the best New Zealand amateur bred rose gaining a Certificate of Merit.

In recent years, there has been a vote from those in attendance at the awards presentation to vote for the best looking rose on the day. This is presented with the World Federation of Rose Societies People's Choice Award. The trial ground's top award, the Gold Star of the South Pacific, is awarded to the highest pointed rose on trial gaining a Certificate of Merit. This is a highly sought after award, particularly by New Zealand rose breeders, and often leads to the rose becoming a top seller commercially.

A full list of award winning roses trial by trial can be found at https://nzroses.org.nz/nzrs-trials/

The trials are free to visit and are located within the Dugald Mackenzie Rose Garden which is part of the Victoria Esplanade Gardens. The best time to visit and see the roses at their peak is mid-November onwards although there are blooms all the way through until May.

Figure 4: The International Rose Trials are located within the Dugald Mackenzie Rose Garden.

Combined Proceedings IPPS 73:188-194. 2023.

Plant Breeding at Auckland Botanic Gardens and Beyond

Jack Hobbs

Auckland Botanic Gardens, 102 Hill Road, Manurewa, Auckland 2105

Jack.Hobbs@aucklandcouncil.govt.nz

Keywords: *Veronica, Leptospermum, Dahlia, Canna, Hemerocallis, Camellia,* septoria leaf spot, daylily rust, camellia flower blight

Summary

Plant breeding programmes at Auckland Botanic Gardens have produced ornamental plants that perform well under New Zealand conditions. Trials are conducted to evaluate plants for flowering, foliage and habit, and general plant health. Examples of breeding efforts related to selected plants will be described.

INTRODUCTION

In plant breeding, Auckland Botanic Gardens (ABG) is best known for developing the 'Wiri' series of *Veronica* (*Hebe*) and *Leptospermum*. Other crops that have been developed include *Dahlia, Canna, Camellia* and *Hemerocallis*.

All plant breeding programmes at ABG primarily aim to produce plants that perform to a high standard and remain healthy in Auckland conditions without applications of insecticides and fungicides. Seedlings are also evaluated for their ornamental qualities and garden performance with particular focus on high health. Trials are conducted to ascertain the best performing plants using criteria that consider flowering periods, foliage and habit, and general plant health. Our niche is to fill some of the gaps that are understandably unrealistic for most commercial plant breeders.

Numerous popular garden plants such as dahlias, camellias, azaleas, and chrysanthemums were not primarily bred for garden performance. Many were bred to produce exhibition-quality flowers for the show bench, while in recent times most ornamental commercial crops have been bred to have high aesthetic appeal at point of sale. Compact, precocious plants that flower heavily while still young have understandable appeal at point of sale, but often they do not perform sustainably well when planted into a garden or container.

Plant breeding is closely aligned with and ultimately dependent upon plant selection. In my view a programme will only succeed if the quality of the parent material is of the highest quality.

I have collaborated on several breeding ventures with Dr Keith Hammett, particularly with *Dahlia*. From Keith I learned the importance of having a plan that includes clear objectives, accurate documentation, and scouring every possible avenue to secure the best possible germplasm to execute the plan successfully.

Others who have made substantial contributions to the ABG breeding programme are Terry Hatch and the late Graeme Platt. Terry Hatch has a contagious passion for plants and is always on the lookout for anything new, different, and ultimately superior. I learned from him the importance of assembling a wide array of germplasm and having an intimate knowledge of the plant groups that are being worked with.

Graeme Platt taught me that all plants are not created equal, and that there is a vast pool of plant material with unrealised potential. Although he was not a plant breeder, he introduced many of our most important garden plants, especially natives that he sourced mainly from the wild. Several of these were used as parents in my plant breeding programmes.

All three consistently exhibited a deep passion for plants and a dogged determination to make improved options available to our gardeners.

Veronica (*Hebe*) breeding

The *Veronica* breeding programme started at ABG in1982. The first objective (prior to eco-sourcing) was to produce a form of koromiko (*Veronica stricta*) that was resistant to septoria leaf spot. I was nursery manager at the time, and for revegetation programmes we were growing large numbers of koromiko that invariably became riddled with spots. Ultimately, I found one plant that remained relatively free of disease and named it 'Wiri Spears' which we produced for plantings in Regional Parks.

Shortly afterwards, and for similar reasons, I raised 'Wiri Jewel', a form of *V. speciosa* that was relatively resistant to septoria leaf spot and downy mildew. 'Wiri Jewel' is the parent of several hybrids including 'Wiri Gem' and 'Wiri Charm'.

We decided to use a common prefix for all our cultivar names to associate them with

ABG and settled on 'Wiri' which was applied to all the *Veronica* (*Hebe*) and *Leptospermum* cultivars produced at the Gardens.

The objective of producing healthy hybrids was never lost, but over time more focus was put on improving aesthetic and performance characteristics.

Parent plants were grown in containers to facilitate hand pollination of flowers rather than just collecting open-pollinated seed as I had done previously. The laborious pollination process involved removing the corolla from the seed parent with tweezers which simultaneously removed the stamens and prevented self-pollination. Then pollen from the pollen parent was transferred to the stigmas of the seed parents.

The breeding process was simply to produce numerous seedlings and cull any that developed disease symptoms. Large numbers were produced, with only about 2% of the seedlings that were pricked out making it past the tube stage for subsequent pot and field trials. Most of these seedlings were later culled owing to unsatisfactory performance or appearance. I was initially looking for healthy high-performance garden subjects, and it was later that I realised that for a cultivar to be commercially viable it had to perform well in nursery production systems and look good at point of sale. To date, just 15 Wiri cultivars have been named by ABG from tens of thousands of seedlings raised. A few cultivars were given the Wiri prefix by others looking to take advantage of the commercial appeal of the brand at the time.

Veronica diosmifolia was used as one of the foundation parents as it invariably remains clean and healthy and passes this characteristic on to its offspring. Crossing a pink-flowered form of *V. diosmifolia* with *H. speciosa* 'Wiri Jewel' resulted in 'Wiri Gem' and 'Wiri Charm' which have both proved popular with gardeners. 'Wiri Charm' is particularly popular in the UK.

'Wiri Mist' is a cross between *V. diosmifolia* and *V. albicans* and is invariably healthy and especially attractive when smothered with white flowers in late spring.

'Wiri Splash' deserves in my view to be more widely grown. It forms a healthy compact mound of green foliage tinged yellow and carries masses lilac flowers in early summer.

'Wiri Image' is another that could be grown more widely, although it is larger and not suitable for small gardens.

F_1 hybrid seedlings proved to be consistently uniform in appearance, but this was not the case with hybrids containing large numbers of species in their pedigree. Such hybrids often bear little resemblance to wild species, increasing opportunities to produce novel new hybrids.

A turning point in my understanding of the commercial viability of hebes came with invitations in 1990 and 1993 by Danish hebe growers to visit their growing operations. There I encountered hundreds of thousands of hebes in 10 cm pots, with 2 million or more produced annually for the European market, largely as house plants.

My plant breeding days were interrupted when I became manager of ABG in 1997 and most of the plant breeding activity at ABG paused.

Recently I have begun once again to dabble in plant breeding, especially with hebes, which I believe have untapped potential as garden subjects and in commercial production. My main initial objective is to produce cultivars with performance and health attributes similar to 'Wiri Mist' but in a wider range of flower colours. The specific objectives of the programme are to produce cultivars with the following characteristics:

- remain healthy without pesticides
- are high performing garden subjects
- have buyer appeal in small pots at point of sale
- have attractive foliage
- remain compact
- have attractive flowers with significant peak flowering period
- ideally flower when grown in small pots

Many of the most colourful seedlings have *V. speciosa* in their pedigree and have inherited its disease susceptibility. Many also flower over a relatively prolonged period although not all produce a peak flush. Some of the most prolific flowering *V. speciosa* hybrids I have produced flower progressively as their branches extend, but the downside is they do not remain compact and therefore require annual pruning.

A large proportion of seedlings derived from 'Wiri Mist' produce healthy attractive healthy foliage but fail to flower sufficiently and therefore are culled. Those that flower sporadically without producing a significant main flush are generally culled. Rare exceptions are those with especially impressive foliage.

It has proved particularly challenging to develop hybrids that make outstanding garden subjects and also look attractive at point of sale. Based on extensive trials at Auckland Botanic Gardens it appears that many recently released hybrids have been selected primarily for their appearance at point of sale as few have performed well when trialed as garden subjects.

The key plant breeding lessons I have learned are:

- start with a plan
- stick to the plan
- assemble the best possible array of germplasm
- research genetics (for compatibility)
- keep detailed records
- grow large numbers
- cull ruthlessly
- do not release anything until extensively trialed.

BREEDING PROCESS

Seedlings are germinated and pricked out. Seedlings grown in greenhouse for several weeks. Even in these relatively controlled conditions some die.

When large enough they are shifted outdoors. The attrition rate is high as any seedlings that develop downy mildew are culled. On average just 5% make it through this stage and are potted into 10 cm pots.

Seedlings that have attractive foliage and good form and still remain healthy are propagated by cuttings.

Cutting grown plants are further tested as pot plants and some are planted into garden trials.

Selected individuals are evaluated over the next 2 or 3 years to determine their suitability for container and garden use.

Some still develop disease symptoms and are culled. Those that remain healthy continue to be assessed for health, compactness and flowering performance. Those with a tendency to become woody are also culled. The majority of remaining seedlings do not flower well enough to persevere with.

Hemerocallis

The daylily breeding programme at ABG commenced shortly after daylily rust (*Puccinia hemerocallidis*) was first reported in Auckland in 2011. The breeding programme was initiated in 2014 and has been led by Jack Hobbs and Emma Simpkins (née Bodley). The aim is to produce evergreen cultivars that remain healthy and produce prolonged displays of attractive flowers in a wide range of colours.

Daylily rust spread rapidly and soon most cultivars grown in Auckland became so debilitated they were no longer suitable for garden use. Typical symptoms are yellow pustules on the underside of the leaves that eventually cover much of the foliage. The rust is particularly severe in warm humid climates such as Auckland's.

Prior to the incursion of daylily rust large drifts of daylilies were planted at ABG. The first major trial was some 30 years ago when more than 400 cultivars were evaluated. From that trial we selected 40 cultivars that we used and recommended.

In December 2009 55 cultivars were planted in ABG trial beds and evaluated for five years. Just 12 cultivars were assessed as being suitable for ornamental use in Auckland gardens.

Daylily rust rendered many of the high performing cultivars identified in our trials unusable as garden subjects. Today few daylilies are rust free in Auckland, and those that do remain healthy can be hard to source. However, we have identified some cultivars that are still worth growing and the best of these have been included in our breeding programme.

'Squeaky' was chosen as the foundation parent as it was an outstanding performer in ABG trials and proved to be particularly resistance to rust disease. The yellow-orange flowers have a distinctive crinkle and although not the most impressive they appear for many months. The attractive evergreen foliage is narrow and spreads densely, making it an effective groundcover, and the plants invariably look healthy. In essence the ABG breeding programme aims to replicate the best attributes of this daylily in a range of different flower colours.

During the summer of 2014/2015 a selection of 23 rust-resistant cultivars was used to pollinate 30 plants of *H.* 'Squeaky' which was the sole seed parent. In the summer of 2015/2016 the number of pollinators was reduced to 12 rust-resistant cultivars used to pollinate 30 plants of *H.* 'Squeaky' which again was the sole seed parent.

Cultivars used as pollinators in the breeding programme include:

'Nashville' (evergreen, reddish flowers with yellow centres)

'Chicago Apache' (deciduous, pinkish red)

'Glitter' (yellow)

'Moon Goddess' (yellow)

'Chosen One' (lemon).

'Peek a Boo Eyes' (evergreen, soft yellow with crimson eye)

'Baby Betsy' (pink with yellow eye)

'Little Grapette' (reddish purple)

'Lullaby Baby' (soft peach)

'Mini Pearl' (peach).

In recent years several hybrids derived from 'Squeaky' have been introduced into the breeding programme including being used in sibling crosses.

Canna

By the early 1980s large numbers of *Canna* cultivars were collected and trialled at ABG. Many showed symptoms of virus infection such as vein clearing and stunted growth and some developed bacterial soft rot (*Pectobacterium carotovorum*) which was considered likely to be secondary to the virus infection. The worst affected cultivars were culled, and those that remained were evaluated over several years, with the best performers being included in our display gardens.

Subsequently a *Canna* breeding programme was instigated at ABG to produce virus-free cultivars that performed well as garden subjects. Seed from seven high-performing healthy cultivars was sown on 4 May 1987 in a heated greenhouse with timed lighting providing 16 hours light per day. The resulting seedlings were planted in trial beds in early October 1987.

Seed of *Canna* 'Cupid' was sown on 17 May 1988 with resultant seedlings planted in the trial ground on 13 October 1988.

On 16 June 1988, the following pairs of cultivars were planted to act as parent blocks with the objective of producing new virus-free cultivars in the main colour range groups:

- red-flowered pair: *C.* 'America' (no seed produced) and *C.* 'Assault'
- pink-flowered pair: *C.* 'Cupid' and *C.* 'La Boheme'
- yellow-flowered pair: *C.* 'Banner' and *C.* 'Felix Ragout'.

Seed was sown on 26 May 1989, and the resultant seedlings planted out on 10 October 1989. Seedlings that demonstrated desirable characteristics and high performance were initially given a code. Seedlings that did not reach the required standards were culled.

ABG trialled *Canna* cultivars from 2017–2020 including those available on the market, with ABG-bred selections as a comparison.

Each *Canna* was given an overall rating according to ABG star performer criteria (1 = poor performer to 10 = excellent performer). Cultivars that scored 8 or more were considered top performers and are recommended for Auckland based on the results of these trials. Overall ratings took into consideration flowering period, quality of flowers, absence of pests and diseases, habit and vigour. 'Gabriel' (coral pink) received the highest rating and is promoted as a Star Performer. 'Hampton' was named in 2019 by Emma Simpkins and plants have been made available to the market.

Camellia

ABG holds an extensive collection of some 500 different camellias including around 60 species. When camellia flower blight (*Ciborinia camelliae*) arrived ABG initiated a breeding programme to develop attractive garden hybrids resistant to this debilitating disease. The first crosses were made in 2015. The aim is to also produce seedlings with handsome glossy foliage and attractive flowers (preferably scented) over long flowering periods.

Matt Denton-Giles (Massey University) tested 39 *Camellia* species in the ABG collection for susceptibility to camellia flower blight and in 2013 reported variable degrees of susceptibility, with *C. lutchuensis, C. transnokoensis, C. yunnanensis* and *C. yuhsienensis* as having flower blight resistance. *C. lutchuensis* and *C. transnokoensis* have been primarily used in the ABG breeding programme, *C. yunnanensis* has been used sparingly. To inform the genetic compatibility of planned crosses the chromosome counts of these species were researched: *C. yunnanensis* ($2n = 30$), *C. lutchuensis* ($2n = 30$), *C. transnokoensis* ($2n = 90$)

The selected species were initially crossed mainly with a selection of larger-flowered japonicas and reticulata hybrids identified as being petal blight resistant. Parents used in recent times include 'Wildfire', 'Bob Hope', 'Fairy Blush' ''Transpink' and 'Transtasman'.

Seedlings are currently being grown in field trials to ascertain their suitability of gardens and subsequent commercial production.

SUMMARY

The primary aim of ABG breeding programmes has been developing disease- resistant garden plants through breeding for genetic resistance. This aligns with ABG's pesticide minimisation programme that precludes the use of pesticides on ornamental plants. Although emphasis is on developing outstanding garden subjects, increasingly the priority is also placed on commercially viable hybrids. The *Hemerocallis* and *Camellia* programmes are currently active at ABG, the latter being a particularly long-term project. Active breeding of *Canna* is not currently undertaken at ABG. The *Veronica* programme has resumed in recent years on my own property.

LITERATURE CITED

Hobbs, J. (1991). *Hebe* breeding at the Auckland Regional Botanic Gardens. Horticulture in New Zealand, Journal of the Royal NZ Institute of Horticulture, *2(1)*: 20-23.

Hobbs, J. and Bodley E. (2020). Canna breeding at Auckland Botanic Gardens. New Zealand Garden Journal *24(2)*: 2-3.

Bodley, E., Green, R., Jones, J. and Hobbs, J. (2015). Auckland Botanic Gardens *Hemerocallis* rust trial. New Zealand Garden Journal, *18(2)*: 2-4.

Denton-Giles, M., Bradshaw, R.E. and Dijkwel, P.P. (2013). *Ciborinia camelliae* (Sclerotiniaceae) induces variable plant resistance responses in selected species of *Camellia*, Phytopathology *103(7)*: 725-732.

Combined Proceedings IPPS 73:195-201. 2023.

Rhododendron Potting Mix – Our Trials and Results using Coir

Ian Swan

Tawa Glen Nursery, 100 Richmond Road, RD3, New Plymouth 4373

tawa_glen@xtra.co.nz

Keywords: capillary watering, rhododendron dieback, heavy and light potting mix

Summary

Potting mixes are integral to superior plant production systems. Creating a suitable mix can be challenging given the changing available substrates. A new mix using coir was developed for *Rhododendron* production.

INTRODUCTION

This was a journey of trials, frustrations, successes, and disappointments. We grow rhododendrons in 6L pots on capillary beds with 60% overhead shade. Most of our watering is capillary watering, with overhead sprinklers used in dry spells, which in Taranaki is about 2 to 3 weeks without rain.

What is a good potting mix? I am a grower, not a potting mix manufacturer,

IPPS Vol. 73 – 2023

and like a lot of growers we rely on a supplier to tell us what we need and manufacture a mix to suit our conditions.

What constitutes a good potting mix for me:

- Grows good strong plants of high health
- Retains enough moisture without being too wet
- Contains enough air space for drainage
- Light enough for staff to handle plants easily
- Heavy enough that plants don't blow over too easily
- Cost effective. If you get 100% of your plants saleable cost is then not so crucial

We were having some issues with the potting mix that we were using at the time. The plants were very wet at the base with no roots in the bottom 80-100ml of the pot. We were getting quite bad dieback around the leaf margins of certain varieties and spent far too many hours trimming foliage to make plants saleable. Some were unsaleable.

TRIALLING MIXES WITH COIR

We were on an IPPS field trip to the Napier area and saw a big bin of medium coir at one of the places visited. I had not seen this grade in bulk and it looked like it would make a great medium. We set about getting a few half cubes bags of coarse coir, medium coir, and a 50/50 mix of medium coir and CAN fines, all with the usual fertiliser and 12-month slow release.

(CAN fines are produced from crushed and screened *Pinus radiata* bark blended with trace elements and Calcium Ammonium Nitrate, pH adjusted and composted).

Our initial trials were with Rhododendrons 'Kaponga' and 'Harry Tagg'. These were 2.5L plants potted into a 6L, and these results after 6 months looked pretty promising. The only problem was that the mix was too light, and we were constantly standing these plants up and giving them extra top watering. We had root systems that went right to the bottom of the pot in all 3 mixes.

We had our standard rhododendron mix, a new coir trial mix, a new rhododendron mix, and we mixed up a trial (trial A) 50/50 standard rhododendron mix with medium coir to try to open it and drain it better, and trial B 60/40 new rhododendron mix with medium coir.

It was about this time that IPPS along with New Zealand Plant Producers Incorporated ran the Dirt, Fert and Squirt workshop with Professor Paul Fisher. One of the mixes there had a high coir percentage and Paul mentioned that it had no capillary action. Some sales reps have told me that with that long fibre in the mix it would be great for capillary action. I bought a moisture probe and found that we could not get the mix above 20% moisture even after heavy Taranaki rain.

Our mix at present is 50% bark CAN, 10% coir, 20% NZ peat, 20% pumice. It's a bit heavier than I would like as after carrying 6L pots around for the day it feels like the arms have stretched a bit, but they are not blowing over in our windy conditions and the plants are looking great.

I think that it could be a great product with the right irrigation systems, but it is not for our capillary watering beds. It has almost priced itself off the market with the huge increases in shipping costs.

Standard Rhododendron mix (heavy)

- 60% bark fibre (pine bark)
- 20% coir
- 20% 7mm pumice

New coir trial mix

- 60% medium coir
- 30% CAN fines
- 10% pumice

Trial mix A

- 50% standard rhododendron mix
- 50% medium coir

New Rhododendron mix

- 65% CAN fines
- 25% coir
- 10% 7mm pumice

Trial mix B

- 60% new rhododendron mix
- 40% medium coir

The following figures graphically shows root system development in various Rhododendron cultivars related to container mixes.

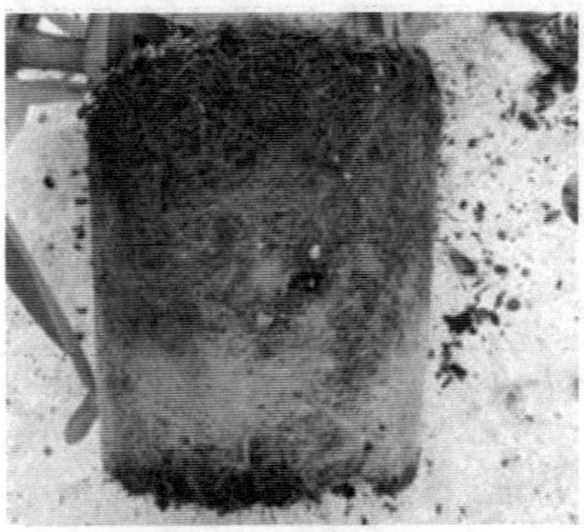

R. 'Kaponga' 100% coarse coir Excellent root development. Too light, blows over and more watering required.

R. 'Kaponga' 100% medium coir Excellent root development. Too light, blows over and more watering required.

R. 'Kaponga' 50% medium coir, 50% CAN fines. Heavier mix, doesn't blow over, not as good root development, roots to the bottom of the pot.

Standard rhododendron mix. Roots not to the bottom of the pot.

Two varieties in the new rhododendron mix. 65% CAN fines, 25% coir, 10% pumice. First variety here, showing good roots.

Two varieties in the new rhododendron mix. 65% CAN fines, 25% coir, 10% pumice. Second variety here, showing roots that are still okay.

R. 'Rubicon' new rhododendron mix. 65% CAN fines, 25% coir, 10% pumice. Roots okay, not quite as good as with the coir.

Trial mix B
65% new rhododendron mix, 35% medium coir. Good root formation, healthy roots, roots to the bottom.

R. 'Rubicon' Trial mix B
65% new rhododendron mix, 35% medium coir. Not bad root system

R. 'Rubicon' New coir mix
60% medium coir, 30% CAN fines, 10% pumice. A better root system, healthy roots to the bottom of the pot.

R. 'Cherry Cheesecake' Standard rhododendron mix. 60% bark fibre, 20% coir, 20% pumice. Poor root development, plant not saleable.

R. 'Cherry Cheesecake' Trial mix A 65% standard rhododendron mix, 35% medium coir.. Average root development, fungus growth in mix, can become hydrophobic.

R. 'Dad's Indian Summer' Trial mix B 60% standard rhododendron mix, 40% medium coir. Much better root system, right to the bottom of the pot.

R. 'Princess Alice' New rhododendron mix. 65% CAN fines, 25% coir, 10% pumice. Roots healthy but not a lot of them.

R. 'Princess Alice' Trial mix B 60% new rhododendron mix, 40% medium coir. Healthy roots but not much volume, not to the bottom of the pot.

R. 'Princess Alice' New coir trial mix 60% medium coir, 30% CAN fines, 10% pumice. Healthy roots, right to the bottom of the pot.

R. 'Rubicon' Trial mix B 60% new rhododendron mix, 40% medium coir. Healthy roots, not much volume, not to the bottom of the pot.

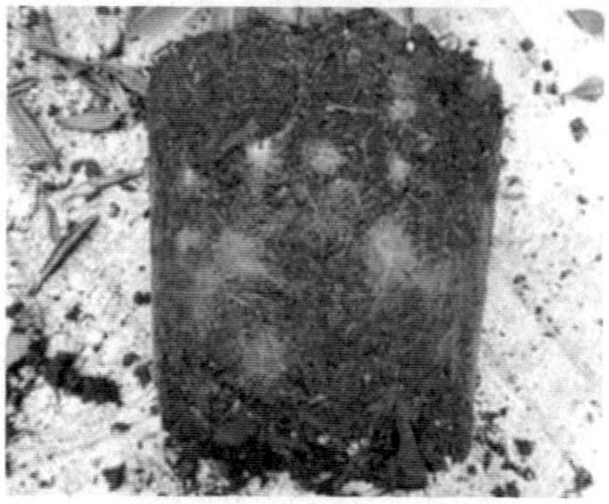

R. 'Rubicon' New coir trial mix 60% medium coir, 30% CAN fines, 10% pumice. Healthy roots, good volume, roots to the bottom of the pot.

R. 'Rubicon' comparison. Plants on the left have been grown in the New rhododendron mix. Plants on the right have been grown in the New coir mix. No difference in the tops but consistently different root volumes.

CONCLUSION

Is coir better than bark? In our experience, sometimes. We are getting a better root system with some varieties but they need overhead watering. Practically all of the yakushimanum hybrids hate it. We lost most of these varieties. Vireya rhododendrons like it a lot. They are predominantly epiphytes and like to be well drained.

The foliage on this year's crop was far better with no burning on leaf edges.

But - there is always a but: it depends on your watering system. Does the coir mix suit our growing and watering system? No, coir has no capillary action at all and must be watered from above. Coir is certainly worth trying. Just make sure that your watering system suits it. The price of freight has almost prohibited the use of this product in any quantity.

Combined Proceedings IPPS 73:202-203. 2023.

Old Dogs Teaching Young Chicks Old Tricks with a Twist

Terry Hatch

Joy Plants, 78 Jericho Road, RD2, Pukekohe 2677

hatch.clan@xtra.co.nz

Keywords: *Vitex*, *Metrosideros*, hammer cuttings, horseshoe cuttings, king fern

Summary

Several old propagation methods have been successful on a small number of native species; two-metre-long trimmed branches of *Metrosideros* have rooted, hammer cuttings are recommended for some species where there is limited material, and horseshoe cuttings of king fern produce fernlets within a few months.

INTRODUCTION

In the late 1970's, I was helping to eliminate feral cats on Hauturu Island (Little Barrier Island). The dense bush was magic, huge trees of many species some rarely seen on the mainland. Outstanding was a very large *Vitex* possibly hundreds of years old, where large branches wept down to the earth. Some had produced roots and new trees had arisen. These had again grown, and later branches had done what the original tree had and rooted, producing more large trees, eventually into three generations of very amazing trees.

IPPS Vol. 73 – 2023

In our nursery area are stands of *Vitex* with many other species of native trees. One still day one of the old *Vitex* fell across the main track. The top 3 metres was sawn off and the 4-metre trunk pushed to one side. A large hole was dug for the top to be planted so climbers and stag horn ferns could be planted on it and around it. Steel waratahs and re-bar with concrete were placed around the outside to hold the top in place. To our amazement this top would grow for several years and finally came to its demise in a summer drought. We have since found large young cuttings are much better.

On a trip to the Coromandel, I was shown a vast *Metrosideros* that had done a similar thing as the *Vitex* on Hauturu. It also had produced many young trees from where the large branches had touched the earth, so I figured we would give this a try. I pruned some large branches of my old *Metrosideros* in the garden, these I cut into 2 metre lengths and proceeded to plant them with a post rammer. They are now large trees 25 years old. From this we have figured many other natives have the potential to be propagated in this way.

Hammer cuttings

Sixty years ago, when working at John Pettit's Nursery I was asked to propagate a variegated *Pisonia brunoniana* by cuttings. They always took months to grow roots, so I tried hammer cuttings. These are single buds with a small part of the leaf stem and only a tiny part of the main stem. These seem to grow roots quite fast and produce shoots within 4-6 months later. We have since also produced plants of *Pennantia baylisiana* and *Hedycarya arborea* by this method. When there is limited material, this may be a way of bulking up stock from other thick, juicy stem species.

Horseshoe cuttings

The major species is the *Ptisana salicina,* king fern (similar to the tropical *Angiopteris evecta).* The whole stipe is wrenched from the base, there should be a horseshoe shaped end. The stipe is shortened to 30cm and planted into sand. In a few months young fernlets are produced and can be detached when large enough. Other ferns can also be grown by this method.

Combined Proceedings IPPS 73: 204-206. 2023.

Fifty Years of Change in a Family Run Production Nursery and a Brief Overview of the Industry as We Have Seen It

Lindsey Hatch

Joy Plants, 78 Jericho Road, RD2, Pukekohe 2677

hatch.clan@xtra.co.nz

Keywords: Joy Plants, founding member, Bevlynn bulbs

Summary

Joy Plants is a well-known New Zealand nursery started by Terry and Pamela Hatch in the 1960's. A wide range of plants has been produced by Joy Plants throughout the years, with plant trends leading to diversification in main production lines. Landscaping and even beehive supply to kiwifruit orchards has shown how this successful family nursery has adapted to change over 50 years.

INTRODUCTION

Joy Plants was established in 1965 in the backyard of a rental in Mangere while working at Pettit's Nursery.

We set up in Manurewa in 1967. Site 3 was purchased and established in 1972 in Runciman Road, Pukekohe. Site 4 starts set up in 1992 in Jericho Road, Pukekohe. Site 5

IPPS Vol. 73 – 2023

was purchased in 2018 in Parker Lane, Pukekohe. A new plan.

Manurewa was a small nursery with various lines being grown for landscaping purposes, which was an extra revenue while Terry was foreman for Pettit's Nursery. Growing mainly alpines and small rockery species and anything that could be salvaged from gardens that were being worked on.

Joy Plants would take full swing with the purchase of land in a rural part of Pukekohe known as Pukekohe East on Runciman Road. Starting off small and slowly growing on its four-acre land parcel from 1972 to 1997, established by Terence and Pamela Hatch. Staff numbers ranged from zero to three at any one given time not including themselves. Terry would be a part of the first ever New Zealand meetings and a founder member of IPPS New Zealand, a major part of his nursery production life. Plants were produced by cutting, seed, and division, and grown in PBs. Soil mix was made by hand with spade and barrow from pumice, sand and peat, lime, and various other forms of fertiliser. A computer would be bought in 1982 because this was the way of the future - a Commodore 64 so we could print labels of plants at shows and do plant lists for mail orders.

Production would be mainly a range of perennials rock plants and small bulbs and over the years an increase of other lines - a selection of small Australian shrubs and perennials and a selection of small rockery natives. This included a range of small growing flaxes that would be named and grown by Terry, that would continue to be a popular line for many in the industry for years to come. Sale of plants would be through plant/pet shops and retail nurseries around the North Island with road trips in a little blue Morris Minor to nurseries around Auckland and Hamilton by Pam and longer road trips by Terry as far north as Kaitaia and south to New Plymouth. Bus parties and garden groups would be a regular occurrence as the nursery became more known after displays and sales at various flower and garden shows.

Bulbs would become a big part of the nursery, especially African species. Two such varieties would become another string to the nursery bow and the little but short-lived cut flower production business would start up Bevlynn Bulbs growing *Nerine* and *Zantedeschia,* or calla lily, for cut flowers from an extensive breeding programme - one even featuring on a New Zealand stamp. There was also a dry bulb mail order list boasting over two hundred species of bulbs, many rare and endangered at the time. Over summer months most of these would need lifting, cleaning, and packaging before stock and excess bulbs were replanted. The use of tissue cultures, some of the first in New Zealand, would play a major part in increase of calla corms to increase some of our selected colour forms. This was also going to be a life changer for plant production and our cut flower business - or was it? If this was not enough, by the 1980's we would be doing landscaping again, and as a sideline provision of beehives for local kiwi fruit orchards, reaching almost one hundred hives for pollination purposes.

Production of natives for revegetation projects would also start in the 1980's due to the lack of suppliers. We would soon have one of the largest selections on the market. These were mainly for one project that would continue for thirty-two years, the growing and planting of more than 650,000 trees on Mercury Island. This would see the need for expansion once again, the purchase of 24 acres in Jericho Road in 1992 and the increase of the Hatch clan. Perennials would still play a part in the nursery, but the bulb market hit a downturn, and many would not make the transition to the new site due to various reasons - vermin being one, time of work required and lack of interest by gardeners. Also, the need to increase native plant production with more local farmers requiring natives, a few more projects further afield, and an increase of natives in private gardens would see us increase our native ornamental production for landscape use.

The late 1990's to the present day 2020's has not seen much major change in our growing style, just an increase in volume. We are still a family business, now with three partners and two extra workers of which one is extended family. We grow what we like, and we sell what we can to those we like and those who pay their bills. Plants are now grown in hard plastic pots in a potting mix made of pumice, sand and pine bark, dolomite lime and long-life composite fertilisers. Not mixed by hand now but by rotary hoe on a tractor - simple but it works. 2018 has seen us purchase 44 acres with the intention to increase contract grown production. Who knows what is around the corner? One would hope more plant breeding and production of worthy plants for the masses, but one step at a time.

Combined Proceedings IPPS 73:207-210. 2023.

Technical Sessions of the International Plant Propagator's Society – Southern Region of North America (SNRA) Annual Meeting

Judson LeCompte

Spring Meadow Nursery
12601 120th Ave, Grand Haven, Michigan 49417, USA

judson@springmeadownursery.com

Keywords: IPPS-SR, SRNA, awards, scholarships

Summary

The 47th Annual Meeting of the International Plant Propagators' Society-Southern Region of North America (SRNA) convened at 8:00 am on 30 October 2023 at the DoubleTree Hotel, Raleigh-Durham Airport Research Triangle Park, Durham, North Carolina with President Judson LeCompte presiding.

INTRODUCTION

President LeCompte welcomed everyone to Durham, North Carolina for the 47th Annual Meeting of the SRNA. It is so awesome to be here in Athens to "seek and share" with one another! The relationships and experiences forged at our meetings are what distinguishes the IPPS. LeCompte came to his first SRNA meeting in 2009 in Biloxi, Mississippi - with eight fellow students who shared one hotel room. He has not missed a meeting since! He remarked that the SRNA has been so important for his career – and professional relationships developed.

He thanked Local Site Committee Chair, Dr. Elizabeth Riley, Co-Chair, Dana

Massey and their committee and volunteers for their outstanding work in arranging the excellent tours, hotel, other planning activities, and all their attention to detail. The SRNA remains financially strong, with the due diligence of Sec-Tres, Donna Foster, and is the largest international region with 245 active, paying members, and 67 student/military veteran members for a total of 312 members.

LeCompte thanked the Executive Committee, and the Sponsorship Committee: Chair, Michael Row, Dr. Elizabeth Riley, Dr. Anthony Witcher and Matt Sawyer - who raised $56,667, which is outstanding! LeCompte encouraged the membership to thank, visit and show their support of our sponsors during the meeting. The SRNA is deeply indebted to our loyal sponsors who make our annual meeting financially possible. He encouraged all members to make new members and first-time attendees feel welcome — share with them and seek from them.

LeCompte announced that the SRNA is in its sixth year of the Southern Region Educational Endowment, with a base donation of $20,000 from an anonymous donor. The Education Endowment balance is now at $127,020 – and growing. It will greatly enhance our region's ability to support students and early career professionals – and ensure continued quality of the outstanding educational, out-reach programs our region is known for. All of this year's contributions to the silent and live auction are to go to the Endowment Fund – so please contribute! He thanked Kevin Gantt for leading the endowment effort.

The SRNA currently supports 5-scholarship programs – that cost around $20,000 per year. This includes: Vivian Munday Young Horticultural Research Scholarship Work Program, Charlie Parkerson Student Research Competition, Early-Career Exchange Program, Margie Jenkins Industry Scholarship, and new Vince Dooley Scholarship.

Three years ago - the IPPS-SR initiated the **Margie Jenkens Industry Scholarship** to support industry professionals attending our conference for the 1st time. Margie Jenkens had a major impact on the Louisiana Nursery industry with her plant selections. LeCompte recognized this year's recipient: Emily Ellis. He recognized the 2023 Vince Dooley scholarship recipient: Kayla Morrison. Coach Dooley was a big proponent of landscape horticulture, taking a plant materials course with Dr. Michael Dirr at the University of Georgia – and developing great interest in new landscape plant selections.

For the **Early-Career Exchange Program** between the SRNA and the European Region, Teagan Young of the University of Florida, represented the Southern Region at the European Conference and toured the European nursery industry. LeCompte remarked that he was the first recipient – and the wonderful professional and personal experience he had visiting the European Region. Thomas McDonald of the European Region was their professional nominee for the program; he accompanied the International Tour, hosted by the SRNA. LeCompte asked McDonald to come forward and presented him with a gift. This is a wonderful exchange program for an early career professional to go to Europe and visit the European Region. He also recognized the International Board Chair, Tim Lawrence, from the European Region – who hosted LeCompte in England as part of the exchange program.

This is the twelfth year the SRNA is doing the **Vivian Munday Young Horticultural Professional Scholarship Work Program** (formerly Vivian Munday Scholarship). He introduced the four interns: Jerry Yu of North Carolina State University, Grace Carapezza of the University of Florida, Katelyn Reffler of Stephen F. Austin State University, and Kaitlyn Swaintek of the University of Georgia. These young professionals are making a strong contribution to this year's program. LeCompte remarked that our future is young people!

Next year at the 2024 mid-year board meeting in Orlando, Florida – the SRNA will be doing a strategic planning session – and developing a 5-year strategic plan for the organization.

LeCompte encouraged the membership to fill out questions for the Question Box – and attend the Tuesday night Question Box/Ice Cream Social.–The Question Box was to be moderated by Dr. Mengmeng Gu and Richard May.

LeCompte thanked Program Chair and 1st Vice-President, Dr. Cheryl Boyer for the excellent program and incredible group of speakers she assembled!

PROGRAM CHAIR, DR. CHERYL BOYER

Program Chair Boyer welcomed all members, guests and students. She acknowledged President LeCompte for his leadership and very capably serving as President. She thanked the membership for the opportunity to serve them, and then reviewed the scheduled program.

There were eight outstanding paper submissions for the Charlie Parkerson Student Research Competition. There will be four students competing in the oral competition, and poster presentations for the membership to visit with student presenters during the meeting. She acknowledged International Delegate, Laura Miller, and Director, Dr. Anthony Witcher – for seamlessly running the audio-visuals for our speakers and meeting. She then introduced the first speaker, Dr. Andrew King of King's Nursery and Stephen F. Austin State University.

Figure 1. President Dr. Judson LeCompte (left) with Dr. Cheryl Boyer (right), Program Chair of the 2023 Durham, North Carolina, 47th annual conference.

Combined Proceedings IPPS 73:210-216. 2023.

Myrica rubra, a New Ornamental with Edible Fruit and its Propagation Challenges

Zachary Hutzell[a] and Donglin Zhang

Department of Horticulture, University of Georgia, Athens, GA 30602, USA.

zacharyhutzell@uga.edu, donglin@uga.edu

[a]First Place – Charlie Parkerson Graduate Student Research Paper Competition

Keywords: adventitious rooting, edible ornamental, cutting propagation, indole-3-butyric acid (IBA), Yangmei, yumberry.

Summary

The successful introduction and adoption of new taxa in the ornamental market depends on developing adequate propagation protocols. If a new plant cannot be reliably and quickly propagated, nursery producers may not be able to justify large scale production of the taxon while wasting resources, labor, and time. *Myrica rubra* is a new species and its selected clones are worthy of introduction for their ornamental value and edible fruits. However, these new clones are difficult to clonally regenerate with stem cuttings. This study evaluated hormone applications and timing of cuttings to optimize the protocols for stem cutting propagation for *Myrica rubra.* Some success was achieved in rooting semi-hardwood stem cuttings taken between May 25 and September 5, with an application of 8,000 ppm indole-3-butyric acid (IBA) talc powder. Further studies should focus on the best-found protocol for rooting stem cuttings of *Myrica rubra* for large-scale clonal propagation for nursery production.

INTRODUCTION

Yangmei (*Myrica rubra*), or yumberry, is an evergreen shrub or small tree native to southeast Asia (Zheng-Yi Wu, 1999). Maturing at a height of 4.5-6.0 meters, this plant stands out as an option for use as a screening shrub and small shade tree in the southern United States (**Fig. 1**). Its fresh fruit is sold in Chinese markets at a premium price, while dried fruit can be ground and incorporated into culinary dishes, cosmetics, and traditional medicines - hence the common name "yumberry". Breeding programs in Asia have produced cultivars for fruit production and are typically propagated asexually via grafting onto seedling rootstock, which aids to maintain desirable fruit characteristics (Davies et al., 2018).

Figure 1. A) *Myrica rubra* is a 4.5-6.0m evergreen shrub or small tree. B) Fruit, approximately 3 cm (1.2 in) in diameter. C) Purple-bronze new growth.

The plant is dioecious, so male plants are separate from fruit-producing females (**Fig. 2**). This supplies a valuable opportunity, as consumers who want to produce fruit can plant females, while consumers who do not want fruit (which can be messy) can plant males. It is important to note that females need a male pollinizer to produce fruit; this can be achieved with grafting a prolific, pollen producing male branch onto a female plant.

Myrica rubra and its cultivars have not yet been introduced to the United States, which may be in part due to the plant's propagation challenges. Clonal propagation is essential for the widespread distribution of woody cultivars, as growing plants from seed can take years and produces genetic variation in offspring. For yumberry, this can result in plants of varied sizes, shapes, foliage characteristics, and fruit qualities. While grafting desirable scion onto seedling rootstock provides a viable solution, propagation via stem cuttings according to a best-practices protocol can be faster, less labor intensive, and more reliable (Davies et al., 2018).

Figure 2. Male flowers (A) and female flowers (B) of *Myrica rubra.*

Developing a protocol for repeatable success in stem cutting propagation of *Myrica rubra* has proven to be a challenge for the Woody Ornamental Lab at the University of Georgia. Many factors can affect the rooting percentage and root quality of stem cuttings, including the genetics of the mother plant, the time at which cuttings are collected, nutrient cofactors, media formulation, irrigation schedule, hormone type, hormone application method, and hormone concentration (Davies et al., 2018).

Objectives of this study were to establish the best window of time to take cuttings, explore different hormone concentrations and types, application of nutrient cofactors to increase rooting percentage and quality of *Myrica rubra* – for its commercial introduction in the U.S. as a woody ornamental and edible landscape plant.

MATERIALS AND METHODS

Over the course of the last decade, a total of 75 *Myrica rubra* seedlings have been evaluated at the University of Georgia Durham Horticulture Farm. Among them, 35 have been identified as female, and 21 have been identified as male (we have not observed flowers on the remaining plants). Many of the seedlings are, as of 2023, 4.5-6.0 m tall and have displayed beautiful, purplish bronze new growth (**Fig. 1**). Individual plants were selected for cutting propagation based on ornamental qualities and fruit production, including shape, foliage color, pollen production (males), fruit size, fruit color, and fruit taste.

From 2016 to 2018, the Atlanta Botanical Garden and Woody Ornamental Lab at the University of Georgia set out to root

Myrica rubra stem cuttings. Based on prior knowledge of rooting evergreen stem cuttings, hardwood stem cuttings were taken from September to December during these three years. From May 2019 to February 2020, cuttings were taken on the 11th of every month from the same seedling to establish the best time to collect cuttings. From May 2021 to September 2022, semi-hardwood cuttings were targeted, which were collected from the first flush of growth from late spring to early summer, and the second flush of growth from mid to late summer.

To prepare the cuttings, leaves were removed from the bottom of stem cuttings. The top of stem cuttings retained leaves 2.5-5cm (1-2in) from the apical end of the stem, leaving approximately 5-7.6 cm (2-3 in) of exposed stem (**Fig. 3**). Leaves left on the stems were trimmed to reduce transpiration. Prior to hormone applications, the base of the cuttings were abraded lightly with the blade of pruning shears for wounding to better induce adventitious root formation (Davies et al., 2018).

Figure 3. Semi-hardwood stem cuttings are approximately 11cm (about 4.3 in) in length. A) *Myrica rubra* cutting prior to processing. B) *Myrica rubra cutting* after processing with basal leaves removed, basal end wounded, and remaining leaves trimmed.

For cutting treatments, prepared stem cuttings were soaked in a diluted solution of SUPERthrive https://en.wikipedia.org/wiki/SUPERthrive in water (approximately 10 ml SUPERthrive per 1 liter of water) for 20 minutes. SUPERthrive is marketed as a "vitamin solution" (nitrogen derived from kelp). We treated this soaking as a co-factor that could help improve the following rooting hormone application. Soaked cuttings were then dipped in Hormodin 3 talc [8000 ppm indole-3-butyric acid (IBA)].

Cuttings were then placed in 32-cell propagation trays filled with 1:1 (by volume) mix of Pro-mix, a peat-based substrate, and perlite. Cell inserts were 2 in x 2 in and 3.5 in deep. Trays with cuttings were placed under a controlled mist system in a temperature-controlled greenhouse at 24° C (75° F). Misting was scheduled to 10-sec every 10 min for the first two weeks, then 10 sec every 20 min for the second set of two weeks, and then 10 sec every 30 min until rooting data was collected.

After rooting, individual cuttings were rated on a scale of 0-5, based on quantity and length of roots. This value was referred to as "root quality". If root emergence was visible on a cutting, the cutting was considered successful. The ratio of successful cuttings to all cuttings under the treatment for each replicate was referred to as "rooting percentage". A randomized complete block design was employed for all experiments. There were eight cuttings per replication with 3, 4 or 5 replicates per treatment. Statistical analysis was conducted in R; a one-way ANOVA test at a significance level of $P = 0.05$ was used to find impacts of timing and hormone treatment on rooting of *Myrica rubra* stem cuttings.

RESULTS

From 2016 to 2018, the Atlanta Botanical Garden and Woody Ornamental Lab at the University of Georgia were unsuccessful in rooting any *Myrica rubra* (hardwood) stem cuttings. From May 2019 to February 2020, the University of Georgia found some success with semi-hardwood cuttings taken in late spring and early summer, with a highest rooting percentage of 33% (Zou et al., 2022).

Based on the success found from May 2019 to February 2020, semi-hardwood cuttings were taken in 2021 and 2022 beginning once new growth began to lignify (late May) and ceased once the second flush of growth had fully lignified (early September). For these years, increasing IBA hormone applications from 8,000ppm to 16,000ppm did not significantly affect rooting percentage (P-value = 0.879) or root quality (P-value = 0.270). The highest success found from 2021-2022 was a 45% rooting percentage and a rooting quality of 4.

The most successful treatment in our study was with semi-hardwood cuttings collected on June 16, 2020, which were soaked for 20 minutes in the SUPERthrive solution and dipped into Hormodin 3. This resulted in 79% rooting and a root quality of 3.5.

DISCUSSION

The propagation of *Myrica rubra* via stem cuttings has proven challenging. However, we can now see a clear path towards developing a successful and reliable propagation protocol (**Fig. 4**). From 2021-2022, we attempted to repeat the success found in June 2020 to reach a 79% rooting percentage. We were unsuccessful, which may be due in part to clonal differences and excess water.

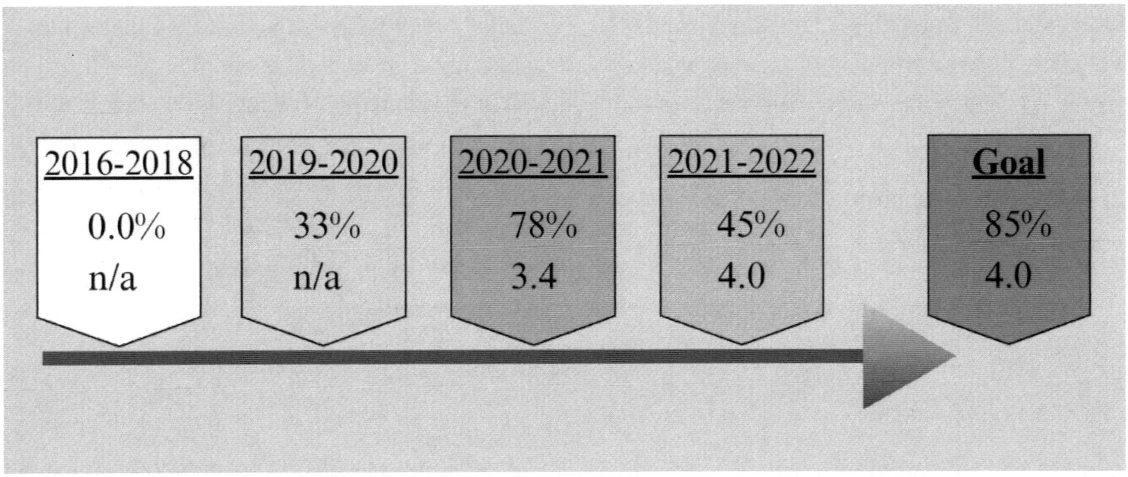

Figure 4. Timeline of progress in stem cutting propagation of *Myrica rubra*. In 2016-2018 there was no success in rooting cuttings. In 2019-2020 was 33% rooting, while the greatest success was in 2020-2021, with a maximum 78% rooting and 3.4 root quality. In 2021-2022, we were unable to repeat the success of 2020-2021. The goal of future studies is to achieve an 85% rooting success with a 4.0 root quality.

Due to a lack of significant variance in rooting percentage and root quality between periods of time within the optimal window as identified from 2021-2022 (Zou et al., 2022), semi-hardwood cuttings taken any time between May 25 and September 9 are the most viable for the clonal propagation of *Myrica rubra*. In addition, Hormodin 3 (8,000 ppm talc IBA) was found to be sufficient for hormone applications; higher concentration of IBA may not be advantageous.

One crucial factor not addressed in this study was mist scheduling. Weeks after cuttings had been placed under the mist benches with the previously described schedule, many of the cuttings deteriorated, which could be due to excess water. This may have resulted in lower rooting percentages and lower root quality. Additionally, the media of 1:1 mix of Pro-mix to perlite may compound the effects of excessive misting by retaining too much water. Optimizing a misting schedule can be tedious and taxa specific for cutting propagation, but may improve propagation success (Davies et al., 2018).

CONCLUSION

Based on our findings to date, the best protocol for clonal stem cutting propagation of *Myrica rubra* is to utilize semi-hardwood cuttings - collected between May 25 and September 5. The cuttings should be soaked in a SUPERthrive solution (10 ml of SUPERthrive into 1 L of water) for 10-20 min and dipped in Hormodin 3 IBA talc powder (8,000ppm) before sticking.

The Woody Ornamental Lab at the University of Georgia plans to conduct further studies with a factorial design in 2024. These will address mist scheduling and media water retention, to see if less water will increase rooting percentages and root quality. These studies will also explore alternate hormone formulations, such as Clonex Rooting Gel, and hormone concentrations. The Woody Ornamentals Lab also hopes to identify differences in the success of cuttings collected from different clones (seedlings), to find the impact of genetic variation on rooting percentages and root quality.

The propagation of *Myrica rubra* via stem cuttings has proven challenging. However, within six years, rooting percentages have increased from 0% to a high of 79% (**Fig. 4**). In the future, the successful and reliable propagation of *Myrica rubra* may enable the introduction of a new species and its cultivars to consumers across the United States with high ornamental and culinary value.

LITERATURE CITED

Davies, F. T., Geneve, R. L., and Wilson, S. B. (2018). Hartmann and Kester's Plant Propagation: Principles and Practices. Ninth ed. Pearson, NY, NY.

R Core Team (2021). R: A language and environment for statistical computing. R Foundation for Statistical Computing, Vienna, Austria. <https://www.R-project.org/>.

RStudio Team (2020). RStudio: Integrated Development for R. RStudio, PBC, Boston, MA. <http://www.rstudio.com/>.

Wickham, H. (2016). ggplot2: Elegant Graphics for Data Analysis. Spri.-Verl. New York. ISBN 978-3-319-24277-4. <https://ggplot2.tidyverse.org>.

Wu, Z. Y. and Raven, P. H. (1999). Flora of China, Vol. 4, Cycadaceae through Fagaceae. (p.276). Miss. Bot. Gar. Press.

Zou, Y. P., Zhang, D. L., and Hutzell, Z. (2022). Windows of opportunity for rooting woody stem cuttings. Comb. Proc. Intl. Plant Prop. Soc. *72*:207-214.

Combined Proceedings IPPS 73:217-224. 2023.

Evaluation of Sawdust Derived from Three Different Softwood Tree Species as Substrate Amendments

Amanda Mizell[a], Jeb S. Fields, and Maureen Theissen

Hammond Research Station, LSU Agricultural Center, 21549 Old Covington Hwy, Hammond, LA 70403, USA

amizel5@lsu.edu

[a]Second Place – Charlie Parkerson Graduate Student Research Paper Competition

Keywords: soilless media, wood fiber, peat moss, alternative potting media, greenhouse production

Summary

Peat moss alternatives are needed as the use of soilless substrates has increased. Wood as a peat moss amendment has been used for decades. In this study, three different softwood tree species: Douglas Fir (*Pseudotsuga*), Hemlock (*Tsuga*), and Southern Yellow Pine (*Pinus*) were blended with peat and perlite at 20- and 40% (by vol.) to create six unique soilless substrate blends. Plugs of Marigold (*Tagetes patula*) 'Janie Yellow', *Zinnia* 'Preciosa Yellow', and *Helianthus* 'Busy Bee' were grown in the sawdust substrates. Static physical properties, chemical properties, and plant health were evaluated. Overall, findings were similar to other studies that show sawdust having low bulk density, high air space and container capacity, and can grow crops comparable to a standard greenhouse growing media.

INTRODUCTION

Sphagnum peat moss is the primary component of most greenhouse production growing media. However, as the horticultural industry continues to grow, peat moss alternatives are needed to keep up with demand and lower costs for growers. Wood fibers are currently some of the leading peat alternatives that have promise (Bilderback et al., 2013; Durand et al., 2021; Jackson, 2016).

Wood as a component in soilless media has been used since the 1980's (Laiche and Nash, 1986). Its use in greenhouse production has increased due to its world-wide availability and because it can be processed into different particle sizes and textures to achieve desirable physical properties (Jackson et al., 2009; Jackson, 2018). There are many methods for processing wood components including disc-refining, hammermilling, and screw extruding (Poleatewich et al., 2022). These are all actively processed materials. Currently, the most popular wood fiber amendment for horticultural use in the U.S. is a thermally refined (i.e. heat is applied as the wood chips are spun into a fiber to reduce any chemical or biological activity that could potentially be harmful to crops) product (Hydrafiber, Profile Products, Buffalo Grove, IL). However, sawdusts are another wood product that is ubiquitous across the country and are considered waste products.

Sawdust has been utilized for plant cultivation due to its low cost, high availability, moisture retention, and adequate root-aeration for decades (Bowen, 1983; Jung et al., 2017; Yasin et al., 2023). Sawdust is easily available and widely used in places that have wood processing industries (Jung et al., 2017). Utilizing an amendment that is readily available can be more cost-effective than having components such as peat or coconut coir shipped (Yasin et al., 2023).

The primary challenge with sawdust, and other wood amendments, is nitrogen immobilization. This is when plant available nitrogen, such as nitrate ($NO3^-$) and ammonium ($NH4^+$), is converted to unavailable nitrogen by microorganisms. For this reason, soilless substrates amended with wood are often composted or supplemented with additional nitrogen through fertilization to prevent nutrient deficiencies to the crop (Jackson et al., 2009). If not treated or composted, sawdust can contain phenols and toxins which can harm plants. Additionally, the differences between properties among different species of wood makes the use of sawdust in soilless substrates variable (Jung et al., 2023). While various sources of sawdust such as Douglas Fir (*Pseudotsuga*), Red Cotton Tree (*Bombax ceiba*), and White Spruce (*Picea glauca*) have proven to yield adequate plant growth with proper irrigation and supplemental fertilizer (Depardieu et al., 2016; Yasin et al., 2022), there have not been many studies comparing sawdust from different tree species. Therefore, the objective of this study was to compare sawdust from three tree species, each harvested from a different geographic region, as a component in soilless growing media, and to evaluate the sawdust species effects on crop productivity and health. The three tree species used in this study include Douglas Fir (DF; *Pseudotsuga*), Hemlock (H; *Tsuga*), and Southern Yellow Pine (SYP; *Pinus*).

MATERIALS AND METHODS

Preparation of Substrate Blends. Sawdust was collected from an industrial lumber mill and allowed to age for one year. Substrate blends consisted of peat:perlite:sawdust at rates of 80:20:0 (control), 60:20:20, or 40:20:40 (v/v/v) for a total of seven substrate blends. Substrate treatments will be referred to as CTL, SYP20, SYP40, DF20, DF40, H20 and H40 for the remainder of this paper. Components were hydrated and slowly incorporated using an electric concrete mixer (Yardmax, Roselle, IL).

Greenhouse experiment. Plugs of Marigold (*Tagetes patula*)'Janie Yellow', Zinnia 'Preciosa Yellow', and *Helianthus* 'Busy Bee' (**Fig. 1**) were each planted in a 2.5 L container containing one of the seven substrate blends with five replicates each for a total of 105 units (7 substrate treatments x 3 plant species x 5 replicates). Crops were grown on a greenhouse bench for 63 days and hand-fertigated weekly with 200 mL of 20-10-20 water-soluble fertilizer solution adjusted to 200 ppm N (Peters Professional, Dublin, OH). Measurements including growth index, leaf chlorophyll content, flower count, substrate shrinkage, and pH and electrical conductivity (via pour-through analysis; LeBude and Bilderback, 2009) were collected bi-weekly. Crops were destructively harvested at the conclusion of the study by cutting the shoot at the substrate line and removing the substrate from the roots. Shoots and roots were dried in an oven at 70°C for five days and weighed for accumulated biomass.

Figure 1. Representative plants of each treatment with corresponding R:S ratios. In each photograph the plant on the left is the control (CTL), the center plant has 20% sawdust (by vol.), and the right plant has 40% (by vol.) sawdust.

Static Physical Properties. Physical properties including container capacity (CC), air space (AS), bulk density (D_b), and total porosity (TP) were determined on all substrate blends via NCSU porometer analysis of three replicates as described by Fonteno and Bilderback (1993). Particle size distribution was determined on all substrate blends by shaking 100 g of oven dried substrate through sieves consisting of 6.3, 2.0, 0.7, 0.5, 0.3, and 0.1 mm with a catch pan at the bottom using a Ro-Tap shaker (Rx-29; W.S. Tyler, Mentor, OH, U.S.) for five minutes. The contents of each tray were weighed and classified into four size classifications: extra-large (>6.30 mm), large (2.00–6.30 mm), medium (2.00–0.71 mm), and fine (<0.71 mm).

Data analysis. All data presented in tables and figures with corresponding statistical analysis was analyzed in JMP Pro (17.0; SAS Institute, Inc.; Cary, NC, U.S.) utilizing Analysis of variance (ANOVA) and Tukey's Honestly Significant Difference at the $\alpha = 0.05$ significance level.

RESULTS AND DISCUSSION

Static Physical Properties. There were significant differences in CC across substrate blends

($P < .0001$). The SYP20 blend had the highest CC (0.66 $g \cdot g^{-1}$) and the CTL had the lowest (0.56 $g \cdot g^{-1}$). Research has shown that sawdust substrates contain relatively high water storage capacity (Marinou, 2013). The SYP20 blend also had the lowest AS (0.10 $g \cdot g^{-1}$); however, the SYP40 blend had the highest AS (0.19 $g \cdot g^{-1}$; **Table 1**). The AS and TP across all substrate blends were significant and were all in the recommended range of 10%-30% AS and 50%-85% TP (Yeager et al., 2007; $P < .0001$). The SYP40 blend had the highest TP (0.76 $g \cdot g^{-1}$) and the CTL had the lowest (0.68 $g \cdot g^{-1}$; Table 1). The SYP blends seem to have the most suitable physical properties compared to the other species. The differences in bulk density were not as significant ($P = 0.0273$), considering sawdust is well-known for having low bulk density (Haidar and Rishmany, 2021).

Particle size distribution was significantly different across all substrate blends (**Table 1**). The H40 blend had the greatest amount of extra-large particles (2.26 $g \cdot g^{-1}$) and the lowest amount of fine particles (38.8 $g \cdot g^{-1}$; $P < .0001$; **Table 1**). In contrast, CTL had the greatest proportion of fine particles (45.1 $g \cdot g^{-1}$) and the lowest amount of extra-large particles (0.26 $g \cdot g^{-1}$; **Table 1**). All substrate blends exhibited greater quantities of fine particle proportions (**Table 1**). There was significant substrate shrinkage across all substrate blends (**Fig. 2).** The greater amount of substrate shrinkage may be due to the inherent fine particle percentages of sawdust substrates. Research has shown that smaller particle size wood substrates tend to have more shrinkage than substrates that contain larger particle sizes (Jackson, 2008; Wang, 1994)

Growth Trial. There were no significant differences in growth index across both marigolds or zinnias ($P = 0.7039$ and $P = 0.5515$, respectively); however, the sunflowers grown in 40% sawdust incorporation across all wood species exhibited significantly greater growth than the other treatments ($P = 0.0007$; **Fig. 2**). Overall, all plants were considered salable at the end of this study (**Fig. 1**). When using sawdust as a substrate amendment, it usually does not make up more than 50% of the substrate to avoid nutrient loss (Marinou, 2013). Therefore, using a smaller amount of sawdust, like in this study, with added nitrogen can lead to comparable growth to that of a traditional greenhouse media of peat and perlite. The greater growth could also be due to the adequate air space that sawdust can provide to roots after drainage (Jackson, 2008). The plants grown in CTL had the greatest chlorophyll content across all species and were significantly highest in marigolds and sunflowers (P = 0.0394 and P = 0.0001, respectively; **Fig. 2**).

Table 1. Physical properties of substrate substrates comprised of blends of peat, perlite and sawdust from three tree species.

	Static physical properties				Particle size distribution ($g \cdot g^{-1}$)			
Substrate	Container capacity ($cm^3 \cdot cm^{-3}$)	Air space ($cm^3 \cdot cm^{-3}$)	Total porosity ($cm^3 \cdot cm^{-3}$)	Bulk density ($g \cdot cm^{-3}$)	Extra-large (>6.3mm) ($g \cdot g^{-1}$)	Large (6.3mm-2.00mm)	Medium (2.00mm-0.71mm)	Fines (<0.71 mm)
80:20 peat:perlite[a]	0.56 d[b]	0.12 bcd	0.68 c	0.12 ab	0.26 d	24.9 c	29.9 ab	45.1 a
40:20:40 peat:perlite:DF[c]	0.61 bc	0.12 cd	0.73 b	0.11 ab	2.13 ab	28.1 ab	30.6 ab	40.3 cd
60:20:20 peat:perlite:DF	0.60 bcd	0.16 ab	0.76 ab	0.12 a	0.76 cd	26.6 bc	31.5 a	42.4 b
40:20:40 peat:perlite:H[d]	0.63 ab	0.12 cd	0.76 ab	0.11 b	2.26 a	28.2 ab	30.6 ab	38.8 d
60:20:20 peat:perlite:H	0.61 bc	0.15 bc	0.76 ab	0.12 ab	1.5 abc	30.1 a	29.0 b	40.0 cd
40:20:40 peat:perlite:SYP[e]	0.58 cd	0.19 a	0.78 a	0.11 ab	1.16 bcd	28.9 a	29.2 b	41.8 bc
60:20:20 peat:perlite:SYP	0.66 a	0.10 d	0.76 ab	0.12 a	1.26 bc	27.9 ab	29.2 b	43.0 b
P Value	<.0001	<.0001	<.0001	0.0273	<.0001	<.0001	0.0047	<.0001

[a]Standard 80:20 peat:perlite substrate used as a control. [b]Letters down columns represent similarities and differences according to Tukey's Honest Significant Different $\alpha = 0.05$. [c]Douglas fir (*Pseudotsuga*). [d]Hemlock (T*suga*). [e]Southern Yellow Pine (*Pinus*)

Considering the nitrogen drawdown effect that sawdust tends to have on crops, the low chlorophyll content of the sawdust blends was hypothesized (Jackson, 2009). Large amounts of fertilizer are typically needed to compensate for the nutrient loss associated with using sawdust as a substrate amendment, especially if the sawdust is not treated or composted (Jackson, 2008).

Substrate Chemical Properties. The pH across all substrate blends were significantly different. However, there were no significant differences in electrical conductivity (EC; **Fig. 2**). Initial pH was lower in all substrates and increased over time. Sawdust has been shown to increase in pH over time (Davis, 2022), like what was observed in this study. The opposite was true regarding EC, which decreased over time.

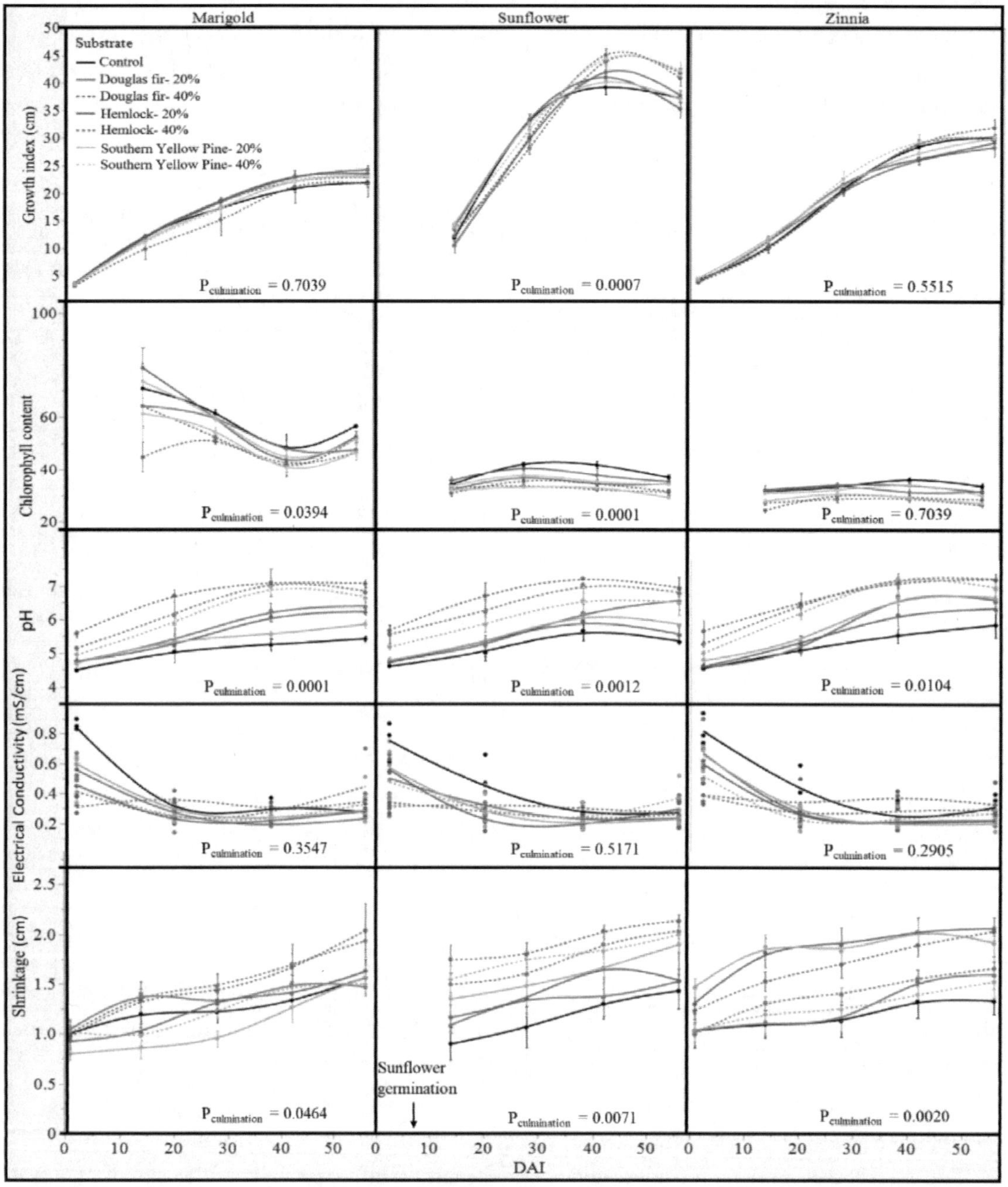

Figure 2. Growth index (cm), chlorophyll content, pH, electrical conductivity (mS/cm), and substrate shrinkage (cm) of marigold, sunflower, and zinnia crops grown in substrates developed made from peat:perlite:sawdust blends.

CONCLUSIONS

In conclusion, there seems to be very little measured differences in these short-term crops grown with and without the presence of 20- and 40% sawdust (by vol.), nor did the sawdust tree species influence crop growth. There is still more research that needs to be done to understand the stability of sawdust as a substrate amendment to better manage its fertility and irrigation requirements in a greenhouse setting. However, considering the low bulk density of sawdust and its ubiquity, it makes an excellent alternative to many peat moss alternatives that may not be as easily available or as inexpensive. This study, in combination with others that exhibit the successful use of sawdust in soilless substrates, further supports the use of sawdust across the country as the industry continues to grow and seek alternative substrate components to extend peat supplies.

LITERATURE CITED

Bilderback, T.E., Riley, E.D., Jackson, B.E., Owen, J.S., Kraus, H.T., Fonteno, W.C., Altland, J.E., and Fain, G.B. (2013). Strategies for developing sustainable substrates in nursery crop production. Acta Hortic. *013*:43–56.

Bowen, P.A. (1983). The effect of oxygen fumigation of sawdust medium on the yield and yield-components of greenhouse cucumbers. Sci. Hortic. *20*:131—136.

Carlile, W.R., Cattivello, C., and Zaccheo, P. (2015). Organic growing Media: Constituents and Properties. Vadose Zone Journal *14*:0. https://doi.org/10.2136/vzj2014.09.0125

Carlile, W.R. (2004). Changes in organic growing media during storage. Acta Hortic. *648*:153–159.

Davis, A and Strik, B.C. (2022). Long-term effects of pre-plant incorporation with sawdust, sawdust mulch, and nitrogen fertilizer rate on 'Elliott' highbush blueberry. HortScience *57*:414-421. https://doi.org/10.21273/HORTSCI16359-21.

Depardieu, C., Prémont, V., Boily, C., and Caron, J. (2016). Sawdust and bark-based substrates for soilless strawberry production: irrigation and electrical conductivity management. PLoS ONE *11*:e0154104. doi:10.1371/journal.pone.0154104.

Durand, S., Jackson, B.E., W.C. Fonteno, W.C., and Michel, J-C. (2021). The Use of wood fiber for reducing risks of hydrophobicity in peat-based substrates. Agronomy *11*:907. https://doi.org/10.3390/ agronomy11050907.

Haidar, AA and Rishmany, J. (2021). Investigation of bulk density and friction coefficient of olive residues and sawdust prior to pelletizing. Biomass Convers. Biorefin.*13*:10513–10525 https://doi.org/10.1007/s13399-021-02063-y.

Jackson, B.E., (2008). Chemical, physical, and biological factors influencing nutrient availability and plant growth in a pine tree substrate (Doctoral Dissertation). Virgina Polytechnic Institute.

Jackson, B.E. (2016). The Evolution and Revolution of Wood Substrates. Greenhouse Grower 36–40.

Jackson, B.E. (2018). Challenges and considerations of using wood substrates: physical properties. Greenhouse Grower.

Jackson, B.E., Wright, R.D., and Alley, M.M. (2009). Comparison of fertilizer nitrogen availability, nitrogen immobilization, substrate carbon dioxide efflux, and nutrient leaching in peat-lite, pine bark, and pine tree substrates. HortScience *44*:781–790.

Jung, Ji., Ha, S., Yang, J.-K. (2017). Steam treated sawdust as soilless growing media for germination and growth of horticulture plants. J. Korean Wood Sci. Technol. *45*:857-871. https://doi.org/10.5658/WOOD.2017.45.6.857.

Laiche A.J. and Nash, V.E. (1986). Evaluation of pine bark with wood, and pine tree chips as components of a container plant growing media. J. Environ. Hortic. *4*:22–25.

LeBude, A.V. and Bilderback, T.E. (2009). The pour-through extraction procedure: A nutrient management tool for nursery crops. North Carolina Coop. Ext. Bul. Ag-171-W.

Marinou, E., Chrysargyris, A., and Tzortzakis, N. (2013). Use of sawdust, coco soil and pumice in hydroponically grown strawberry. Plant, Soil Environ. *10*:452-459.

Poleatewich, A., Michaud, I., Jackson, B., Krause, M., and DeGenring, L. (2022). The effect of peat moss amended with three engineered wood substrate components on suppression of damping-of caused by *Rhizoctonia solani*. Agriculture *12*:2092. https://doi.org/10.3390/ agriculture12122092.

Wang, Y.T. (1994). Using ground kenaf stem core as a major component of container media. J. Amer. Soc. Hort. Sci. *119*:931–935.

Yasin, M., Anwar, F., Noorani, H., Muhammad,S., Mahmood, A., Javed, T., Ali, B., Alharbi, K., Saleh, I., and Abu-Harirah, H. (2023). Evaluating non-composted red cotton tree (*Bombax ceiba*) sawdust mixtures for raising okra [*Abelmoschus esculentus* (L.) Moench] in pots. Agronomy *13*:97. https://doi.org/10.3390/agronomy13010097

Yeager, T.H., Fare, D.C., Lea-Cox, J., Ruter, J., Bilderback, T.E., Gilliam, C.H., Niemiera, A.X., Warren, S.L., Whitwell, T.E., Wright, R.D., and Tilt, K.M. (2007). Best management practices: guide for producing container-grown plants. 2nd Ed. Southern Nurserymen's Assoc, Marietta, GA.

Combined Proceedings IPPS 73:225-234. 2023.

Landscape Performance of Native and Non-native Ornamentals Grown Under Two Different Irrigation Regimes in North and Northcentral Florida

Joanna J. Silva[1a], Sandra B. Wilson[2], Gary W. Knox[2], Rachel E. Mallinger[1]

[1]Department of Entomology and Nematology, University of Florida, P.O. Box 110620, Gainesville, FL 32611, USA; [2]Department of Environmental Horticulture, University of Florida, PO Box 110670, Gainesville, FL 32611, USA

joanna.jaramillo@ufl.edu

[a]Fourth Place – Charlie Parkerson Graduate Student Research Paper Competition

Keywords: pollinator plants, bee pollinators, drought tolerance

Summary

Pollinator gardening has gained momentum in recent years with an increased consumer interest in selecting native over non-native plant species to reduce water dependence and maximize biodiversity value in both public greenspaces and domestic gardens. A two-year study was conducted to determine the main effects of plant provenance (native or non-native) and moisture availability (full or partial irrigation) on landscape performance and flowering of twenty ornamental species planted in two geographic locations (north or northcentral Florida). Represented genera of paired native and non-native species included *Bidens*, *Conradina*, *Coreopsis*, *Gaillardia*, *Hibiscus*, *Ilex*, *Monarda*, *Salvia*, *Scutellaria*, and

Viburnum. A positive response of plant size was observed for native provenance and full irrigation treatments. Floral abundance of native species was also greater than non-native species at both planting locations. Across both irrigation regimes and locations, both native and non-native plants attracted a diverse population of pollinator groups. Notably there was a positive association where 2.3 times more native bees were collected from native species compared to non-native species.

INTRODUCTION

Much attention over the last decade has been directed towards ecologically friendly landscaping where plants not only require less water but bring aesthetic value and biodiversity to our gardens. Among these efforts bee-friendly gardening awareness is paramount to help mitigate the discernible global issue of bee decline. The informed selection of native and non-native plants also plays a major role in creating attractive landscapes that provide floral resources for diverse pollinators (Anderson et al., 2022; Kalaman et al., 2020). Native plants, defined as species existing in the U.S. prior to European contact, are particularly known for their resiliency in gardens, as they are locally adapted to the climate, soil conditions, and natural pests of a given region (Matrazzo and Bissett, 2020). However, the effects of plant provenance and drought tolerance of ornamental plants on pollinator preference remains unclear. For example, Salisbury et al. (2015) reported increased pollinator preference for native over non-native plants, yet Martins et al. (2017) observed no such effect. Likewise, Kalaman et al. (2020) found pollinator visitation to vary by species and planting sites while Descamps et al. (2021) showed that environmental conditions were the primary factor affecting the attractiveness, vegetative and floral traits, and resource value of plants. Thus, the overall goal of this paper was to ascertain landscape performance of a broad range of species as influenced by plant provenance, geographic planting location, and moisture availability. Specific objectives were to: 1) determine the effects of native and non-native provenance on plant growth and flowering in common garden plots, 2) determine the effects of full or reduced irrigation on these same traits, and 3) to characterize overall bee community composition visiting native and non-native plants.

MATERIALS AND METHODS

Twenty ornamental plant species were selected for use in this study based on the following criteria: 1) commercially available in nurseries and appropriate for ornamental use in landscapes, 2) able to flower prolifically and attract pollinators (and bees in particular), and 3) capable of surviving a two-year landscape trial in Florida. Resultant plants represented ten congeneric pairs of Florida native and non-native species, to analyze the effect of provenance on bee attractiveness while controlling for large variation in leaf and floral morphologies, flower colors, growth habits, and blooming periods - a key and novel component of our study design (**Table 1, Fig. 1**). Plants were sourced from as few nurseries as possible and obtained in finished one-gallon pots prior to planting.

Table 1. Natural range and cultivar origin of twenty ornamental species (paired by native and non-native genera) that were evaluated for landscape performance under partial and full irrigation regimes in north and northcentral Florida.

Species	Common name	Native range and cultivar origin
Bidens alba	Spanish needles	Native to Florida and the southern U.S., South America and the West Indies. Naturally found in disturbed sites.
Bidens ferulifolia 'BID 16101'	Goldilocks Rocks® bidens	Native range of parent species is from Arizona and New Mexico to northern Mexico. This cultivar originated from a cross-pollination made by the inventor in Bozen, Italy. USPP 32,646.
Coreopsis leavenworthii	Tickseed coreopsis	Endemic throughout Florida and two counties in Alabama. Naturally found in wet flatwoods and disturbed sites.
Coreopsis× 'Jethro Tull'	Jethro Tull coreopsis	Native range of the female parent species, *C. auriculata* extends from Virginia, Kentucky to Georgia and Louisiana. Native range of male parent species, *C. lanceolata,* includes Florida and most of the U.S. This cultivar originated from crossing *C. auriculata* 'Samfir' and *C. lanceolata* 'Early Sunrise'. USPP 18,789.
Gaillardia pulchella	Blanket flower	Native to northern Mexico, and the southern and central U.S. No longer believed to be native to Florida. Naturally found in dry open spaces with sandy soils.
Gaillardia × *grandiflora* 'Arizona Sun'	Arizona sun blanket flower	Parent species of this hybrid are *G. aristata* (Native from North Dakota to Colorado west to California and British Columbia) and *G. pulchella*. This cultivar was released in 2005 from Benary Co., The Netherlands.
Hibiscus grandiflorus	Swamp rosemallow	Native to swamps and marshes of the southeast U.S. including Florida, Alabama, Georgia, and Mississippi.
Hibiscus syriacus 'SHIMCR1'	Ruffled Satin® rose of Sharon	Native range of the parent species is Asia. This cultivar is a product of a planned breeding program, originating among the progeny of a cross pollination between *H. syriacus* 'Kwangmyung' and *H. syriacus* 'Samchulli'. USPP 26,222.
Ilex glabra	Inkberry; Gallberry	Native to the Eastern coastal plain from Nova Scotia to Florida and West to Louisiana. Naturally found in peripheries of swamps and bogs.
Ilex cornuta 'Dwarf Burford'	Dwarf Burford holly	Native range of the parent species is China and Korea. The cultivar was discovered in 1947 among vegetatively propagated Burford hollies.

Monarda punctata	Spotted beebalm	Native to Florida and the Eastern U.S. where it naturally occurs in flatwoods, dry disturbed sites and sandy sites.
Monarda didyma 'Pardon My Pink'	Pardon my pink beebalm	This species is native to bottomlands, thickets and moist woods from Maine to Minnesota south to Missouri and Georgia. The cultivar was hybridized using *M. didyma* 'ACrade' and *M. didyma* 'AChall'. USPP 24,244.
Salvia azurea	Azure blue sage	Native to flatwoods, hammocks sandhills and prairies of Florida and the central and eastern U.S.
S. longispicata × *S. farinacea* 'PAS1246577'	Big blue salvia	This interspecific cross of Indigo Spires was part of a planned breeding program by PanAmerican Seed. US PVP201700218.
Salvia rosmarinus	Rosemary	Native to dry, rocky area along the Mediterranean, Portugal, and northwest Spain, northern Africa, western Asia, southern Europe.
Conradina grandiflora	False rosemary	Endemic to Florida occurring on the central and southern Atlantic coastal ridge. Naturally found in scrub areas.
Scutellaria arenicola	Florida scrub skullcap	Nearly endemic to well-drained sandhills and scrub of Florida.
Scutellaria javanica	Malaysian skullcap	Native to wet tropical biome of Hainan, Jawa, Maluku, New Guinea, Philippines, Sulawesi, Sumatera, and Vietnam.
Viburnum obovatum	Walter's viburnum	Native to floodplain forests of Florida, Alabama, Georgia, and South Carolina.
Viburnum suspensum	Sandankwa viburnum	Native to subtropical Ryukyu Islands of southwestern Japan.

Field plots were prepared similarly in two locations. The first site was located at the UF North Florida Research and Education Center (NFREC) in north Florida (Quincy, FL, USDA cold hardiness zone 8b) and the second site was located at the UF Plant Science Research and Education Unit (PSREU) in northcentral Florida (Citra, FL, USDA cold hardiness zone 9a) and Two months prior to early spring installation, slightly raised beds were disked and treated with an herbicide prior to covering with a commercial-grade black landscape fabric. Each of 40 plots at each site measured 3 m in length and 0.9 m in width, with 1 m of spacing between each row. To create full floral coverage, a minimum of two (mostly herbaceous) and a maximum of three (mostly woody) plants of each species were assigned to each plot, determined by their predicted size at full maturity. Once established (after 4 weeks), half of the plots were drip-irrigated for 2h per day, while the other half were irrigated at 10% volumetric soil moisture using a SMRT-Y- Soil Moisture Sensor Kit (Rainbird Inc., Tucson, AZ). Plants were top-dressed with a 20N–4P–7K slow-release fertilizer (Osmocote Pro, 8–9-months) upon planting and between years.

Figure 1. Floral representation of ten native (N) and ten non-native ornamental plants selected for this study.

Standard soil analysis using a Mehlich-3 extraction method revealed macro and micronutrients at both locations were within normal limits with a 6.7 and 7.0 soil pH in north and northcentral FL, respectively (UF Extension Soil Testing Laboratory, Gainesville, FL). As reported by an automated weather network system

(https://fawn.ifas.ufl.edu), field conditions in north FL were as follows: avg. monthly rainfall 0.42 cm, mean minimum and maximum temperatures 13.56 and 26.30 °C, respectively, and 78.76 % relative humidity. In northcentral FL field conditions were as follows: avg. monthly rainfall 0.35 cm, mean minimum and maximum temperatures 15.55 and 28.27 °C, respectively, and 81.5% relative humidity.

Each month, plant height and perpendicular widths were measured for each plant at both locations to generate a maximum growth index ([height + (avg. width1 + width2)]/2 for the first year of the study. Also, a floral survey was conducted monthly for the entirety of the two-year study where the total number of flowers were counted for each plot across all treatments. Capitulate inflorescences (*Bidens*, *Coreopsis* and *Gaillardia*) were notated as a single flower (**Fig. 1**). To characterize the insect community composition (bees in particular) among species from native and non-native provenances, active sampling techniques were deployed within plots where each observer (consisting of 2- 4 people) walked down each row collecting foraging insects for a period of one to three minutes per plot. Specimens were placed in vials and stored in the freezer for subsequent identification.

To evaluate the main effects of provenance and irrigation on growth and flowering, this study utilized eight rows (blocks) per planting location, with four rows receiving partial irrigation and four rows receiving full irrigation. Twenty plots containing congeneric native and non-native species were assigned to each block using a completely randomized design. Generalized linear models (GLM) were used with plant 'provenance' (native versus non-native) and 'irrigation' treatment (full versus partial) as fixed effects and 'plot' as the random effect. Data were subjected to an analysis of variance (ANOVA) using statistical software RStudio (Version 2023.06.2+561, Boston, MA) with significance determined at P=0.05.

RESULTS

The maximum growth indices measured for each species are presented in **Table 2**. A significant effect of irrigation (P=0.0124), provenance (P<0.0001), and planting location (P<0.0001) was observed with a non-significant 3-way interaction (P=0.9566). In north FL, six of the 20 species (Spanish needles, Arizona sun blanket-flower, spotted beebalm, pardon my pink beebalm, Walter's viburnum, and sandankwa viburnum) grew larger under full irrigation than partial irrigation regimes. In northcentral FL, Spanish needles, Arizona sun blanket-flower and spotted beebalm also grew larger under full than partial irrigation, as well as Jethro Tull coreopsis. Collectively, native and non-native plants grown in north FL had 1.2 times greater plant size than plants grown in northcentral FL. Across both irrigation treatments and planting locations, native plants were 1.4 times larger than non-natives plants.

Table 2. Maximum growth index (cm) [plant height (cm)+ average of two perpendicular widths (cm)/2] for each of twenty ornamental species, grown under two irrigation treatments (partial and full) at each location (north and northcentral Florida) during the first year of the study. For each planting location, means are presented ±SD with different letters indicating significant responses among full and partial irrigation treatments (*P*=0.05). Native species are indicated with a (N).

	North Florida growth index		Northcentral Florida growth index	
Species	Full (cm)	Partial (cm)	Full (cm)	Partial (cm)
Spanish needles-N	204.0 ± 12.8a	181.6 ± 14.8b	151.3 ± 7.0a	136.1 ± 13.5b
Goldilocks Rocks® bidens	83.7 ± 6.2a	76.6 ± 5.4a	44.8 ± 7.9 a	47.7 ± 7.8a
False Rosemary-N	61.5 ± 12.5a	68.2 ± 25.6a	85.6 ± 4.2a	79.2 ± 5.9a
Rosemary	78.4 ± 8.6a	84.4 ± 4.8a	64.2 ± 8.6a	63.3 ± 5.1a
Tickseed coreopsis-N	162.2 ± 47.4a	178.1 ± 16.9a	144.1 ± 15.6a	143.7 ± 16.6a
Jethro Tull coreopsis	95.5 ± 3.8a	86.5 ± 13.7a	83.1 ± 6.0a	69.7 ± 7.1b
Blanket-flower-N	193.7 ± 14.5a	173.3 ± 14.2a	150.7 ± 3.3a	154.3 ± 16.9a
Arizona sun blanket-flower	98.8 ± 7.3a	85.7 ± 7.5b	72.8 ± 13.4 a	65.4 ± 5.4b
Swamp rosemallow-N	184.8 ± 25.7a	186.3 ± 9.7a	143.6 ±12.7a	150.8 ± 12.8a
Ruffled Satin® rose of Sharon	68.6 ± 7.3a	60.3 ± 8.3a	39.6 ± 4.1a	39.9 ± 5.6a
Inkberry; Gallberry-N	106.4 ± 3.3a	105.2 ± 27.1a	90.0 ± 20.51a	78.9 ± 5.4b
Dwarf Burford holly	161.8 ± 5.7a	164.8 ± 43.1a	116.5 ± 12.7a	102.2 ± 12.3b
Spotted beebalm-N	224.5 ± 59.4a	161.5 ± 17.8b	199.7 ± 49.3a	180.0 ± 49.1b
Pardon my pink beebalm	64.9 ± 4.5a	57.2 ± 1.1b	54.3 ± 9.7a	45.9 ± 4.9a
Azure blue sage-N	159.1 ± 15.3a	167.3 ± 11.1a	144.8 ± 8.9a	143.1 ± 11.3a
Big blue salvia	162.6 ± 27.4a	150.4 ± 9.2a	142.7 ± 34.4a	131.18 ± 17.5a
Florida scrub skullcap-N	112.6 ± 14.2a	114.3 ± 6.8a	89.4 ± 6.0a	80.4 ± 15.0a
Malaysian skullcap	77.9 ± 5.8a	75.8 ± 5.5a	59.9 ± 7.1a	54.0 ± 6.1 a
Walter's viburnum-N	145.6 ± 10.0a	165.9 ± 61.8b	107.3 ± 9.4a	116.6 ± 11.8a
Sandankwa viburnum	114.3 ± 6.2a	102.9 ± 7.1b	91.6 ± 7.8a	87.3 ± 7.9a

The number of flowers counted for each species plot for the entirety of the two-year study are presented in **Figure 2**. Similar to growth responses, a significant effect of provenance ($P<0.0001$) was observed for floral abundance. However, the effects of irrigation ($P=0.2844$), planting location ($P=0.1528$), and their interaction ($P=0.6206$) were non-significant. An impressive 974,143 and 34,517 floral counts were recorded among species grown in north and northcentral FL, respectively (**Fig. 2**). In north FL, regardless of irrigation treatment, six of the ten native species (Bidens, Coreopsis, Gaillardia, Ilex, Monarda, and Rosemary) had greater floral abundance than their respective non-native congener. This same species response was observed in northcentral FL, but additionally, the native Walter's viburnum produced more flowers than the non-native Sandankwa viburnum. Interestingly, at both locations, the non-native big blue salvia was the only species to have greater floral abundance than its native congener, azure blue sage. Also, at both locations, flowering responses to provenance were not observed for *Scutellaria* and *Hibiscus*.

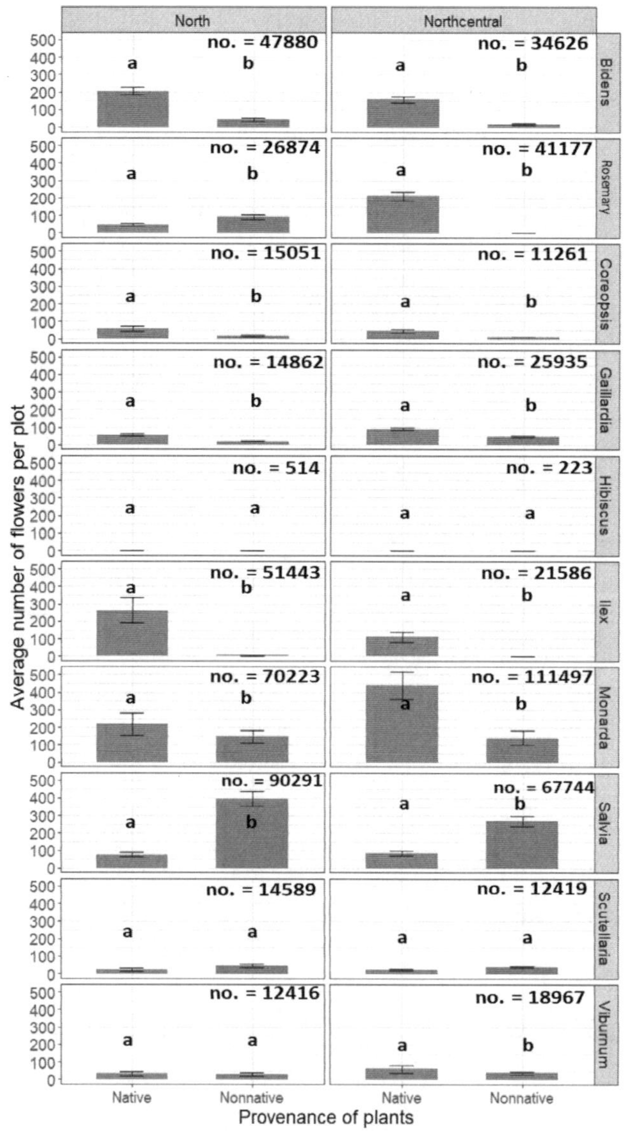

Figure 2. Average number of flowers counted for native (green bars) and non-native (orange bars) species of each genus grown in north and northcentral Florida (across combined nonsignificant irrigation regimes). Flowers were counted in each plot each month during a two-year period. Error bars denote standard deviation of the mean (n=20). Within each graph, different letters indicate response is significant at $P=0.05$. The total number of flowers per congener was included at the top of each graph as a reference to floral abundance.

We collected 937 insects visiting flowers comprising of 599 specimens visiting the native species and 338 specimens visiting the non-native plants (**Fig. 3**). Of the total insects sampled from native plants: 45.4%, 17.7%, 13.4% and 4.5% were, respectively, wild native bees, wasps, honeybees, and bumblebees. Of the total insects sampled from non-native species: 35.2%, 23.4%, 16.9% and 5.5% were, respectively, wild native bees, honeybees, bumblebees, and wasps.

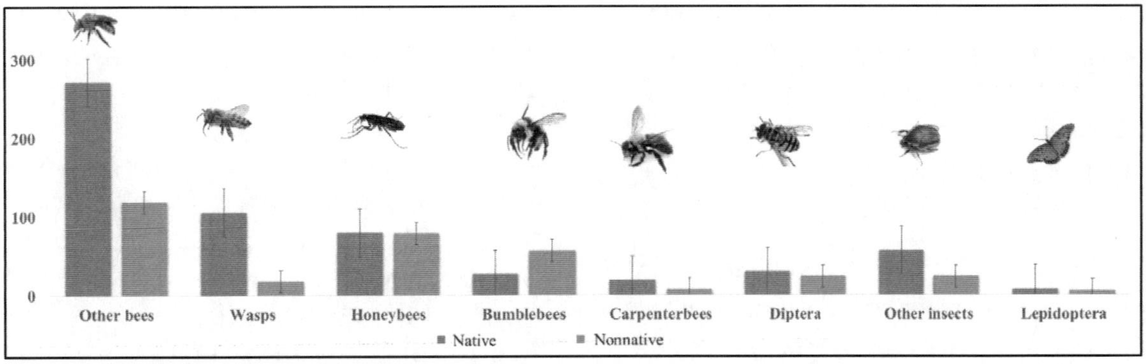

Figure 3. Number of different groups of pollinators actively collected from flowers on ten native (green bars) and ten non-native (orange bars) ornamental species across both irrigation treatments and locations.

DISCUSSION

Unique to this study, pollinator friendly, native and non-native ornamental species were used to assess the effects of moisture availability on plant growth and flowering. Of the twenty species evaluated, a positive growth response to full rather than partial irrigation was observed for less than a third of all native and non-native plants. This nominal effect of provenance on plant growth is consistent with results from Schober et al. (2010) who found irrigation frequency positively affected plant size of only 20% of the woody plants evaluated.

In addition to plant growth in the present study, floral abundance was also measured for native and non-native species. While flowering response to irrigation was non-significant, a 1.8-fold increase in floral abundance was observed for native plants compared to non-native plants. This is consistent with a 2.0-fold increase reported in an earlier study (Kalaman et al., 2020). Likewise, the type of pollinators attracted to flowers was also influenced by provenance, where native plants attracted 2.3 times more native bees than non-native plants. In their review of global pollinator decline, Potts et al. (2010) point out that non-native plants may be primarily attractive to generalist and non-native bee species. In the current study, the number of honeybees (generalists) collected were similar for native and non-native plants. Yet, more bumblebees (also generalists) were sampled from non-native rather than to native species.

CONCLUSIONS

Results presented herein show that provenance has more of an effect on plant performance than irrigation for the species evaluated in this study. Selecting plants that are both attractive, tolerant of varying environmental conditions, and rich in floral resources may be a priority for effective pollinator gardening. Further work is underway to determine the responses of pollinators to changes in floral resources (nectar and pollen), floral traits, visual signals, and volatile organic compounds (VOCs).

LITERATURE CITED

Anderson, A.G., Costner, L., Best, L., and Langellotto, G.A. (2022). The bee fauna associated with Pacific Northwest (USA) native plants for gardens. Conserv. Sci. Pract. *4*(10): 1-18. doi:10.1111/csp2.12801.

Baldock, K. (2020). Opportunities and threats for pollinator conservation in global towns and cities. Curr. Opin. Insect Sci. *38*:63-71. doi:10.1016/j.cois.2020.01.006.

Garbuzov, M., Alton, K., and Ratnieks, F.L.W. (2017). Most ornamental plants on sale in garden centers are unattractive to flower-visiting insects. Peer J. *5*, e3066. https://peerj.com/articles/3066/

Kalaman, H., Wilson, S.B., Mallinger, R.E., Knox, G.W., and van Santen, E. (2020). Evaluating native and non-native ornamentals as pollinator plants in Florida: I. Floral abundance and insect visitation. HortScience *57*:126-136. doi:10.21273/HORTSCI16123-21.

Martins, K.T., Gonzalez, A. and Lechowicz, M.J. (2017). Patterns of pollinator turnover and increasing diversity associated with urban habitats. Urban Ecosyst. *20*:1359–1371. doi.org:10.1007/s11252-017-0688-8.

Matrazzo, S. and Bissett, N. (2020). Native Plants for Florida Gardens. Palm Beach (FL) Pineapple Press, Inc.

Potts, S. G., Biesmeijer, J.C., Kremen, C., Neumann, P., Schweiger, O. and Kunin, W.E. (2010). Global pollinator declines: Trends, impacts and drivers. Trends Ecol. Evol. *25*(6): 345-353. doi:10.1016/j.tree.2010.01.007.

Salisbury, A., Armitage, J., Bostock, H., Perry, J., Tatchell, M., Thompson, K. (2015). Editor's Choice: Enhancing gardens as habitats for flower-visiting aerial insects (pollinators): should we plant native or exotic species? J. Appl. Ecol. *52*: 1156-1164. doi: 10.1111/1365-2664.12499.

Shober, A.L., Moore, K.A., Wiese, C., Scheiber, S.M., Gilman, E.F., and Paz, M. (2010). Native and non-native shrub post-transplant performance under low-volume irrigation in three hardiness zones. HortTechnology. *20*:751-757. doi:10.21273/HORTTECH.20.4.751.

Combined Proceedings IPPS 73:235-245. 2023.

Cultivating Sustainability: Biodegradable Containers in Horticultural Production

Melanie Hill[a], Emily Stamm, and Paul C. Bartley

101 Funchess Hall, Department of Horticulture, Auburn University, Auburn, AL 36849, USA

mch0097@auburn.edu

[a]Third Place – Charlie Parkerson Graduate Student Research Paper Competition

Keywords: Basil, biodegradable containers, commercial production

Keywords: basil, biodegradable containers, commercial production

Summary

Plastic containers are the standard in the horticulture industry for the production of plants for retail and landscape use. Sustainability has become a popular topic because the pandemic affected the way people think about sustainability and their impact. The pandemic also created a boom in the plant industry because people were home and had more time to explore their current interests as well as new interests, such as indoor and outdoor gardening. Six containers made from materials other than petroleum-based plastic were used in this study to look at their performance in relation to plastic containers, as well as their degradation and marketability. Plant size, container gravimetric differences from trial initiation to harvest week, container wet and dry tensile strength from trial initiation to harvest week were evaluated; and a consumer opinion survey was conducted at each harvest week. The largest plants were grown in peat and BioPax containers. BioPax containers also had the highest tensile strength at all testing intervals - as wet and dry test pieces.

INTRODUCTION

In the U.S. Horticulture Industry, plastic containers used in the greenhouse and nursery industries amount to more than 750,000 metric tons of plastic (Shrader et al., 2015). Plastic is the standard container type for the horticulture industry because they have a durability that withstands automated production and the strains of shipping, the ease of acquiring many shapes and sizes of containers, and a relatively low cost (Kratsch et al., 2015). However, it is estimated that ~98% of plastic containers are disposed of in landfills due to contamination risks and the cost of cleaning and sanitizing containers (Fuentes et al., 2021; Shrader et al., 2015). In addition to cost-prohibitive factors, containers used in greenhouse and nursery production systems experience degradation due to the light and heat conditions, further disincentivizing container reuse (Fuentes et al., 2021).

Sustainable growing containers fall into one of three categories: plantable, compostable, and bioplastic (Soulliere-Chieppo., 2020). Containers that fall into the plantable category are typically constructed from materials such as coconut coir, manure, peat, paper, and wood pulp and are intended, as the category name implies, to be planted in the soil with the plants still inside them (Soulliere-Chieppo., 2020). Compostable containers have been produced from rice hulls, poultry feathers, recycled paper or cardboard, bamboo, or other fibers and require a home or industrial compost system to be broken down (Soulliere-Chieppo., 2020). Bioplastic containers are made of plastic that started as a plant constituent or has plant constituents instead of petroleum (Soulliere-Chieppo., 2020). Growers prefer bioplastic containers due to their consistency, stability, and durability during handling, processing, and shipping, and similarities to traditional plastic containers. Several sustainable containers are already on the market and have been used in research conducted in greenhouse and landscape settings. Containers previously evaluated include rice straw, rice hulls, paper, peat, coconut coir, composted cow manure, and wood fiber (Conneway, 2013; Kuehny et al., 2011). Results from these investigations varied, but the consensus was that most container types produced plants of marketable size and quality (Kuehny et al., 2011) and that low container strength can be attributed to containers made of coir, wood fiber, peat, manure, and straw (Conneway, 2013).

While some of these sustainable containers are commercially available, an online survey conducted in 2020 found that 83% of horticultural growers do not purchase biodegradable containers (Harris, 2020). Potential reasons for this are that biodegradable containers lack the necessary strength for automation processes and the durability to remain structurally sound for the length of production cycles (Kratsch et al., 2015). If these containers do not maintain structural integrity, then producers risk experiencing losses when containers break during production, shipping, or in the retail environment (Harris, 2020). Since the pandemic, demand for horticultural commodities and sustainably sourced products has increased across horticultural consumers. A study by the University of Georgia found that approximately 1,400 of the 4,200 respondents started gardening in 2020 due to being at home more (Campbell, 2022). Recent market research suggests ornamental plant consumers are willing to pay more for nonplastic and recyclable containers, and an increasing number attempt to avoid the use of plastic or opt for products with packaging that is environmentally friendly (Emmert, 2021; Fulcher et al., 2015).

Considering the variety of plants grown in varying production cycles and changes in consumer appetite, a new investigation into sustainable containers was warranted. This study aimed to evaluate commercially available alternatives to traditional plastic containers in bi-weekly production intervals using crop performance, material testing, and consumer evaluations to determine their viability in floriculture production systems.

MATERIALS AND METHODS

Six biodegradable or biobased containers were evaluated with an industry-standard plastic container through an eight-week basil production cycle (**Table 1**).

Table 1. Biodegradable, biobased, and plastic containers which were assessed in an eight-week greenhouse trial.

Product	Composition	Container size	Image
CowPot	Composted cow manure	5"	
FertilPot	Wood fiber	4"	
EverEco	Tapioca starch	3.5"	
PlantBest	Coconut coir	4.5"	
BioPax	Wood pulp and additives	4"	
Jiffy	Peat	5"	
Control	Plastic	4"	

Eighteen units of each container type were labeled, weighed, filled with a peat-based substrate, and received a basil transplant. Planted containers were randomly arranged on a greenhouse bench. All containers were fertilized once per week (250 ppm of 15N-5P-15K; JR Peters, Allentown, PA) and irrigated with clear water every other day. Four replications from each treatment were randomly selected for harvest every two weeks. Basil plants were cut at the substrate surface at each harvest interval, and fresh and dry weights were recorded. The containers were then dried at 65 C for one week. After the containers were dried, the substrate was removed, and the containers were reweighed to calculate the percentage of weight lost from initiation to the date of harvest.

Tensile strength testing was performed on the containers, both wet and dry, to evaluate changes in material characteristics due to degradation. Harvested containers were cut into rectangular strips (1" in width) and conditioned for at least 40 hours at 23 C ± 2.0 C and 50% relative humidity before tests were run. Dry tests were performed immediately following conditioning. Wet tests were conducted after the samples were submerged in water for 105 min and set to drain for 60 min. Tests began by securing the samples with clamps to the load frame (Series 5565; Instron, Norwood, MA) and concluded when the samples failed (i.e., broke or began to stretch). Each test was performed with the utmost care to ensure sample integrity. However, material integrity changed throughout the study, which resulted in some materials being more fragile than others and, thus, some specimens could not be tested accurately.

At each harvest interval, a survey was conducted to gauge consumer appetite for each container type. Each consumer was polled by age and the frequency of purchasing plants. Only the container, with the plant removed, was revealed to the consumer to survey. Consumers were then asked to rank their likeliness to purchase a plant grown in each container type using a 1-5 scale, where "1" was very unlikely and "5" was very likely.

All data were analyzed via ANOVA with the PROC Glimmix procedure, SAS 9.4 (SAS Institute Inc., Cary, NC). Means were separated using Tukey's honest significant difference (HSD) at a 5% alpha level.

RESULTS AND DISCUSSION

Basil Growth. Basil dry weights were similar across all container types two weeks after transplant **(Fig. 1)**. By Week 4, differences in basil dry weight were observed. Basil growing in coir, plastic, and peat produced the largest plants four weeks after transplant.

Basil grown in EverECO containers, produced from tapioca starches, had the lowest dry weight at Week 4. This trend would continue through Week 6. A significant increase in basil dry weight occurred from Week 4 to Week 6. No differences were observed between container types with the exception of EverECO, which produced the smallest basil.

By Week 8, container effects on basil dry weight differentiated broadly. Peat containers grew basil 25% larger than CowPots and 350% larger than those produced in EverECO. BioPax and plastic containers were the most similar in shape and size.

Dry weights for BioPax and plastic were similar until Week 8. At the final harvest interval, basil grown in plastic containers were 26% larger than those produced in BioPax.

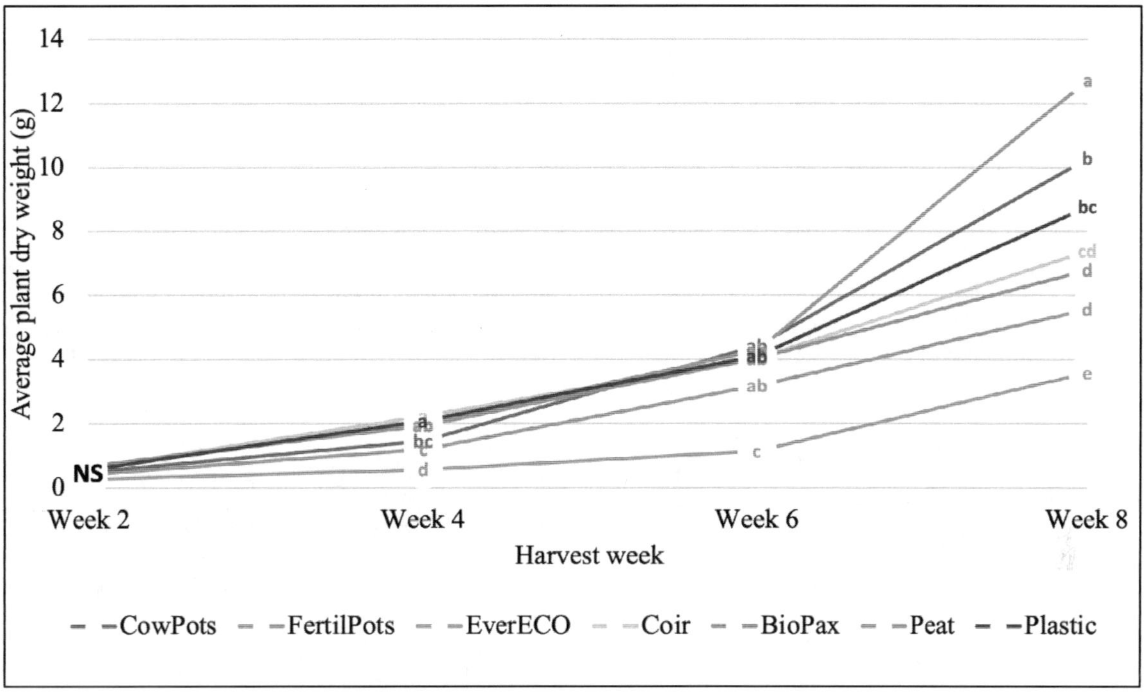

Figure 1: Average plant dry weights for each treatment by harvest date.

Due to volumetric differences between containers, differences in basil growth were likely the result of a more/less restrictive growth environment and may not reflect the characteristics of the materials (**Fig. 2**). Between Week 2 and Week 6, similarities in basil dry weight could be attributed to the containers' stability and volume, which demonstrated only subtle signs of degradation and provided ample space for root growth and development. After Week 6, degradation of the container walls, wetting and drying cycles, and increased resource demand from basil plants could have contributed to the differences observed at the Week 8 harvest interval.

A common challenge with biodegradable plant containers is the evaporation of water through the walls of the container. Some container materials are more porous, reducing plant available water and affecting plant growth. For example, EverECO and wood fiber containers were observed to dry down faster than other containers. Consequently, EverECO and wood fiber containers produced the smallest basil plants on average. The plants grown in peat pots and CowPots were larger than other containers, likely due to the size of the container.

Container Degradation. Degradation can occur rapidly in biodegradable containers (**Fig. 3**). By Week 2, all biodegradable and biobased containers had lost weight. EverECO containers degraded 12.8% by Week 2, but degradation moderately stabilized through Week 6. However, by Week 8, EverECO containers had lost an average of 33.8%. CowPots also degraded quickly and, by the study's conclusion, had lost more than 24% of its initial weight. FertilPots,

Coir, BioPax, and Peat containers demonstrated similar degradation rates at each harvest interval. By the conclusion of the study, all container types had lost at least 5% of their original weight.

Figure 2. Progression of plant size and container state at each harvest week.

Containers of manure and coir fiber were fragile and the most difficult to dry and clean without compromising their integrity. EverECO containers were remarkably pliable and slimy when wet but dried out quickly and became brittle. BioPax containers, feeling and appearing the most plastic-like, degraded nearly three times more than plastic containers. Surprisingly, by Week 8, plastic containers had lost ~5% of their initial weight. Due to the persistence and invasiveness of microplastics in our environment, further investigations may be warranted to understand plastic fate in production systems.

Materials Testing. Material tensile strength, determined by the maximum force withstood before sample failure, was affected by container type, harvest week (Week 0, Week 4, and Week 8), condition (wet or dry), and their interactions (p=<0.0001). Since differences between container materials were distinct, mean comparisons were determined by material across harvest week and condition (**Table 2**).

BioPax containers withstood the greatest force of any container material before failure. As BioPax degraded, tensile strength improved by 44% and 73% by Week 4 and Week 8, respectively. Wetting BioPax reduced its tensile strength by 14-22% at each harvest interval. Plastic containers were able to withstand considerable tensile force without breaking.

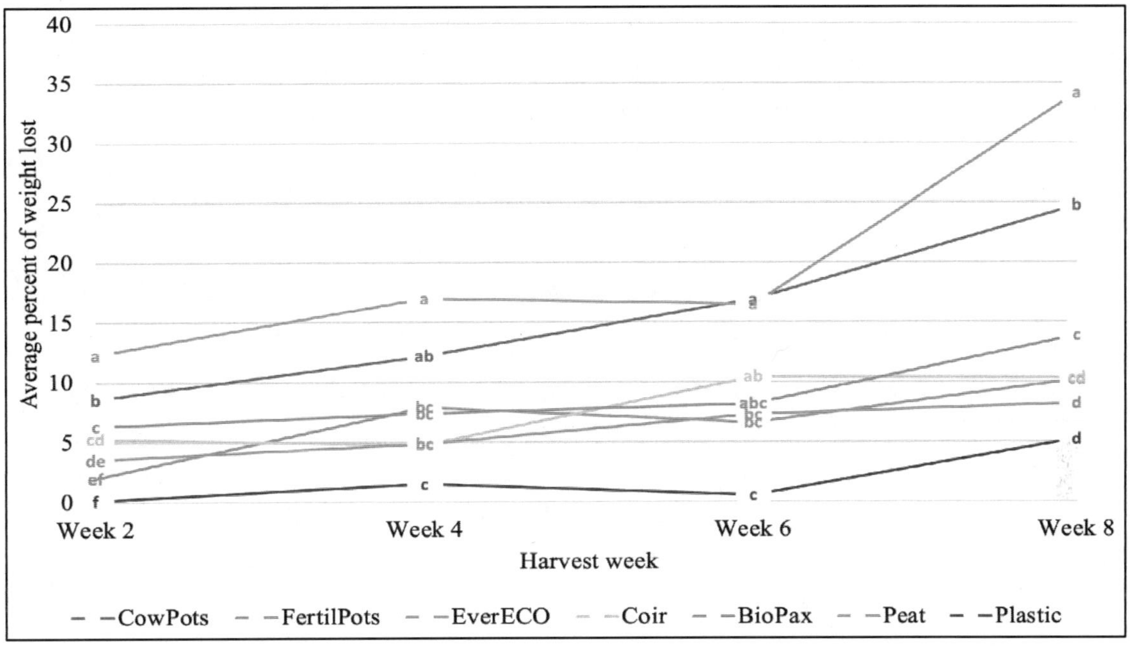

Figure 3. Average percent of weight lost from containers at each harvest date to estimate container degradation.

The elasticity of the plastic allowed the samples to stretch only, and no breakage was recorded. As expected, the tensile strength of plastic was the least impacted by hydration. While the tensile strength of BioPax and plastic increased with age, the tensile strength of other biodegradable containers exhibited decreasing trends as they aged and degraded. Hydrating biodegradable containers significantly reduced the tensile strength of CowPots, FertilPot, coir, and peat containers. Hydration of these samples often resulted in instability in handling.

Delamination occurred in containers made of coir fiber when wet. CowPots, having the lowest tensile strength, were markedly fragile when wet. Wet samples of EverECO (tapioca starch) could not be tested due to a complete loss of sample integrity. These results highlight a significant problem shared by many sustainable container types.

Consumer Opinion. Greater than half of the survey participants fell into the age group of 18-24 years old (**Table 3**). The highest percentage of participants (48%) purchased plants seasonally. Consumer purchasing habits were unaffected by age. Survey participants were most likely to buy a plant grown in a coir, BioPax, peat, or plastic container. Survey participants were least likely to purchase a plant grown in an EverECO container (**Table 4**). Few trends in consumer likeliness to purchase were observed by week. However, initial preferance for FertilPot decreased sharply from Week 2, 3.54, to Week 8, 2.73.

In discussion with survey participants (post submission), many responded in favor of coconut coir containers and in disapproval of plastic containers. Often, participants thought that BioPax containers were petroleum-based and were delighted to learn the product was biobased.

Table 2. Tensile strength, determined by the maximum force (N) withstood before failure, of biodegradable, biobased, and plastic containers after 0, 4, and 8 weeks of production

		Harvest interval		
Container	Condition	Week 0	Week 4	Week 8
CowPot	Dry	9.0a[z]	8.2a	8.4a
	Wet	3.6b	2.5b	2.3b
FertilPot	Dry	40.3a	17.6b	22.9b
	Wet	5.5d	4.2d	8.7c
EverECO	Dry	79.4a	49.3b	23.0c
	Wet	--	--	--
Coir	Dry	23.2a	21.6a	12.2b
	Wet	21.2a	14.0b	8.0c
BioPax	Dry	398.5d	542.7b	622.9a
	Wet	309.1e	446.6c	534.6b
Peat	Dry	57.0a	45.5b	36.5c
	Wet	19.0d	10.8e	10.1e
Plastic	Dry	245.6b	260.2ab	277.8a
	Wet	213.8c	258.8ab	251.9b

[z] Data were analyzed using an ANOVA and subsequent means were compared within container type using the Tukey honest significant difference ($p \leq 0.05$). Means within a container type with the same letter do not significantly differ from each other.

Table 3. Demographics and purchasing habits of survey participants.

Age	% of respondents	Plant purchasing frequency	% of respondents
18-24 years old	68.8%	Once per week	7.8%
25-34 years old	12.5%	Monthly	18.8%
35-44 years old	0%	Seasonally	48.4%
45-54 years old	7.8%	Once per year	14.1%
55 or older	10.9%	Less than once per year	10.9%

Table 4. The likeliness to purchase each container changed over time.

	Container						
Harvest interval	CowPot	FertilPot	EverECO	Coir	BioPax	Peat	Plastic
Week 2	3.5[z]	3.5	2.7	3.9	3.8	3.7	3.7
Week 4	3.2	3.1	1.9	3.5	4.0	3.5	4.2
Week 6	2.9	3.3	2.2	3.8	3.9	3.8	3.5
Week 8	3.5	2.7	2.7	4.4	3.8	3.7	3.9

[z] Participants were asked to rank their likeliness to purchase a plant grown in each container type using a 1-5 scale, where "1" was very unlikely and "5" was very likely.

CONCLUSIONS

The adoption of biodegradable containers by growers in the horticulture industry is a topic with many points of concern, including the quality of plants able to be grown in them, if the containers can withstand the rigors of a typical production setting, and whether they would be marketable by the end of lengthy production cycles. This study has shown that the biodegradable containers included in this study produce plants of marketable size and quality, with the exception of EverECO tapioca starch containers. BioPax containers maintained the greatest tensile strength throughout the trial and are most likely to withstand stresses incurred in commercial production systems. With respect to marketability, EverECO and FertilPot containers were ranked lowest by survey participants after the eight-week trial. However, other sustainable container products invoked similar or greater consumer enthusiasm as traditional plastic containers. Given the consumer enthusiasm for sustainable cultivation - considerable research and product development is needed to improve the industry's trust in biodegradable and biobased containers.

LITERATURE CITED

Conneway, R.L. (2013). Evaluating biodegradable containers as alternatives to plastic pots, *Graduate Theses, Dissertations, and Problem Reports*. 541. https://researchrepository.wvu.edu/etd/541 (Accessed: 22 March 2023).

Emmert, A. (2021). The rise of the eco-friendly consumer, strategy and business. https://www.strategy-business.com/article/The-rise-of-the-eco-friendly-consumer (Accessed: 22May 2023).

Fuentes, R.A. et al. (2021). Development of biodegradable pots from different agroindustrial wastes and byproducts. Sustain. Materials Tech. https://www.sciencedirect.com/science/article/pii/S2214993721000932 (Accessed: 15 March2023).

Harris, B.A., Pennisi, S.V. and Florkowski, W.J. (2020). Horticulture industry adoption of biodegradable containers. HortTech. *30*(3): 1-13.

Kratsch, H.A. *et al.* (2015). Performance and biodegradation in soil of novel horticulture containers made from bioplastics and biocomposites. HortTech. *25*(1): 119-131.

Kuehny, J.S., Taylor, M., and Evans, M.R. (2011). Greenhouse and landscape performance of bedding plants in biocontainers, HortTech. *21*(2): 155-161.

Schrader, J.A. *et al.* (2015). Development and evaluation of bioplastic containers for sustainable greenhouse and nursery production, Acta Hortic. *1104*: 79-88.

Soulliere-Chieppo, M. (2020). Plastic pots and the green industry - production, use, disposal and environmental impacts. https://cdn.ymaws.com/www.mnla.com/resource/resmgr/docs/apld-plastic-pots-white-pape.pdf (Accessed: 22 March 2023).

Poleatewich, A., Michaud, I., Jackson, B., Krause, M., and DeGenring, L. (2022). The effect of peat moss amended with three engineered wood substrate components on suppression of damping-of caused by *Rhizoctonia solani*. Agriculture *12*:2092. https://doi.org/10.3390/ agriculture12122092.

Wang, Y.T. (1994). Using ground kenaf stem core as a major component of container media. J. Amer. Soc. Hort. Sci. *119*:931–935.

Yasin, M., Anwar, F., Noorani, H., Muhammad,S., Mahmood, A., Javed, T., Ali, B., Alharbi, K., Saleh, I., and Abu-Harirah, H. (2023). Evaluating non-composted red cotton tree (*Bombax ceiba*) sawdust mixtures for raising okra [*Abelmoschus esculentus* (L.) Moench] in pots. Agronomy *13*:97. https://doi.org/10.3390/ agronomy13010097.

Yeager, T.H., Fare, D.C., Lea-Cox, J., Ruter, J., Bilderback, T.E., Gilliam, C.H., Niemiera, A.X., Warren, S.L., Whitwell, T.E., Wright, R.D., and Tilt, K.M. (2007). Best management practices: guide for producing container-grown plants. 2nd Ed. Southern Nurserymen's Assoc, Marietta, GA.

Characterization and Efficacy of a Novel Poultry-Derived Fertilizer for Container Production

Austin Lindquist[1], Paul Bartley[1] and Rishi Prasad[2]

[1]101 Funchess Hall, Department of Horticulture, Auburn University, Auburn, AL 36849; [2]201 Funchess Hall, Department of Crop, Soil, and Environmental Sciences, Auburn University, Auburn, AL 36849, USA

all0091@auburn.edu

Keywords: poultry litter, fertilizer, nutrient release, plant assays

Summary

The poultry industry is a major rural economic industry within the Southern US. Raw poultry litter, a waste product of the industry, is facing increasing regulation for its use as a broadcast fertilizer. Alternatively, poultry litter can be processed utilizing aerobic digestion to produce a new product with potentially fewer negative environmental and health consequences. A novel digested litter-based fertilizer produced by Cleaned and Green, LLC (C&G) was assessed for nutrient release characteristics and plant responses to gauge its efficacy for container production. Rapid water incubation tests demonstrated that the majority of C&G nutrients released within 10 min, similar to traditional synthetic fertilizers. In soil-based incubation tests, ammonium release occurred in two phases, at Day 1 and Day 35. Nitrate concentration remained low through Day 15 but increased dramatically through Day 55. Potassium was immediately available upon application. Phosphorus concentrations were not significantly higher than control soil, indicating this product may help alleviate some environmental concerns. Tomatoes grown with C&G and C&G blended fertilizers showed improved vitality at higher N rates.

However, tomatoes growth with poultry litter at higher rates produced larger plants. Growth assays on tomatoes was recorded. Due to its rapid release, this novel fertilizer appears best fit for short-term floriculture crops.

INTRODUCTION

Extended-release fertilizers fall into two categories: slow-release and controlled-release. Controlled-release products often contain synthetic, plastic-coated prills that rely on environmental factors to mediate release (Vejan *et al.*, 2021). Slow-release fertilizers often lack a synthetic coating and rely on microbial processes to release nutrients (Fu *et al.*, 2018). Poultry litter falls into the rapid-release fertilizer category; nutrient release and availability begin upon release (Wang *et al.*, 2015). Controlled-release fertilizers are common in container production to meet plant requirements over time and limit excess leaching (Chen and Wei, 2018). The rapid release of high nutrient amounts may lead to environmental concerns and require multiple applications for desired plant nutrition in the growing season (Vejan *et al.*, 2021).

The United States Department of Agriculture estimates the value of poultry products at $46 billion, with Georgia (1.3 billion birds) and Alabama (1.2 billion) overtaking Arkansas (1 billion) in total birds (USDA, 2022). Poultry litter (PL) applications can enhance plant growth, demonstrating generally positive relationships within soils (Wang *et al.*, 2015). However, concerns pertaining to the eutrophication of aquatic environments have brought on renewed legislation and regulation of its applications. Raw PL may undergo the process of aerobic digestion to slow nutrient release and reduce the risk of disease. Aerobic digestion utilizes microbial activity or thermophilic heat from an acid to break down the litter (Zhang *et al.*, 2022). Aerobic digestion may be combined with additional processes, such as that utilized by Cleaned & Green (C&G), to produce a PL-based fertilizer with N-P-K ratios similar to fertilizers in container production (USPO, 10,723,665 B1).

The objectives of this study were two-fold: first, to characterize the nutrient release patterns of C&G, and second, to conduct a growth assay with C&G and other nutrient sources. Physical prill characterization, chemical composition, and microbial viability were also tested (data not reported).

MATERIALS AND METHODS

Incubation Test. Water-based incubation assessments were conducted to determine the nutrient release patterns of a novel, PL-derived fertilizer, C&G (12 N - 2 P_2O_5 - 2 K_2O - 12 S), a synthetic fertilizer (herein referred to as "Synthetic") blended to contain similar nutrient values, and a polymeric resin-coated fertilizer (17-5-11; Osmocote, Scotts Miracle-Gro, Marysville, OH). One gram of each fertilizer product was added to 100 mL DI water (n=3). Solutions were stirred at 60 rpm and maintained a temp of 20 C. Electrical conductivity (EC) measurements occurred at set time intervals over a 24-hour period.

Soil-based incubation tests were performed with a clay soil (pH=6.6). The soil was homogenized by drying, pulverizing, and sieving to exclude particles >2 mm. Incubation jars (n=80) were filled with 100 g of dry soil. C&G was incorporated at a rate of 1.5 lbs. N/yd^3 in half of the jars. Soil samples were maintained at a volumetric water content of 0.3 cm^3/cm^3 and 30 C. On each sample date, four replications per treatment (C&G and Control) were removed, air-dried, homogenized, and partitioned for nutrient extraction. Ammonium and nitrate concentrations were determined using 2M KCL extraction, fluorimetry, and colorimetry on a multimode plate reader. Total P and K were determined using a Mehlich 1 extraction (Mehlich, 1953).

Plant Assays. Tomatoes (*Solanum lycopersicum* 'Celebrity') were transplanted into 1 gal containers (Classic 400; Nursery Supplies Inc., Fairless Hills, PA) filled with a pine bark: peat (3:1 v:v) amended substrate. Fertilizer treatments included the following: Synthetic, C&G, Blend (Synthetic+C&G), PL, and a control receiving no fertilization. Fertilizer treatments were applied at three rates: 0.75 lbs. N/yd^3, 1.5 lbs. N/yd^3, and 3 lbs. N/yd^3. Each treatment-rate combination included six replicants. Substrate pH and EC were monitored weekly using the Pour-Through method (Wright, 1986). Plant dry weight and foliar nutrient concentration were determined eight weeks after transplant.

Soil-based nutrient release data were analyzed via PROC Reg procedure of SAS 9.4 (SAS Institute Inc., Cary, NC). Effects of fertilizer, rate, and the fertilzer*rate interaction on dry weight and foliar nutrient concentrations were analyzed via ANOVA with the PROC Glimmix procedure. Means were separated using Tukey's honest significant difference (HSD) at a 5% alpha level.

RESULTS

Incubation Tests. In rapid water release incubation, Synthetic released nutrients both quickest and in the greatest quantities (**Fig. 1**).

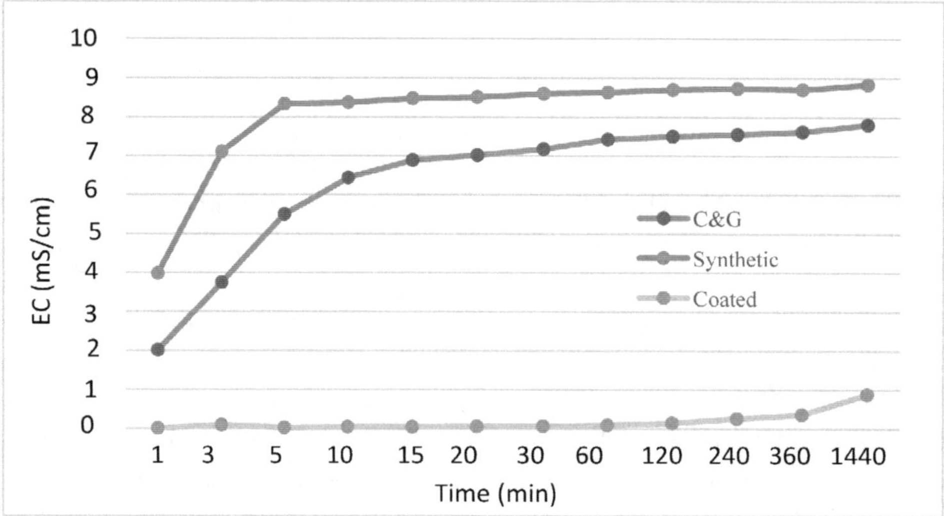

Figure 1. Nutrient release rates, determined by EC, of a novel poultry litter-derived fertilizer (C&G), a synthetic granular fertilizer (Synthetic), and a resin-coated fertilizer (Coated) over 24 hours in a water incubation.

Within 10 minutes, Synthetic released 95% of recorded nutrients compared to 82% for C&G. As expected, little to no change in EC was recorded in the resin-coated product after 24 hrs. C&G performed similarly to quick-release fertilizers and should be considered a quick-release product.

In soil, C&G immediately released ammonium at 65 ppm as compared to 14 ppm for the Control (**Fig. 2**). A secondary release of ammonium, nearly doubling the initial release, occurred on Day 45. By Day 55, ammonium concentration had declined from its peak to 119 ppm. Background ammonium concentrations were 6 ppm at Day 55. Nitrate concentrations increased slowly over the first two weeks before sharply increasing through Day 55. On Day 15, nitrate comprised 25% of the total nitrogen released from C&G (**Fig. 3**).

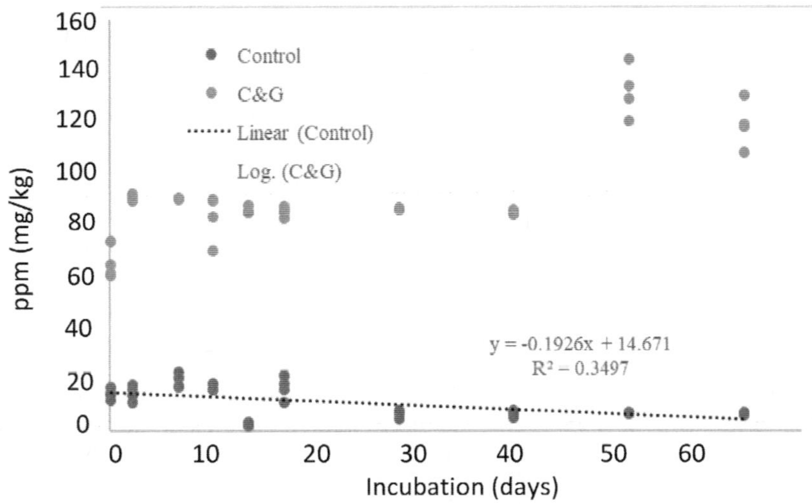

Figure 2. Recorded ammonium concentrations within soil after applying a novel poultry litter-derived fertilizer (C&G) over 55 days.

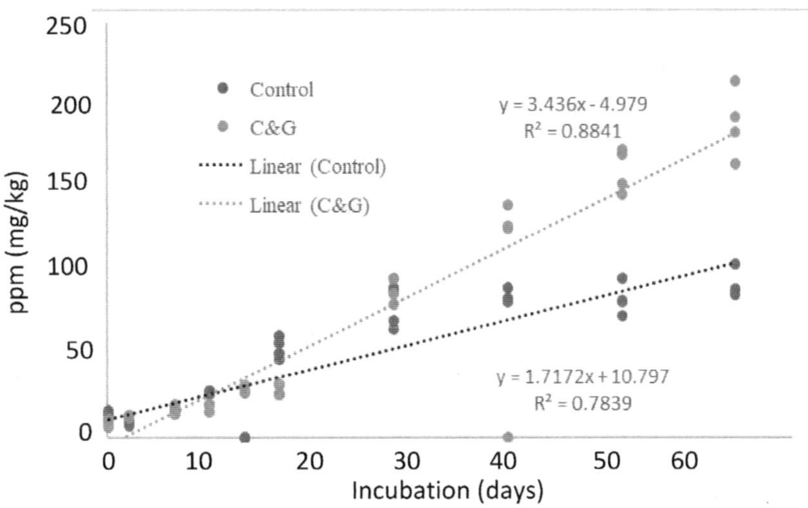

Figure 3. Recorded nitrate concentrations within soil after applying a novel poultry litter-derived fertilizer (C&G) over 55 days.

Nitrate concentrations surpassed ammonium on Day 35 and had a final concentration of 189 ppm. In contrast, the Control nitrate concentration on Day 55 was 92 ppm. Total nitrogen release at the conclusion of the trial was 308 ppm for C&G and 178 ppm for the Control (**Fig. 4**). Potassium was immediately released from C&G. By Day 9, 94% of the total K had been released (**Fig. 5**). Although C&G increased the mean soil-extractable P, concentrations of P were not significantly higher ($p = 0.44$) than the control soil (**Fig. 6**).

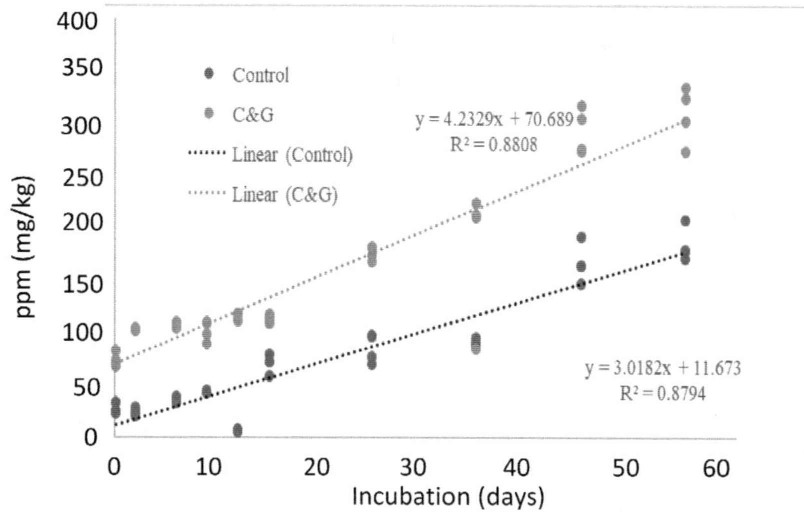

Figure 4. Total nitrogen (sum of nitrate and ammonium concentrations) within soil after applying a novel poultry litter-derived fertilizer (C&G) over 55 days.

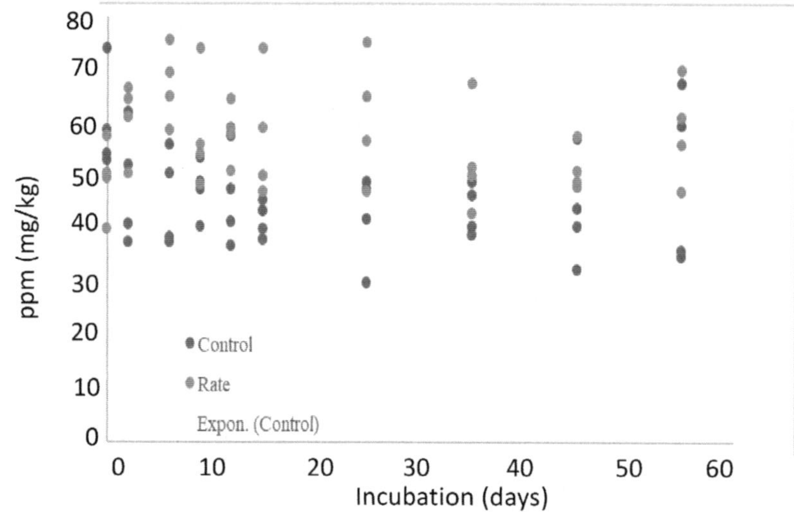

Figure 5. Recorded phosphorus concentrations within soil after applying a novel poultry litter-derived fertilizer (C&G) over 55 days.

Plant Assays. Electrical conductivity levels rapidly dropped from a high of ~8 mS/cm for all fertilizer applications within the first three weeks after transplant (**Fig. 7**). Poultry litter was the only treatment to maintain EC levels >1 mS/cm for eight weeks. By the conclusion of the study, Synthetic and Control recorded similar EC readings at 0.2 mS/cm. Containers with C&G and PL applied maintained higher EC levels by the conclusion of the study. Across all fertilizer treatments, substrate pH was reduced by approximately 1 unit but increased steadily to 6.5-7 (**Fig. 8**).

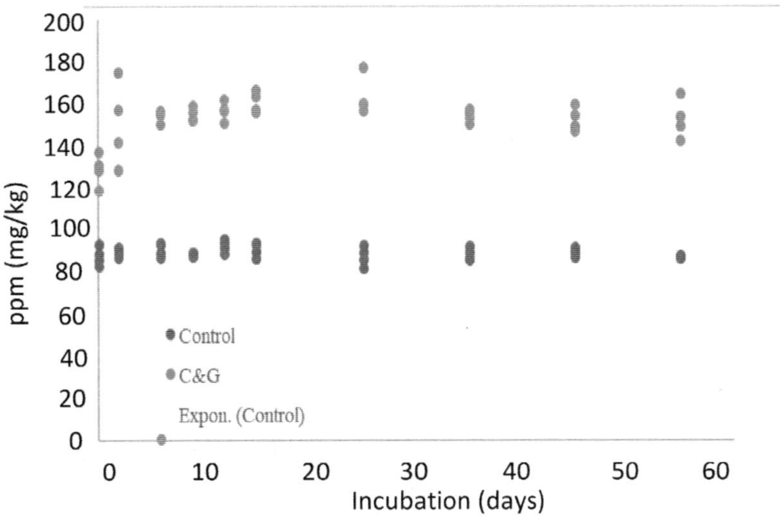

Figure 6. Recorded potassium concentrations within soil after applying a novel poultry litter-derived fertilizer (C&G) over 55 days.

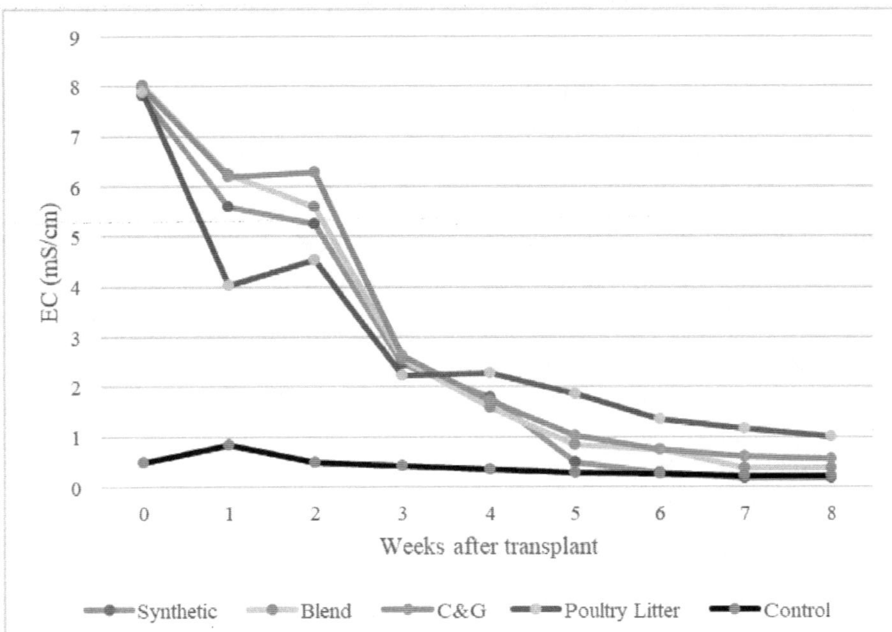

Figure 7. Measured electrical conductivity (EC) of container-grown tomatoes fertilized with a synthetic fertilizer (Synthetic), a novel poultry litter-derived fertilizer (C&G), Blend (Synthetic+C&G), poultry litter, and a control receiving no fertilization. All fertilizers were applied at a rate of 1.5 lbs. N/yd^3.

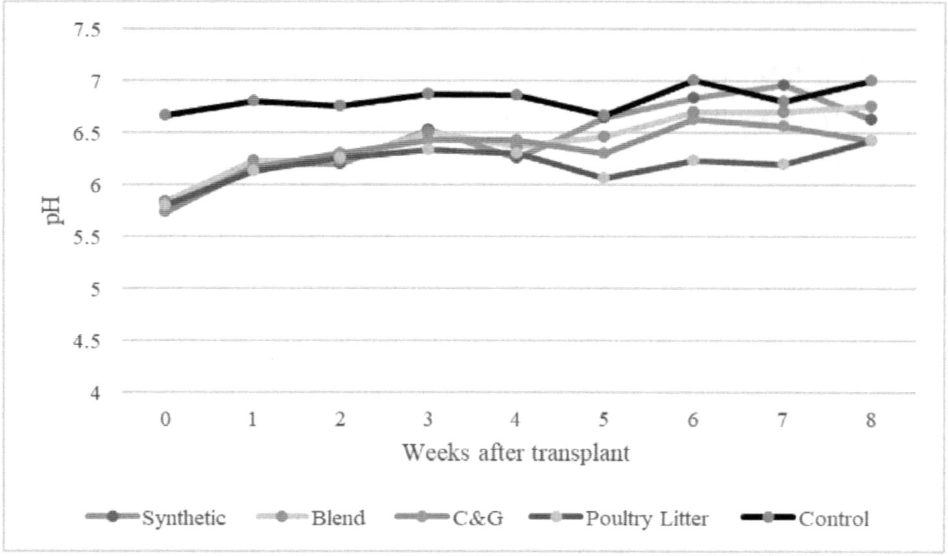

Figure 8. Effect of fertilizer treatments on substrate pH. Fertilizer treatments consisted of a synthetic fertilizer (Synthetic), a novel poultry litter-derived fertilizer (C&G), Blend (Synthetic+C&G), poultry litter, and a control receiving no fertilization.

Tomato dry weight was affected by rate, fertilizer type, and the interaction between rate and fertilizer type (p=< 0.0001). PL at 1.5 lbs N/yd^3 and 3 lbs. N/yd^3 produced the largest tomato plants (**Table 1**). PL applied at 0.75 lbs. N/yd^3 produced plants similar in size to C&G or Blend applied at rates of 0.75 lbs. N/yd^3 and 1.5 lbs. N/yd^3. Increased mortality was observed in all fertilizers applied at 3 lbs. N/yd^3 except PL. Tomatoes receiving Synthetic at 3 lbs. N/yd^3 had an 83% mortality rate. Tomatoes receiving C&G or Blend at 3 lbs. N/yd^3 demonstrated improved vitality, only losing one specimen each. However, Blend applied at 3 lbs. N/yd^3 produced the smallest plants after eight weeks. Visually, tomatoes fertilized with PL exhibited greater nutrient deficiencies (**Fig. 9**).

Foliar N concentrations were affected by nutrient source, rate, and their interaction (p=< 0.0001). Products containing C&G contained the highest concentrations of nitrogen (**Table 2**). At 3 lbs. N/yd^3, C&G and Blend produced plants with foliar N of 2.47% and 2.56%, respectively. Synthetic applied at 0.75 lbs. N/ yd^3 produced plants with foliar N of 1.97%. Although PL tomatoes were the largest, foliar N was lowest at 0.86%, 0.96%, and 1.39% for rates of 0.75, 1.5, and 3 lbs. N/yd^3, respectively.

Foliar P concentrations were affected by nutrient source, rate, and their interaction (p=< 0.0069). Tomatoes fertilized by PL, all rates, Blend at 3 lbs. N/yd^3 contained the highest foliar P concentrations. Few differences were observed in other treatments. Foliar K concentrations were unaffected by rate, but differences were recorded by nutrient source (p=< 0.0001). However, no discernable trends in foliar K were observed.

Table 1. Plant assay dry weights.

Fertilizer	Rate (lbs. N/yd^3)	Dry weight (g) Tomato
	0.75	19.5cd[z]
Synthetic[y]	1.5	29.9bc
	3	--
	0.75	30.4b
Blend[x]	1.5	32.5b
	3	12.6d
	0.75	33.7b
C&G[w]	1.5	32.7b
	3	24.9bc
	0.75	34.8b
Poultry Litter	1.5	49.9a
	3	58.2a
Control	--	0.2e

[z] Data were analyzed using a one-way anova and subsequent means were compared using the Tukey honest significant difference ($p \leq 0.05$). Means within a column with the same letter do not significantly differ from each other.
[y] Synthetic, a rapid-release fertilizer blended to contain the same nutrient ratio and Cleaned and Green minus sulfur

[x] Blend is a 1:1 ratio of Synthetic and Cleaned and Green fertilizer

[w] Cleaned and Green (C&G) is a poultry litter derived product containing a similar N-P-K ratio as blend and Synthetic but with the addition of 11% Sulfur

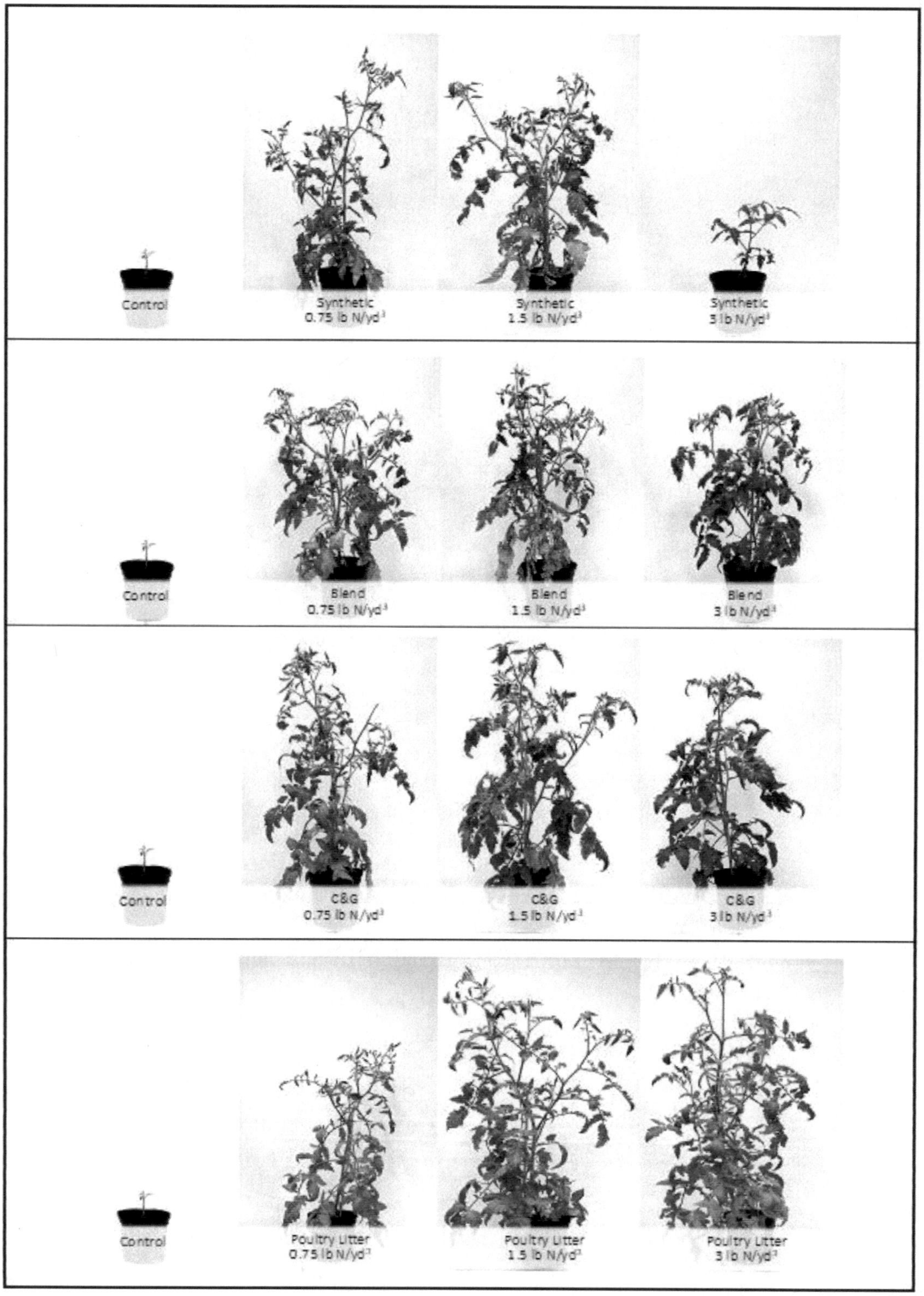

Figure 9. Tomatoes eight weeks after transplanting which received nutrients from a synthetic fertilizer (Synthetic), a novel poultry litter-derived fertilizer (C&G), Blend (Synthetic+C&G), poultry litter, and a control receiving no fertilization.

Table 2. Plant assay of plant macronutrient percentages within tissue.

Fertilizer	Rate (lbs. N/yd³)	Nitrogen	Phosphorus	Potassium
	0.75	1.97abc[z]	0.07d	0.74ab
Synthetic[y]	1.5	1.53dc	0.19cd	0.53b
	3	--	--	--
	0.75	1.55cd	0.12 cd	0.46b
Blend[x]	1.5	1.78cd	0.17cd	0.39b
	3	2.56a	0.37ab	0.63b
	0.75	1.49cde	0.12cd	0.47b
C&G[w]	1.5	1.94bcd	0.19cd	0.34b
	3	2.47ab	0.26bc	0.4b
	0.75	0.86f	0.44a	0.79ab
Poultry Litter	1.5	0.96ef	0.46a	0.71b
	3	1.39def	0.38ab	1.32a

[z] Data were analyzed using a one-way anova and subsequent means were compared using the Tukey honest significant difference ($p \leq 0.05$). Means within a column with the same letter do not significantly differ from each other.

[y] Synthetic, a rapid-release fertilizer blended to contain the same nutrient ratio and Cleaned and Green minus sulfur

[x] Blend is a 1:1 ratio of Synthetic and Cleaned and Green fertilizer

[w] Cleaned and Green (C&G) is a poultry litter derived product containing a similar N-P-K ratio as blend and Synthetic but with the addition of 11% Sulfur

DISCUSSION

Both water- and soil-based incubation tests characterize C&G as a quick-release fertilizer. However, C&G did exhibit a delayed release of ammonium. Between Day 35 and Day 45, ammonium concentration rose 56%. Similarly, nitrate concentrations began to climb rapidly after Day 15. These mechanisms for delayed N release need to be studied further. While P and K were applied in similar concentrations, their releases and availability drastically differed. Potassium and phosphorus were quickly released after incorporation. The mean P concentrations were 61% less than K concentrations. Additional testing is required to determine the speciation of P and C&G potential implications to reduce the environmental burdens of raw PL applications.

Differences were observed between nutrient products and rates in container-grown tomatoes. The Synthetic fertilizer, a custom blend matching the macronutrients of C&G, was volatile at 3 lbs. N/yd^3, resulting in the death of most tomatoes by Week 4. Fatalities were also overserved, but to a lesser extent, in the Blend (Synthetic + C&G) and C&G fertilizers applied at 3 lbs. N/yd^3. Surviving treatments produced smaller plants but mostly recovered by Week 8. No fatalities occurred from PL applications, which produced plants 67% heavier than Synthetic and 53% heavier than C&G. By Week 8, all treatments experienced yellowing, purpling, or a combination of nutrient deficiency symptoms. Significant declines in measured EC occurred across all treatments through Week 8.

Tissue analysis revealed that C&G nutrient levels resemble the Synthetic fertilizer. Plant produced with PL contained higher levels of P and K, but lower N concentrations. Due to the mobility of N within the plant, the rapid growth observed in tomatoes grown in PL would predictably cause N in the most recently mature leaves to be reallocated to meristematic regions. Higher P and K concentrations were expected in tomatoes grown in PL due to its balanced N-P-K ratio.

Overall, C&G, both as a stand-alone and supplemental nutrient source, has shown positive effects on plant growth. It was less volatile than the ammonium sulfate blend of synthetic fertilizer, resulting in fewer plant fatalities. This novel, litter-based product has the greatest potential in short-duration production systems where irrigation and leachate fractions can be carefully controlled. However, further testing in container ornamental crops is needed.

LITERATURE CITED

Chen, J. and Wei, X. (2018). 'Controlled-Release fertilizers as a Means to Reduce Nitrogen Leaching and Runoff in Container-Grown Plant Production', in K. Amanullah and S. Fahad (eds) Nitrogen in Agriculture: Updates - Google Books, pp. 38–39. DOI: 10.5772/intechopen.73055

Chen, Z. and Jiang, X. (2014). 'Microbiological safety of chicken litter or chicken litter-based organic fertilizers: A Review', Agriculture *4*:1-29. https://doi.org/10.3390/AGRICULTURE4010001.

Fu, J. et al. (2018). Classification research and types of slow controlled release fertilizers (SRFs) used - a review. Comm. Soil Sci. Plant Anal. *49*(17): 2219–30. DOI: 10.1080/00103624.2018.1499757

Mehlich, A. (1953). Determination of P, Ca, Mg, K, Na and NH4 by North Carolina soil testing laboratories. University of North Carolina, Raleigh, NC. https://www.ncagr.gov/agronomi/pdffiles/mehlich53.pdf

Poultry - Production and Value 2022 Summary (April 2023) 3 USDA, National Agricultural Statistics Service. usda.library.cornell.edu/usda-esmis/files/m039k491c/wm119387d/5138kw352/plva0423.pdf

Vejan, P. et al. (2021). Controlled release fertilizer: A review on developments, applications and potential in agriculture. J. Control. Release. *339*: 321-334. https://doi.org/10.1016/j.jconrel.2021.10.003.

Wang, Y. et al. (2015). Phosphorus release behaviors of poultry litter biochar as a soil amendment. Sci. Total Environ. *512–513*: 454–463. https://doi.org/10.1016/j.scitotenv.2015.01.093

Zhang, M. et al. (2022). Application of autothermal thermophilic aerobic digestion as a sustainable recycling process of organic liquid waste: Recent advances and prospects. Sci. Total Environ. *828*:154187. https://doi.org/10.1016/J.SCITOTENV.2022.154187.

Combined Proceedings IPPS 73:258-265. 2023.

Effects of Cytokinin Type and Concentration on Shoot Proliferation in a Novel *Tripidium* Hybrid

Tanner Hamerling, Darren Touchell and Thomas Ranney

Mountain Crop Improvement Lab, Department of Horticultural Science, Mountain Horticultural Crops Research and Extension Center, North Carolina State University, 455 Research Drive, Mills River, NC 28759-3423, USA

tom_ranney@ncsu.edu

Keywords: micropropagation, cytokinin, thidiazuron, zeatin, 6-benzyl aminopurine

Summary

Tripidium, a genus within the sugarcane complex (Andropogoneae tribe of the Poaceae), is used as a landscape plant and more recently has been recognized for its bioenergy potential. Micropropagation protocols were investigated to expedite shoot proliferation. Four cytokinin types including 6-benzylamino purine (BAP), thidiazuron (TDZ), zeatin and kinetin were investigated at concentrations of 0. 2.5, 5.0, 10.0 or 20.0 µM. In a second experiment the effect of BAP or TDZ, alone or a 1:1 ratio combination, at 2.5, 5.0, or 10.0 µM on shoot regeneration was investigated. Media supplemented with either 5 µM TDZ or 20 µM BAP produced 6.05 and 5.75 shoots per explant, respectively. The combination of BAP and TDZ did not significantly improve multiplication rates. This research provides protocols for rapid multiplication and micropropagation of *Triipidium.*

INTRODUCTION

Tripidium is a genus within the Poaceae Subtribe Saccharinae, commonly referred to as the sugarcane complex. New interspecific hybrids between *Tripidium ravennae* and *T. arundinaceum* have considerable potential as biomass crops and have demonstrated high yields and overwintering survival rates in USDA Zone 6b (Maren et al., 2021). Traditionally these bioenergy grasses have been commercially propagated through divisions or rhizomes. However, newly developed high-yielding varieties show dense clumping rhizomatous masses with increased culm densities. While these qualities have led to increased yields, they have also led to reduced efficiency of conventional propagation and field establishment methods using divisions and rhizomes. Development of in vitro propagation methods would provide a valuable option for more rapid multiplication.

While micropropation studies on *Tripidium* have been limited (da Silva et al., 2020), there have been several reports on related genera within the sugarcane complex including *Saccharum* sp. (Jahangar et al. 2010; Ramgareed et al., 2010; Salokhe, 2021) and *Miscanthus* sp. (Zhang et al., 2010). In these studies, 6-benzylamino purine (BAP) alone has been the most predominant cytokinin for effective shoot proliferation, though optimal concentrations varied among genera. For *Saccharum* sp., concentrations between 1 to 5 μM BAP were effective for shoot proliferation (Ajadi et al., 2018; Jahangir et al., 2010), while for *Miscanthus* 10-20 μM BAP were more effective (Zhang et al., 2012). However, for some *Saccharum* sp. low concentrations of kinetin in combination with BAP have been utilized to improve shoot proliferation (Balagalla et al., 2018; Cheong et al., 2009; da Silva et al., 2020).

Zeatin and TDZ have been effectively used to a lesser extent to induce shoot proliferation in species within the sugarcane complex. For example, Vinayak et al. (2009) found zeatin to be the most effective cytokinin for shoot proliferation for *Saccharum spontaneum* hybrids. Vazques-Molina et al. (2005) found TDZ to be effective for shoot proliferation in sugarcane (*Saccharum* spp,) cultivars. Interestingly, Sukendah et al. (2023) found that a combination of TDZ and BAP was most effective for shoot proliferation in several *Saccharum* sp. genotypes.

With the development of *Tripidium* as a bioenergy feedstock it would be beneficial to develop micropropagation protocols to facilitate rapid propagation. Considering the variation in media used for related species, the objective of this research was to investigate the effect of cytokinins type and concentration on shoot proliferation of a novel *Tripidium* hybrid.

MATERIAL AND METHODS

Plant material and culture conditions. Nodal sections with actively growing axillary shoots were used to initiate in vitro cultures. Actively growing shoots were collected from field grown plants and rinsed under tap water for 4 h. Explants were surface-disinfested in 20% (v/v) bleach (6.15% NaOCl) solution containing two to three drops of Tween® 20 for 25 min with periodic mixing followed by three 5-min rinses in sterile distilled water. Explants were cultured on initiation medium consisting of Murashige and Skoog (MS) basal salts and vitamins (Murashige and Skoog, 1962) supplemented with 20 µM BAP, 100 mg/L *myo*-Inositol, 100 mg/L 2-(N-Morpholino) ethanesulfonic acid (MES) monohydrate, and 30 g/L sucrose. Media were solidified with 6.5 g/L agar (Phytotechnology Laboratories, Shawnee Mission, KS) and adjusted to pH of 5.75, and 25 mL was dispensed to 180-mL glass jars. Microshoots were maintained by transfer to fresh culture medium every 4 to 6 weeks and incubated under standard culture conditions [23° ± 2°C and a 16 h photoperiod of 80 $\mu mol \cdot m^{-2} \cdot s^{-1}$ provided by cool-white fluorescent lamps].

Cytokinin Type and Concentration. Effects of cytokinins and their concentration on microshoot growth and proliferation were examined. Media consisted of MS salts and vitamins supplemented with BA, TDZ, zeatin, or kinetin at 0, 2.5, 5.0, 10.0 or 20.0 µM. Cytokinins were added to media prior to autoclaving, except for zeatin, which was filter sterilized and added to cool autoclaved media. All media were supplemented with 100 mg/L *myo*-Inositol, 100 mg/L MES, and 30 g/L sucrose, solidified with 6.5 g/L agar and adjusted to pH 5.75. Media (25 mL) was dispensed into 180-mL glass jars. Three-week- old actively growing stock plants were divided into single shoots and placed onto experimental media. Five microcuttings (10–20 mm long) were placed vertically in each jar. Eight replicates of each media composition were incubated under standard culture conditions, as described previously, in a completely randomized design. After 8 weeks, data were recorded on the number of surviving explants, number of microshoots, and microshoot length (of longest shoot). Data sets were subjected to regression analysis.

BAP and TDZ Combinations. The effects of the cytokinin BAP and TDZ, alone or in combination, on shoot production was investigated. Basal media for all treatments was MS salts and vitamins, 6.5 g/L of agar, 0.1 g/L of Myo-inositol, 0.1 g/L of MES buffer, and 30 g/L of sucrose. To test the effects of cytokinins, media were supplemented with either BAP, TDZ, or a 1:1 ratio of each at 0, 2.5, 5, or 10 µM. Media was dispensed into 180 jars with approximately 25 ml in each jar. Three-week-old actively growing stock plants were divided into single shoots and placed onto experimental media. Each treatment consisted of 8 reps (jars) each containing 5 subsamples (shoots) and were cultured under standard culture conditions as previously described in a completely randomized design. After eight weeks data can be collected on explant survival shoot number and length. Data sets were subjected to regression analysis.

Rooting. Media used for in vitro rooting consisted of half strength MS salts and vitamins supplemented with 5 µM NAA, 30 g/L sucrose, 100 mg/L MES, and 100 mg/L *myo*-inositol. Media were solidified with 6.5 g/L agar and adjusted to a pH of 5.75. Microcuttings, 10 to 20 mm long, were subcultured on 25 mL of media in 180-mL jars. All jars were incubated under standard culture conditions as described previously. Following 2 weeks of growth, microshoots were transferred ex vitro.

RESULTS AND DISCUSSION

Cytokinin Type and Concentration. Shoot multiplication was achieved for all treatments. There were significant interactions between cytokinins and their concentration that affected explant survival ($P < 0.05$) and shoot production ($P < 0.05$). Explant survival was reduced on basal media and increased in the presence of cytokinins. For BAP and kinetin, explant survival also declined at higher concentrations (**Fig. 1a**). In general, shoot production was highest on media containing BAP or TDZ (**Fig. 1b**). For BAP, shoot production followed a quadratic model with the highest number of shoots per explant reaching 5.73 ± 0.75 at 20 μM. Similarly, for TDZ shoot proliferation followed a quadratic model, reaching maximum shoot proliferation at 10 μM and then declines with increasing concentration. There was no effect of cytokinin or their concentration on shoot length.

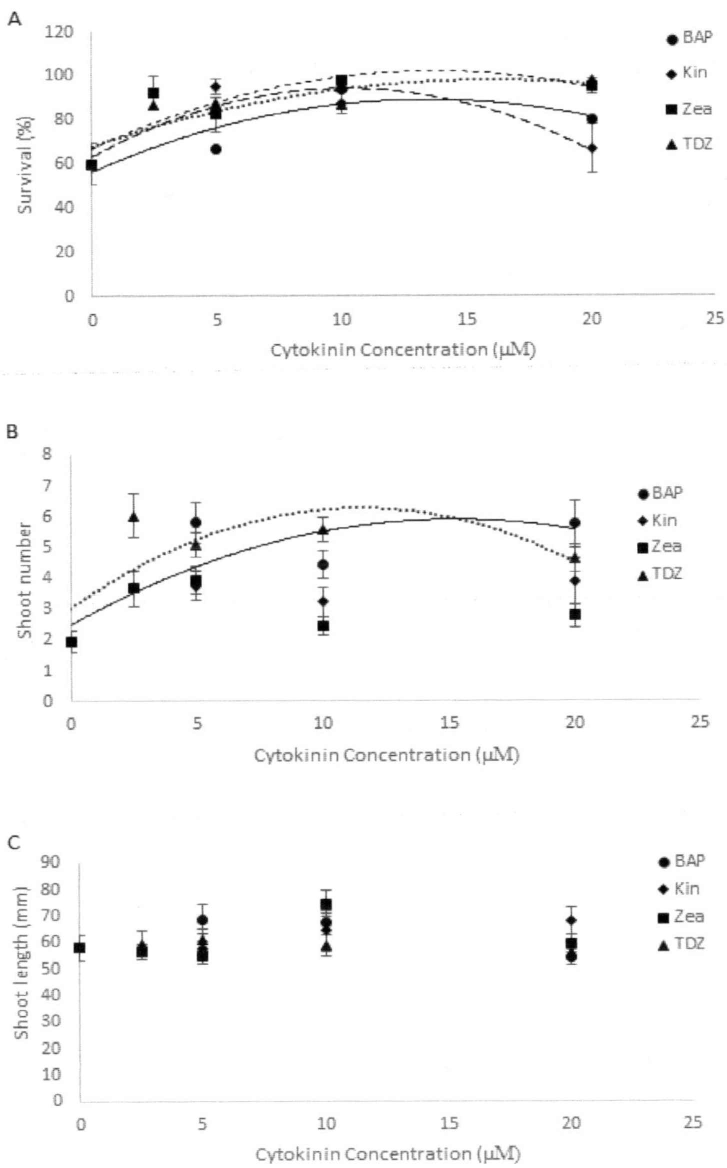

Figure 1. A) Influence of cytokinin type and concentration on explant survival. Symbols represent means, (n=8, ± SEM). BAP, $y = 0.57 + 0.05x - 0.002x^2$; Kinetin, $y = 0.63 + 0.06x - 0.003x^2$; Zeatin, $y = 0.67 + 0.05x - 0.002x^2$; TDZ, $y = 0.66 + 0.04x - 0.001x^2$. B) Shoot number as a function of cytokinin type and concentration. Symbols represent means, (n=8, ± SEM). BAP, $y = 2.38 + 0.44x - 0.01x^2$; Kinetin, NS; Zeatin, NS; TDZ, $y = 2.18 + 0.6x - 0.03x^2$. C) Shoot length as a function of cytokinin type and concentration. Symbols represent means, (n=8, ± SEM).

Interestingly, both zeatin and kinetin resulted in relatively low shoot proliferation rates. Kinetin, in particular, has been widely used for genera within the sugarcane complex. For *Saccharum officinarum,* maximum shoot regeneration was obtained using BAP in combination with kinetin (Salokhe, 2021; Shimelis et al., 2014; Tesfa et al., 2016).

BAP and TDZ Combinations. Shoot multiplication was achieved for all treatments. There were significant interactions between cytokinins and their concentration that affected explant survival ($P < 0.05$) and shoot production ($P < 0.05$), but not shoot length. Explant survival declined in a quadratic model when exposed to increasing concentrations of BAP and TDZ (**Fig. 2a**).

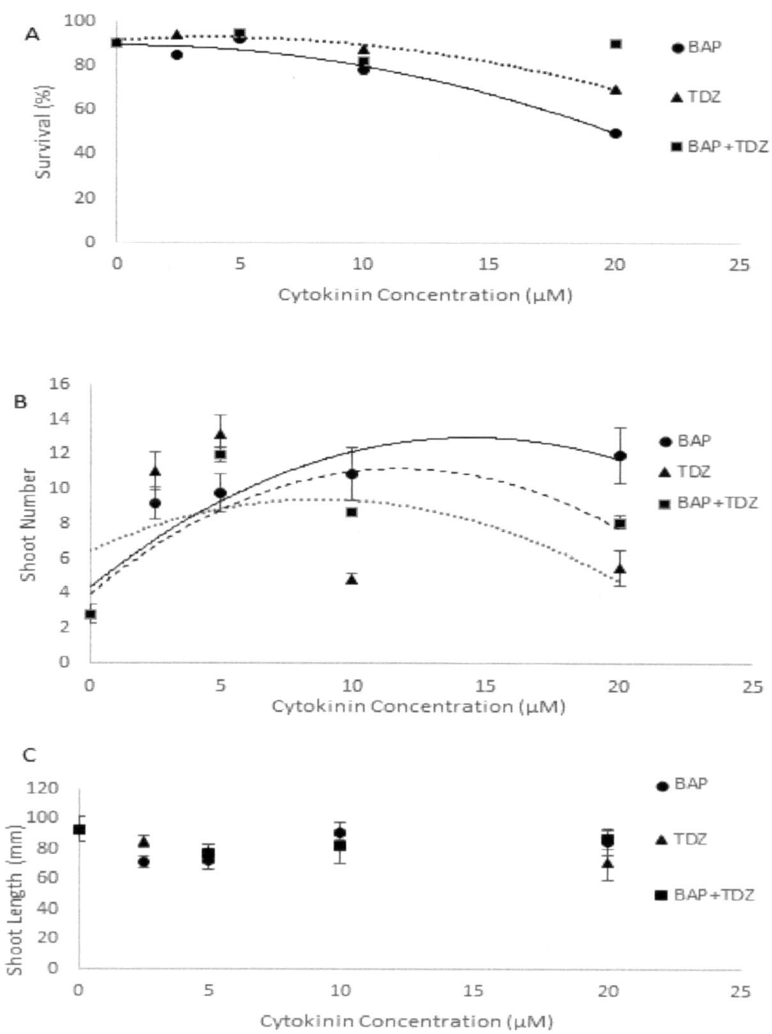

Figure 2. A) Influence of cytokinin type and concentration on explant survival. Symbols represent means, (n=8, ± SEM). BAP, $y = 88.3 + 0.95x - 0.09x^2$; TDZ, $y = 89.3 + 0.105x - 0.104x^2$; BAP+TDZ, NS. B) Shoot number as a function of cytokinin type and concentration. Symbols represent means, (n=8, ± SEM). BAP, $y = 4.4 + 1.2x - 0.04x^2$; TDZ, $y = 6.48 + 0.67x - 0.038x^2$; BAP+TDZ, $y = 4.0 + 1.2x - 0.05x^2$. C) Shoot length as a function of cytokinin type and concentration. Symbols represent means, (n=8, ± SEM).

Explant survival remained high across all concentrations in response to a combination of BAP and TDZ (**Fig. 2a**). High concentrations of cytokinins are known for inducing programmed cell death and could be attributed to the reduction in survival. However, it is interesting that the same response what not observed when using a combination of cytokinins.

Regression analysis showed shoot production followed quadratic responses to concentration. For both TDZ and the combination of BAP and TDZ, shoot production increased until 10 μM before attenuating with increased concentrations. For BAP, shoot production followed a quadratic model and produced the highest number of shoots per explant reaching 11.98 ± 1.6 at 20 μM (**Fig. 2b**). The use of BAP as the sole cytokinin has been successful for shoot proliferation in numerous studies on plants in the sugarcane complex. However, in a more recent study, Sukendah et al. (2023) explored different ratios of BAP and TDZ and found a combination of 15 μM BAP and 2 μM TDZ produced the highest number of shoots. In contrast, in the present study, combinations of BAP and TDZ were not as effective as BAP alone. However, this study used concentrations of BAP and TDZ in a 1:1 ratio (i.e., 2.5 μM BAP: 2.5 μM TDZ). Exploring different ratios of BAP and TDZ may be more beneficial for increasing proliferation rates.

Microcuttings initiated roots within 7 to 10 days after being placed into rooting media. After 2 to 4 weeks plantlets had significant root formation (**Fig. 3**) and were transferred to greenhouse conditions under mist.

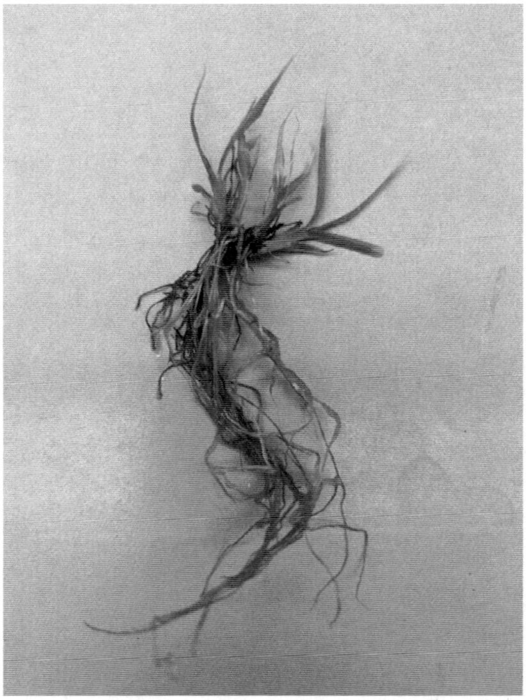

Figure 3. In vitro root formation on *Tripidium* before transferring ex vitro.

CONCLUSIONS

Cytokinins play an important role for in vitro shoot proliferation. Considering the variability in cytokinins used for species and cultivars within the sugarcane complex, this study describes the refinement of cytokinins used for shoot proliferation in a novel *Tripidium* hybrid. Both BAP and TDZ were shown to be effective cytokinins for shoot proliferation. These protocols provide a basis for rapid propagation of *Tripidium* and further enhance the genera as a potential bioenergy feedstock.

AKNOWLEDEMENTS

Thanks to the North Carlonia Department of Agriculture and Consumer Services, Bioenergy Research Initiative for providing funding for this research.

LITERATURE CITED

Balagalla, D.N., Wijesuriya, A., and Ranathunge, N.P., and Perera, A.M.M.S. (2018). A protocol for in-vitro direct plant regeneration from leaf tissues for micropropagation of sugarcane. Tropical Agri. Res. *29*(1):55–64.
https://doi- org /10.4038/tar.v29i1.8297

Cheong, E.J., Mock, R., and Li, R. (2009). Optimizing culture medium for meristem tissue culture of several *Saccharum* species and commercial hybrids. J. Amer. Soc. Sugar Cane Tech. *29*: 149–165. https://doi-org/10.1007/s11240-009-9634-7

da Silva, J.A., Solis-Gracia, N., Jifon, J., Souza, S.C., and Mandadi, K.K. (2020). Use of bioreactors for large-scale multiplication of sugarcane (*Saccharum* spp.), energy cane (*Saccharum* spp.), and related species. In Vitro Cell. Develop. Biol. *56*(3):366–376. https://doi-org /10.1007/s11627-019-10046-y

Maren, N.A, Touchell, D.H, Ranney, TG, Ashrafi, H, Whitfield, M.B., and Chinn, M. (2020). Biomass yields, cytogenetics, fertility, and compositional analyses of novel bioenergy grass hybrids (*Tripidium* spp.). GCB Bioenergy *12*:361-373.
https://doi.org/10.1111/gcbb.12676

Mustafa, G., and Khan, M.S. (2016). Differential role of indolebutyric acid in sugarcane root development. Sugar Tech. *18*(1):55–60.
https://doi.org/10.1007/s12355-014-0362-x

Murashige, T. and F. Skoog. (1962). A revised medium for rapid growth and bioassays with tobacco tissue cultures. Physiol. Plant. *15*:473–497.

Ramgareeb, S., Snyman, S.J., van Antwerpen, T., and Rutherford, R.S. (2010). Elimination of virus and rapid propagation of disease-free sugarcane (*Saccharum* spp. cultivar NCo376) using apical meristem culture. Plant Cell, Tis. Org. Cult. *100*(2): 175–181. https://doi-org /10.1007/s11240-009-9634-7

Shimelis, D. Bantte, K., and Feyissa, T. (2014). Interaction effects of 6-Benzylaminopurine and kinetin on in vitro shoot multiplication of two sugarcane (*Saccharum officinarum* L.) genotypes. Adv. Crop Sci. Tech. *2*:143.

Shubhangi, S. (2021). Development of an efficient protocol for production of healthy sugarcane seed cane through meristem culture. J. Agri. Food Res. Vol. 4. 100126, ISSN 2666-1543.
https://doi.org/10.1016/j.jafr.2021.100126.

Sukendah, S., Purnamingsih, R., and Priyadarshini, R. (2023). Micropropagation of ten genotypes of sugarcane mutants on media containing BA and thidiazuron. Proc. 3rd Inter. Conf. Agri. 20-28.

Tesfa, M., Admassu, B., and Bantte, K. (2016). In vitro shoot multiplication of elite sugarcane (*Saccharum officinarum* L.) genotypes using liquid shake culture system. J. Biol., Agric. Healthcare *6*:35-40.

Vasque-Molina, D.E., De los Santos, A., Lecona Guzman. K.A., Sumano Muniz, O. Velazques Mendez. M., Rosales, R, Olivia LLaven, M.A., Dendooven, L., and Gutierrez-Miceli, A. (2005). Sugar cane buds as an efficient explant for plantlet regeneration. Biol. Plant. *49*:481-485.

Vinayak, V., Dhawan, A.K., and Gupta, V.K. (2009). Efficacy of non-purine and purine cytokinins on shoot regeneraation *in vitro* in sugarcane. Ind. J. Biotech. *8*:227-231.

Jahangir, G.Z., Nasir, I.A., Sial, R.A., Javid, M.A., and Husnain, T. (2010). Various hormonal supplementations activate sugarcane regeneration in-vitro. J. Agric. Sci. *2*: 231-237.

Zhang, Q.X., Sun, Y., Hu, H.K., Chen, B., Hong, C.T., Guo, H.P., Pan, Y.H., and Zheng, B.S. (2012). Micropropagation and plant regeneration from embryogenic callus of *Miscanthus sinensis*. In Vitro Cell. Dev. Biol.-Plant. 48:50–57. https://doi.org/10.1007/s11627-011-9387y

Combined Proceedings IPPS 73:266-272. 2023.

Photosynthetic Performance of *Handroanthus chrysotrichus* Seedlings Grown in Substrate with *Rhizobacteria*

Thiago Souza Campos[1,2], Vania Maria Pereira[2], Antonio Maricélio Borges de Souza[1], Wagner A Vendrame[2], Everlon Cid Rigobelo[1] and Kathia Fernandes Lopes Pivetta[1]

[1]Dept. of Agricultural Sciences - Jaboticabal, SP 14884-900, Brazil; [2]Dept. of Environ. Horticulture, PO Box 110675, Gainesville, FL 32611, USA

t.souzacampos@ufl.edu, Thiago.s.campos@unesp.br

Keywords: *Azospirillum brasilense*, *Bacillus* spp., plant growth, microorganisms, nursery

Summary

Rhizobacteria, regarded as renewable resources, enable a sustainable system for producing vigorous and rapidly growing seedlings. The objective of this study was to evaluate the effect of plant growth-promoting rhizobacteria in the production of *H. chrysotrichus* seedlings. The experimental design was completely randomized. The treatments consisted of microorganisms (*Bacillus subtilis*, *Bacillus megaterium*, *Bacillus amyloliquefaciens* and *Azospirillum brasilense*) plus the absence of microorganisms - control; four repetitions and ten plants per plot. The following were evaluated: leaf number, leaf area, as well as chlorophyll content; minimum and maximum fluorescence; and maximum quantum efficiency of photosystem II. The inoculation with *B. amyloliquefaciens* has been found to enhance leaf area. Moreover, *B. amyloliquefaciens* plays a role in maintaining the functionality of the photosystem reaction center. Consequently, it can be concluded that *B. amyloliquefaciens* stands out as the most effective inoculant for golden trumpet for promoting greater efficiency of the photosystem II.

IPPS Vol. 73 – 2023

INTRODUCTION

Handroanthus chrysotrichus (Mart. ex DC.) Mattos tree belonging to the Bignoniaceae family, meeting the market demands for urban planting, ecosystem restoration, and integrated systems necessitates the reduction of nursery times, achieved through accelerated growth and the development of high-quality seedlings. This enhanced growth and seedling quality can be attained through the application of plant growth-promoting bacteria. In addition to their growth-enhancing properties, rhizobacteria also play a crucial role in mitigating drought and salinity stress, aiding in the phytoextraction of heavy metals, contributing to nutrient supplementation, fixation, or solubilization, facilitating the production of phytohormones, and controlling pathogens (Dias and Santos, 2022).

Among the most prominent rhizobacteria known for their positive impact on seedling growth and quality across various plant species are *Bacillus subtilis* and *Bacillus megaterium* (Guimarães et al., 2021; Santos et al., 2021, Silva et al., 2022), *Bacillus amyloliquefaciens* (Matsumura et al., 2016; Rios et al., 2018; Wang et al., 2020; Ngalimat et al., 2021), and *Azospirillum brasilense* (Gonzalez et al., 2018; Zeffa et al., 2019; Jarquín-Rosales et al., 2023). Given these considerations, the objective of this study was to evaluate the effect of rhizobacteria (*Bacillus subtilis*, *Bacillus megaterium*, *Bacillus amyloliquefaciens* and *Azospirillum brasilense*) on the growth and quality of golden trumpet tree (H. *chrysotrichus*) seedlings.

MATERIALS AND METHODS

The study was carried out between September 2021 and January 2022 in a greenhouse at the College of Agricultural and Veterinary Sciences (UNESP/FCAV), Campus de Jaboticabal, SP, Brazil - under the coordinates 21°15'2" latitude, 48°16'47" longitude and 600 meters of altitude. The Brazil climate of the micro-region by Köppen-Geiger system is tropical savanna Aw type, with dry winter (Andre e Garcia, 2015).

The design of the experiment was entirely randomized. There were five treatments (*Bacillus subtilis, Bacillus megaterium, Bacillus amyloliquefaciens* and *Azospirillum brasilense*, plus the absence of rhizobacteria - control); four repetitions and ten plants per plot. The seeds of golden trumpet tree were collected from existing trees in the Experimental Nursery of Ornamental and Forestry Plants of the Faculty of Agrarian and Veterinary Sciences (UNESP/FCAV) during September 2022. Seeds were sown in tubes with 280 cm³ of volume capacity placed in polypropylene trays for 54 containers, containing Carolina Soil® as commercial substrate, composed of peat, vermiculite, roasted rice husk, calcined dolomite limestone, NPK 14-16-18 fertilizer and micronutrients (information obtained from the packaging). The trays were suspended on metal mesh benches 70 cm from the ground in a covered greenhouse with the sides protected with black screen that allows 50% of the light to pass through and with a clear plastic layer above the screen cover. The irrigation was performed by automatic micro sprinklers, activated three times a day for 15 minutes each, with a flow rate of 30 L h^{-1}.

The microorganisms are part of the collection of the Laboratory of Soil Microbiology of the Department of Vegetal Production of UNESP-FCAV, Campus de Jaboticabal, where they were grown separately, in nutrient broth medium for seven days in flasks kept in B.O.D. (Eletrolab, model 347 F, Brazil), at 25 °C temperature. After the incubation period, the microorganisms were centrifuged separately at 10,000 rpm for 10 minutes at 28 °C (Novatecnica, model MLW K24, Brazil). The inoculum concentration was standardized according to Barry and Thornsberry (1991) and Sahm and Washington II (1991) at 1 x 107 CFU mL-1 using a spectrophotometer (Micronal, model B382, Brazil) at 695 nm absorbance.

The microorganisms were inoculated twice, once at 30 days after the seeds were sown and again at 60 days, by applying 1.0 mL of the solution directly to the substrate near the stem, using a mechanical micropipette (VF-1000, Digipet®). The seedlings belonging to the control treatment were not inoculated. When the roots began to appear at the bottom of the tubes, the following characteristics were evaluated: Leaf number (LN), verified by visual counting of fully expanded leaves; and leaf area (LA, cm^2), measured using an electronic leaf area meter (Li-3100C, LI-COR®, Lincoln, Nebraska, USA). The chlorophyll content was measured with the ChlorofiLOG, model CFL1030, FALKER®; minimum fluorescence (F0), maximum fluorescence (FM) and maximum quantum efficiency of photosystem II (FV/FM) were obtained with a handheld chlorophyll fluorometer (OS30p, Opti Science). The obtained data were submitted to analysis of variance and the means were compared using Tukey's test at 5% probability using the *R* statistical software (R Core Team, 2016).

RESULTS

No significant distinctions were observed among the treatments in terms of leaf number, chlorophyll content, and FV/FM ratio (**Fig. 1A,C,F**). However, concerning leaf area, *B. amyloliquefaciens* exhibited the highest average, while for F0 and FM, the combined application of B. *amyloliquefaciens* and A. *brasilense* yielded the highest mean values (**Fig. 1 B,D,E**).

To visually assess the golden trumpet tree seedlings (**Fig. 2**). Remarkably, at 107 days after sowing, the rhizobacterium B. *amyloliquefaciens* showcased superiority among all treatments.

DISCUSSION

The response of golden trumpet tree seedlings to rhizobacteria inoculation, particularly with *Bacillus amyloliquefaciens*, displayed superior performance across all assessed traits. *Azospirillum brasilense* also exhibited favorable results, underscoring the significant contribution of these two bacteria to the enhancement of seedling growth and quality. The promotion of plant growth by bacteria is influenced by various factors, primarily encompassing nutrient mobilization and solubilization, as well as phytohormone production (Ahemad and Kibret, 2014). While previous studies have highlighted the effective promotion of both plant quality and growth through the inoculation of *Bacillus subtilis* and *Bacillus megaterium* (Guimarães et al., 2021; Santos et al., 2021; Silva et al., 2022), it's worth noting that these species exhibited relatively modest outcomes in the present study. Notably, *Bacillus subtilis*, in particular, demonstrated lower mean values across most of the assessed characteristics.

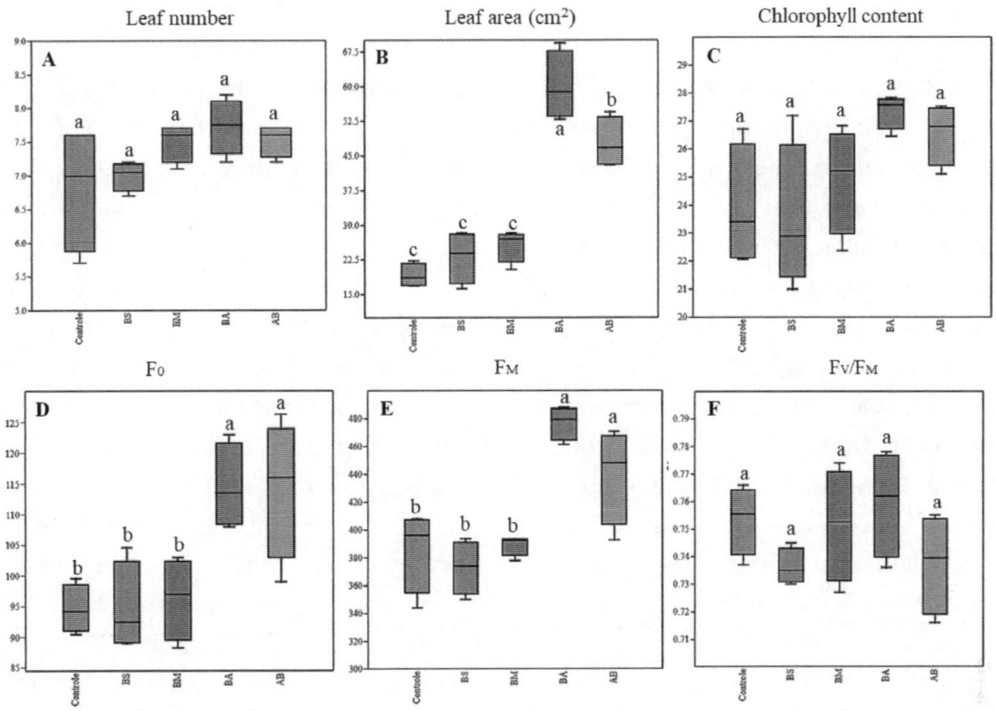

Figure 1. Boxplots of photosynthetic evaluation of *Handroanthus chrysotrichus* seedlings without inoculation (Control) and inoculated *with Bacillus subtilis* (BS), *Bacillus megaterium* (BM), *Bacillus amyloliquefaciens* (BA) and *Azospirillum brasilense* (AB); A) leaf number, B) leaf area, C) chlorophyll content, D) minimum fluorescence (F_O), E) maximum fluorescence (F_M), F) maximum quantum efficiency of photosystem II (F_V/F_M). Means followed by the same lowercase letter do not significantly differ by Tukey's test ($p < 0.05$). Jaboticabal, SP, Brazil, 2022.

Figure 2. *Handroanthus chrysotrichus* seedlings at 107 days after sowing. A) Control, B) *Bacillus subtilis*, C) *Bacillus megaterium*, D) *Bacillus amyloliquefaciens,* and E) *Azospirillum brasilense.* Jaboticabal, SP, Brazil, 2022.

The physiological variables examined in this study provide insights into the growth patterns of golden trumpet tree seedlings. Although there were no significant differences among treatments in terms of leaf number, the inoculation of B. *amyloliquefaciens* resulted in a larger leaf area. This observation suggests an enhanced photosynthetic potential among seedlings with increased surface area for light absorption, thereby facilitating greater carbon assimilation (Taiz et al., 2017). In contrast, the lower mean values for leaf area observed in the control group, as well as with B. subtilis and B. *megaterium* inoculations, contributed to reduced growth and development of the seedlings due to limited accumulation of light energy.

In fluorescence analyses, it's important to note that the minimum (F0) represents the fluorescence when all reaction centers are open, signifying the fluorescence emitted by the chlorophyll a molecules within the Photosystem II light harvesting complex. Conversely, the maximum fluorescence (FM) signifies the complete reduction of the primary quinone as a result of a light pulse incident on its reaction center, resulting in maximum fluorescence (Taiz et al., 2017). Consequently, the superior results achieved by the rhizobacteria B. *amyloliquefaciens* and A. *brasilense* in this study can be attributed to their capacity to establish a greater and more efficient photosynthetic activity under the same climatic conditions, thereby reducing the risk of photoinhibition damage, even though there were no discernible differences among treatments in terms of chlorophyll content. In a study conducted by Samaniego-Gámez et al. (2016), it was observed that the rhizobacterium B. *amyloliquefaciens*, in conjunction with B. *subtilis*, had a positive impact on the photosynthetic performance of *Capsicum chinense* Jacq. Additionally, Gonzalez et al. (2018) that the inoculation of A. *brasilense* with *Prosopis articulata* S. Watson resulted in an increase in chlorophyll levels in 25% of the leaves.

CONCLUSIONS

Based on the results obtained, rhizobacteria exerted a significant influence on the growth, quality, and photosynthetic capacity of the studied golden trumpet tree species. Consequently, the utilization of these beneficial microorganisms in the cultivation of golden trumpet tree seedlings holds great promise.

LITERATURE CITED

Ahemad, M., and Kibret M. (2014). Mechanisms and applications of plant growth promoting rhizobacteria: current perspective. J. King Saud Univ. Sci. *26*(*1*):1-20.

Andre R.G.B., and Garcia A. (2015). Alguns aspectos climáticos do município de Jaboticabal – SP. Nucleus *12*(*2*):263-270.

Barry A.L., and Thornsberry C. (1991). Susceptibility tests: Diffusion test procedures In: Balows A., Hauser W.J., Hermann K.L. et al. Manual of Clinical Microbiology 5th ed. Washington, DC: American Society for Microbiology.

Dias A.S., and Santos C.C. (2022). Bactérias promotoras de crescimento de plantas: conceitos e potencial de uso. Nova Xavantina: Pantanal. ISBN 978-65-81460-63-1.

Gonzalez E.J., Hernandez J.P., De-Bashan L.E., and Bashan, Y. (2018). Dry micropolymeric inoculant of *Azospirillum brasilense* is useful for producing mesquite transplants for reforestation of degraded arid zones. Appl. Soil Ecol. *29*: 84-93.

Guimarães V.F., Klein J., Silva A.S.L., and Klein D.K. (2021). Eficiência de inoculante contendo *Bacillus megaterium* (B119) e *Bacillus subtilis* (B2084) para a cultura do milho, associado à fertilização fosfatada. Resear. Soc. Develop. 10(*12*): e431101220920.

Matsumura, A.T.S., Paz, I.C.P., Guimarães, A.M., Silva, M.E., Ott, A.P., and Duarte, V. (2016). Efeito de três formulações de *Bacillus amyloliquefaciens* ICBB200 sobre o crescimento de plantas de alface (*Lactuca sativa* L.) em cinco condições edafoclimáticas. Scient. Agraria. Paranaensis. *15*(*2*):146-152.

Ngalimat, M.S., Yahaya, R.S.R., Baharudin, M.M.A.A., Yaminudin, S.M., Karim, M., Ahmad, S.A., et al. (2021). A review on the biotechnological applications of the operational group *Bacillus amyloliquefaciens*. Microorganisms. *9*(*3*):614.

Rios, F.A., Freitas, S.K.R., Contiero, R.L., Braz, G.B.P., Roman, R., Belani, R.B., et al. (2018). Agronomic performance of soybean treated with *Bacillus amyloliquefaciens*. Afri. J. Microbiol. Res. *12*(*45*), 1020-1027.

Sahm, D.F., and Washington, J.A. (1991). Antibacterial susceptibility tests: dilution methods. In: Balows, A., Hausler, W.J., Herrmann, K.L., Isenberg, H.D., and Shadomy, H.J. Manual of Clinical Microbiology. 5th ed. Washington, DC: American Society for Microbiology.

Samaniego-Gámez, B.Y., Garruña, R., Tun-Suárez, J.M., Kantun-Can, J., Reyes-Ramírez, A., and Cervantes-Díaz, L. (2016). *Bacillus* spp. inoculation improves photosystem II efficiency and enhances photosynthesis in pepper plants. Chilean J.Agricul. Res *76*(*4*) :409-416.

Santos, A.F., Corrêa, B.O., Klein, J., Bono, J.A.M., Pereira, L.C., Guimarães, V.F., et al. (2021). Biometria e estado nutricional da cultura da aveia branca (*Avena sativa* L.) sob inoculação com *Bacillus subtilis* e *B. megaterium*. Res., Soc. Develop. *10*(*5*): e53410515270.

Silva, K.R.C., Sousa, L.A.M., Silva, F.L.S., Azevedo, J.L.X., Silva, I.A., and Pinto Junior, F.F., Silva, B.G., Andrade, H.A.F., Doihara, I.P., and Silva-Matos, R.R.S. (2022). *Bacillus subtillis* e *Bacillus megaterium* no crescimento inicial de melancia 'Sugar Baby'. Res. Soc. Develop. *11*(*13*): e96111335034.

Souza, A.M.B., Chioda, L.B.; Ferreira, K.B., Vieira, G.R., Campos, T.S., and Pivetta, K.F.L. (2022). Initial growth of *Syagrus romanzoffiana* seedlings in biosolid-based substrate. Pesquisa Agropecu. Trop. *52*: 70577.

Taiz, L., Zeiger, E., Møller, I.M., and Murphy, A. (2017). Fisiologia e desenvolvimento vegetal. Artmed Editora, p. 858.

Wang, C.J., Wang, Y.Z., Chu, Z.H., Wang, P.S., Liu, B.Y., Li, B.Y., et al. (2020). Endophytic *Bacillus amyloliquefaciens* YTB1407 elicits resistance against two fungal pathogens in sweet potato (*Ipomoea batatas* (L.) Lam.). J. Plant Physiol. *253*:153260.

Zeffa, D.M., Perini, L.J., Silva, M.B., Sousa, N.V., Scapim, C.A., Oliveira, A.L.M., et al. (2019). *Azospirillum brasilense* promotes increases in growth and nitrogen use efficiency of maize genotypes. Plos One *4*(4):e0215332.

Combined Proceedings IPPS 73:273-282. 2023.

Germination and Seedling Growth Under Different Sowing Depths for Green and Silver Saw Palmetto (*Serenoa repens*)

Vânia Maria Pereira[1], Patricia Ramalho de Barros[2], Thiago Campos de Souza[1,3], Héctor Pérez[1], Wagner Vendrame[1]

[1]Dept. of Environ. Horticulture, University of Florida, PO Box 110675, Gainesville, FL 32611, USA; [2]Dept. of Plant Science and Landscape Architecture, University of Maryland, College Park, Maryland, Md 20742, USA; [3] Dept. of Agricultural and Veterinary Sciences, Jaboticabal, SP, 14884-900, Brazil

vania.pereira@ufl.edu

Keywords: palm production, seed depth, plant vigor

Summary

Saw palmetto is a native palm of the Southeast United States, present in green and silver forms, of high ornamental value for native landscapes and of potential commercial importance because of its phytotherapeutic properties. This project evaluated the influence of sowing seed depths (1, 2, 4, 6, and 8 cm) in the germination and seedling growth of silver and green forms of saw palmetto - based on nursery growers' assertions that greater depths would yield higher plants. Green saw palmetto achieved 50% maximum germination around 89 days after sowing at any sowing depth, while silver germination occurred within 149 days - and only differed between 1 and 8 cm sowing depth. The green forms had greater seedling height and leaf area one year after sowing – due to earlier germination. As sowing depths increased, seedling height and visual quality increased, but there was no difference in the number of seedling leaves. Root length decreased as sowing depth increased - but there was no effect on the root dry weight; the root biomass was similar at any depth. The green form provided taller plants in a shorter period due to faster germination. However, silver may be preferable due to glaucous coloration of leaves. One year after sowing, seedling height was greater at the deeper sowing depth (8 cm) - confirming nursery growers' observations; however, even with decreased root length - root biomass was unaffected.

INTRODUCTION

Saw palmetto (*Serenoa repens* Bartr.) is a small shrub rhizomatous palm species endemic and widely spread to the southeastern United States, covering coastal areas, pinelands, and prairies (Hilmon, 1985). The species has two forms, the green form found inland and the silver (form glauca), native to the eastern coast of Florida in scrub locations (Moldenke, 1967). Saw palmetto is an essential food source for Florida's fauna, and over 300 insect species visit saw palmetto inflorescences (Carrington et al, 2003; Maehr and Layne, 1996). Besides its food source value, saw palmetto provides medicinal properties (Bennett and Hicklin, 1998; Gilman, 2015).

The landscape uses of saw palmetto started around the 1970s due to the high adaptability (Smith, 1972). Saw palmetto provides excellent soil tolerance from alkaline to acidic soils, drought tolerance, and salt endurance. However, transplanting for landscapes is difficult and becomes more complex as the plants grow older. As transplanting recovery is low, the planting recommendation is twice the final number of plants desired (Gilman, 2015). Due to increased ornamental importance and low transplanting recovery, nursery propagation is essential to meet plant demand. Propagation of saw palmetto occurs by seed germination, with low and slow germination rates and growth (Gilman, 1999). Reports on seed germination studies on saw palmetto have varying results on temperature and pre-germination treatments (Carpenter, 1987; Makus, 2006; Makus).

Seed depth is also essential in seed germination (Murphy et al, 2016). Seed depth varies according to the seed size and species grown. A standard recommendation is to use the seed's diameter as the depth measure to sow (Meerow and Broschat, 2021). As the drying potential increases, seeds should be placed deeper in the soil (Broschat and Donselman, 1986).

In visits to nurseries in the region, a common practice shared was placing saw palmetto seeds deeper in the soil, where it would provide taller plants in shorter periods. Based on this information, this research focused on elucidating the grower's claim and determining the best sowing depths for silver and green saw palmetto for yielding the best quality and taller plants in a shorter period, considering this slow-growth palm. The objectives of this project were to evaluate the influence of seed sowing depth of saw palmetto in germination rate and seedling growth one year after sowing for both forms.

MATERIALS AND METHODS

Green and silver saw palmetto (**Fig. 1A, 1B**) seeds were collected in October 2021. The green form was from Felda, FL, and the silver form was from Gainesville, FL. The endocarp was removed before seed sterilization (2.26% NaCl for 20 min with three rinses in deionized water (DI)) and soaked for 24h, based on Makus, 2006, in 250 ml Erlenmeyer flasks with 150 ml of deionized water with moderate vibration (161 rpm) on a platform shaker. Seeds were placed with operculum downward in contact with the soil (**Fig. 1C**).

Figure 1. A) Green saw palmetto form; **B)** Silver saw palmetto form; **C)** Seed position at containers to facilitate germination reading; **D)** Green form seedlings one year after sowing.

Containers were adapted from clear 50 oz. Drive Thru Bantam Cup. (22.8 x 11.4 cm) with six circular holes (1 cm each) in the bottom, wrapped in two layers of black plastic per plot, filled with general-purpose soilless media (Promix BX, Premier Tech Horticulture, Quebec). Plants were grown under controlled conditions in a greenhouse (30.5 – 22 °C) on the University of Florida campus for one year. Plants were manually irrigated as needed. This study used five blocks in a randomized complete block design; each block contained all treatment combinations in plots, and each plot consisted of 10 containers with two seeds each on opposite sides of the container (500 seeds for each form) (**Fig. 1D**). A 2 x 5 factorial design was used to evaluate the influence of saw palmetto forms (2 levels: green vs. silver) and sowing depth of the seed (5 levels: 1, 2, 4, 6, and 8 cm) on germination and seedling growth.

Germination data collection occurred weekly and seedling growth after a year (November 2021 to 2022). Germination was scored as emerging cotyledonary petiole 2 mm out of the seed coat. Seedling data collection at the end of the experiment included plant height, number of leaves, leaf area (cm^2), SPAD (Minolta Camera Co., Osaka, Japan), root length, visual rate (1-10), canopeo, shoot and root fresh and dry mass (48 h in a forced air 70 °C oven). Statistical analyses were conducted using the software SAS version 9.4 (SAS Institute Inc). For testing of form and/or sowing depth influenced the germination or seedling growth, NLmixed and Glimmix with Anova at (P=0.05) were used.

RESULTS

Considering germination, forms significantly influenced days to achieve 50% of maximum germination (LD50%) (P-value: <.0001) at any depth (**Fig. 2**).

The silver form had a higher LD50%, taking longer to germinate and achieving the 50% of germination, in average 149±10 days. The green form germinated faster within an average of 89±7 days. The estimated difference between both forms was 60±13 days. The maximum germination did not differ between forms or depths, with greater than 70% germination at any treatment combination. The silver form had nearly 90% germination - with no significant differences among sowing depths of 4, 6, or 8 cm (**Fig. 3**).

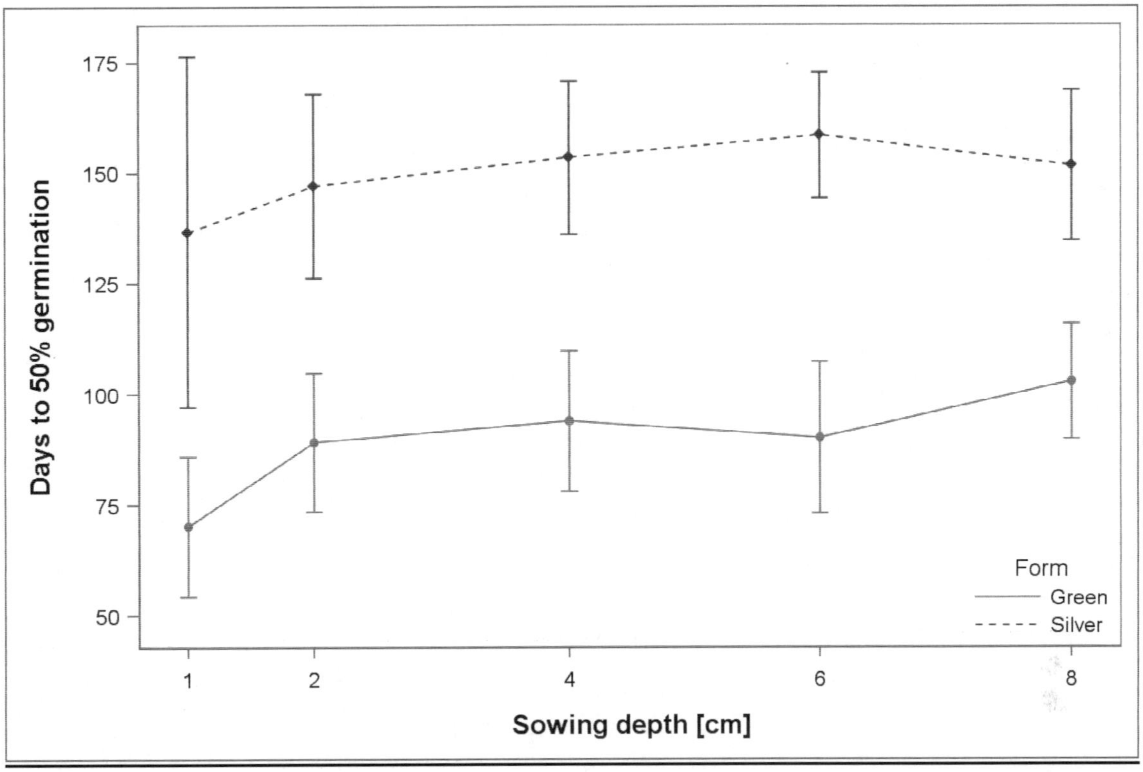

Figure 2. Days to achieve 50% of germination from total germination achieved since sowing. On red, green form, and in blue silver form. The lower and upper 95% confidence intervals are shown as error bars at each sowing depth point. Forms were significantly different while sowing depth was not significantly different for both forms.

Due to germination occurring at different times for both forms, seedling growth data was selected for the seeds that germinated at the peak of germination in March, April, and May (N: 388 seedlings, minimum of three containers (treatment combination) per plot, on each block). Seedling height was significantly affected by sowing depth (P-value: 1.45E-18) and form (P-value: 2.87E-06) with no interaction. Sowing depth differences caused a plant height differential of 10cm between the green and silver saw palmetto forms. For green saw palmetto sown at 8cm depth, seedling height was 31.8±0.5 cm while a 1 cm depth was 21.5±0.8 cm; while for silver saw palmetto, sowing depths of 1- and 8-cm led to seedlings heights of, respectively, 18.9±0.53 cm and 29.8±0.5 cm (**Fig. 4**). Comparing both forms (P-value: 2.87E-06), the green form produced taller seedlings than the silver form, with respectively, plants than silver, 27.0±0.61 cm and 24.48±0.62 cm respectively.

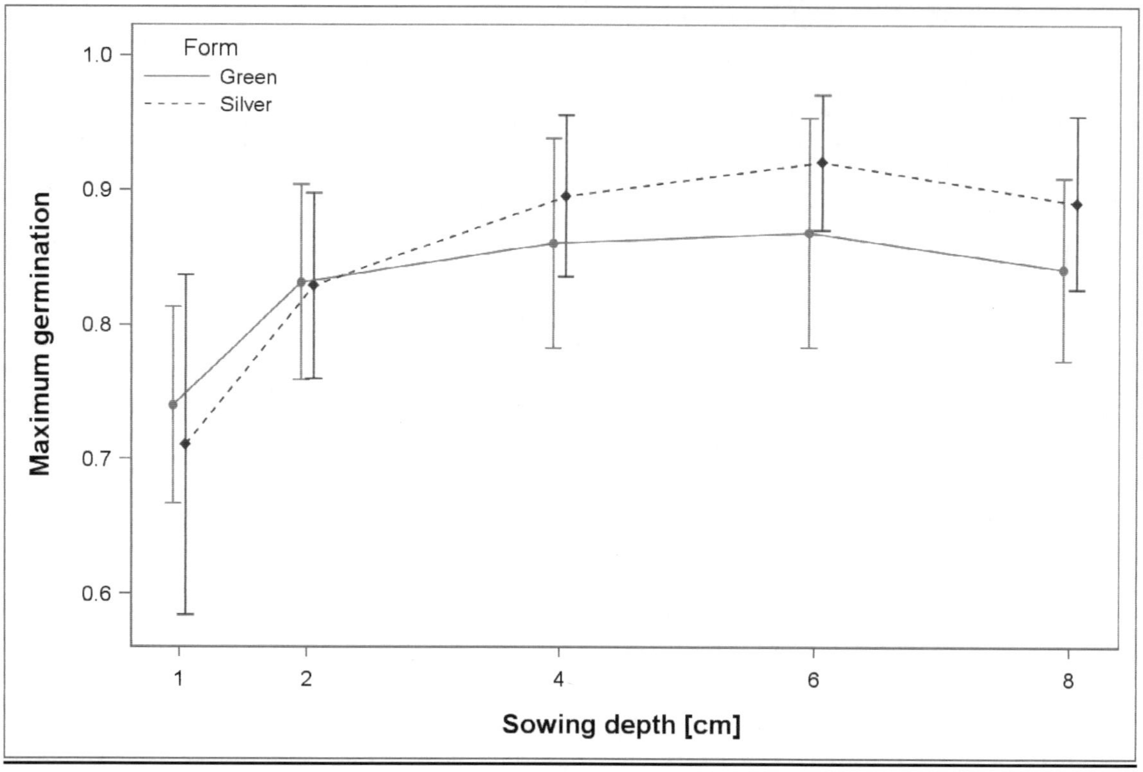

Figure 3. Maximum germination achieved for each form. On red green form, and on blue silver forms. The lower and upper 95% confidence intervals are shown as error bars at each sowing depth point. Even though tough forms or sowing depth were not significantly different, silver form (in blue) presents a trend of increasing maximum germination as the sowing depth increases.

As expected, the root length decreased due to reduced root-growth space in the container as the sowing depth increased from 1- to 8-cm (P-value: 1.45E-18). For green saw palmetto at 1- and 8- cm sowing depths had seedling root lengths of, respectively, 19.2±0.5 cm and 12.4±0.3 cm; for silver saw palmetto, 1- and 8- cm sowing depths had seedling root lengths of, respectively, of 20.3±0.3 cm and 13.4±0.3 cm (**Fig. 5**). However, the dry weight of roots was not affected for the silver form, while differences of the green form at 2, 4, and 8 cm sowing depths were, respectively, 0.60±0.03 g/plant, 0.45±0.03 g/plant, and 0.39±0.02 g/plant (P-value: 0.03). Dry root weight differed between the silver form (0.41±0.03 g/plant) and green form (0.50±0.03 g/plant) (P-value: 0.0006), (**Fig. 6**).

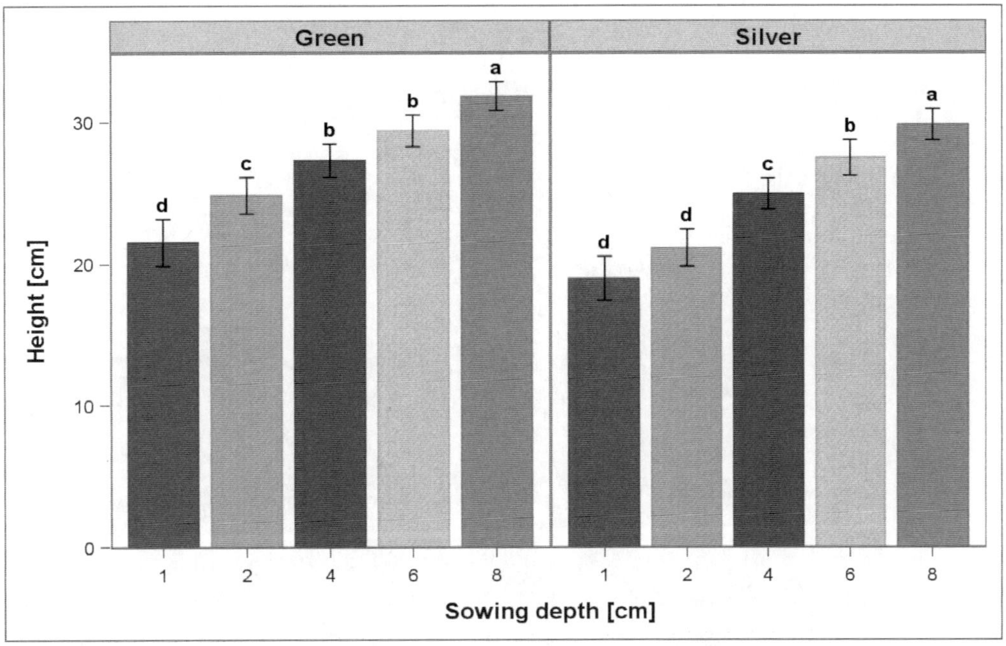

Figure 4. Seedling height (cm) one year after sowing for green and silver saw palmetto forms. The lower and upper 95% confidence intervals are shown as error bars at each sowing depth point. Different letters indicate significant differences between sowing depth based on LSMeans (P=0.05). Increase of sowing depth has increased the seedling height one year after sowing.

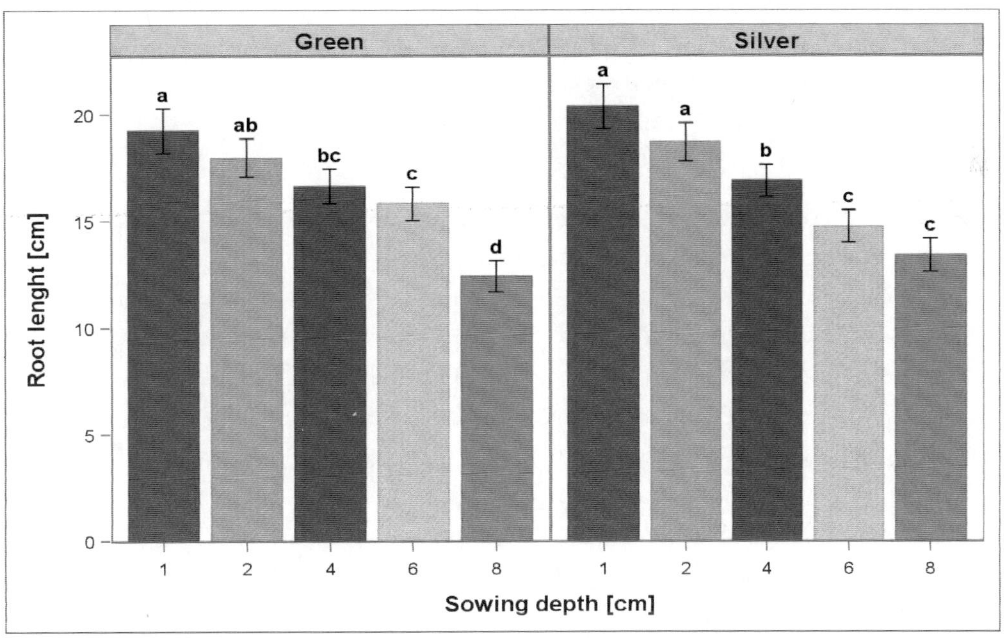

Figure 5. Root length (cm) one year after sowing for green and silver saw palmetto forms. The lower and upper 95% confidence intervals are shown as error bars at each sowing depth point. Different letters indicate significant differences between sowing depth based on LSMeans (P=0.05). Increase of sowing depth has decreased the seedling root length one year after sowing, related to less space in the container.

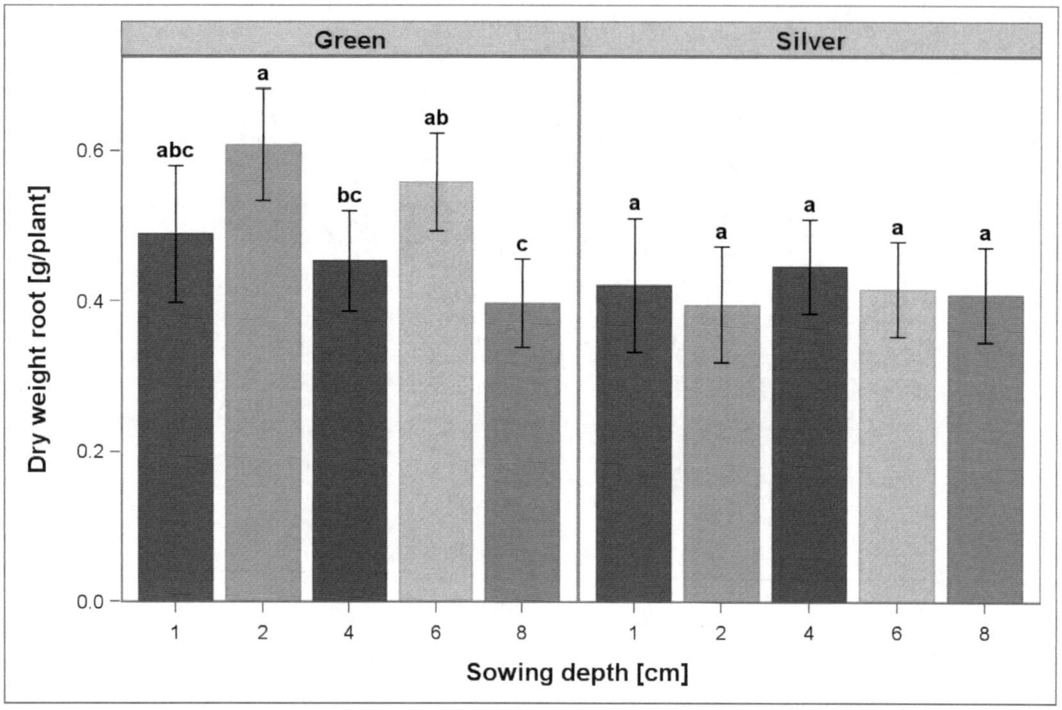

Figure 6. Dry weight root (g/plant) one year after sowing for green and silver saw palmetto forms. The lower and upper 95% confidence intervals are shown as error bars at each sowing depth point. Different letters indicate significant differences between sowing depth based on LSMeans (P=0.05). An increase in sowing depth did not affect root biomass one year after sowing, even though the root length decreased.

DISCUSSION

The results confirmed the grower's claim that increasing sowing depth saw palmetto seedlings will increase plant height. Considering the effects on germination, the sowing depth did not affect the time to achieve 50% of maximum germination for both forms. However, forms were significantly different, with green saw palmetto germinating around 60±13 days before silver saw palmetto at any depth. One hypothesis for this effect could be related to the providence of the seeds and mother plant conditions. Green seeds originated from the south of Florida state, and silver from the north region. This difference in germination response occurs on several species and even in geographically close locations (Baskin and Baskin, 2014). Harper's beauty (*Harperocallis flava*) presented germination variability between populations within 8 km from three seed collection locations (Gardner, 2021). This variability is usually associated with genotypes and environmental heterogeneity in maternal plants (Baskin and Baskin, 2014). Seeds achieved high germination rates (>70%) at any depth for both forms. A standard recommendation is to use the seed's diameter as the depth measure to sown, varying according to the environmental conditions, ranging from 1 to 2 cm for most palm species (Broschat and Meerow, 2000).

Seed depth affected germination percentage and time on areca palm (*Chysalidocarpus lutescens*). In full sun, germination at 1 cm achieved 74%, against 30% in surface or 6 cm. Seed depth effects on germination for areca palm varied according to the environmental conditions, as embryo desiccation is a significant cause of seeds not germinating (Broschat and Donselman, 1986); contrary to their report, sowing depth in our study did not alter germination rates or speed for both forms of saw palmetto.

Seedling growth of the green form produced taller plants one year after sowing, correlated to its earlier germination than the silver form. Increased sowing depth led to taller plants (**Fig. 1D**). Better seedling vigor was also visualized on seeds of Yarey Palm (*Copernicia breweriana*). Seeds sown at 1.27 cm had the greatest germination (79.5%), increase in leaf emergence at 3-month seedling, and higher seedling survival at seven months - when compared to seed sown in the substrate surface (Murphy et al, 2016). On the orchid tree (*Bauhinia retusa* and *Bauhinia variegata*), seeds were sown at 2, 4, and 6 cm, seedling emergence, length, and biomass for *B. retusa* occurred at 4 cm sowing depth. In comparison, *B. variegata* at 2 cm (Yadav et al, 2023). Root length was decreased, associated with the smaller area for root growth in the containers as seeds are set lower in the substrate. However, the biomass (root dry weight) was not affected; seedlings were able to produce a similar amount of root mass even in smaller areas.

CONCLUSION

Green and saw palmetto forms differed in germination timing. However, germination was not affected by sowing depth. Green saw palmetto provided taller plants one year after sowing, which correlated with earlier germination—the deeper sowing depths produced taller seedlings with shorter root lengths - but no effect on root biomass. Future research should evaluate seeds of both forms from other populations for geographical impact and deeper sowing depths. Moreover, the persistence of sowing depths results in increased seedling growth rates.

LITERATURE CITED

Baskin, C. C. and Baskin, J. M. (2014). Seeds Ecology, Biogeography, and Evolution of Dormancy and Germination. Burlington: Elsevier Science.

Bennett, B. and Hicklin, J. (1998). Uses of saw palmetto (*Serenoa repens*, Arecaceae) in Florida. Econ. Bot. *52*(4): 381-393.

Broschat, T. K. and Donselman, H. (1986). Factors affecting storage and germination of Chysalidocarpus lutescens seeds. J. Amer. Soc.Hort. Sci. *111*(6): 872-877.

Broschat, T. K. and Meerow, A. W. (2000). Ornamental Palm Horticulture. Gainesville, FL: University Press of Florida.

Carpenter, W. (1987). Temperature and imbibition effects on seed-germination of sabal-palmetto and serenoa-repens. HortScience *22*(4): 660.

Carrington, M., Gottfried, T., and Mullahey, J. (2003). Pollination biology of saw palmetto (*Serenoa repens:* Palmae) in southwestern Florida. Palms *47*: 95-103.

Gardner, A. G. (2021). Germination Cues, Dormancy, and Ex Situ Conservation of *Harperocallis flava,* a Federally Endangered Endemic Plant of the Florida Panhandle. University of Florida. Dissertation. https://original-ufdc.uflib.ufl.edu/UFE0057749/00001

Gilman, E. (1999). *Serenoa repens*- Saw Palmetto. EDIS: IFAS- University of Florida.Fact Sheet FPS-547. https://hort.ifas.ufl.edu/database/documents/pdf/shrub_fact_sheets/serrepa.pdf

Hilmon, J. B. (1968). Autecology of saw palmetto (*Serenoa repens* (Bartr.) Small). Duke University PhD Dissertation, Duke University, Durham, NC.

Maehr, D. S. and Layne, J. N. (1996). Florida's All-Purpose Plant: the Saw Palmetto. Palmetto. *16*(4): 6-10, 15, 21.

Makus, D. (2006). Seed germination methods and establishment of saw-palmetto, *Serenoa repens*, in South Texas, IV International Symposium on Seed, Transplant and Stand Establishment of Horticultural Crops; Translating Seed and Seedling 782.

Meerow, A. W. and Broschat, T. K. (2021). Palm Seed Germination. EDIS- IFAS. https://edis.ifas.ufl.edu/publication/EP238

Moldenke, H. (1967). *Serenoa repens f. glauca* Moldenke. Phytolog. *14*: 326.

Murphy, V., Clase, T., Rodríguez-Peña, R. A., Jiménez-Rodríguez, F., Jestrow, B., Husby, C. E., and Griffith, M. P. (2016). Conservation horticulture for yarey palm and buccaneer palm: substrate and sowing depth affect germination and early seedling growth. HortTech. *26*(6): 811-815.

Smith, D. (1972). Palm briefs. Fruiting in the saw palmetto. Principes. 16(1): 30-33.

Yadav, N., Khanduri, V. P., Singh, B., Dhanai, C. S., Riyal, M. K., Rawat, D., Ahmad, T., and Kumar, M. (2023). Effect of temperature, seed size, sowing depth, and position on seed germination and seedling growth of *Bauhinia retusa* Roxb. and B*auhinia variegata* L. Forests. *14*(8):1664.

Combined Proceedings IPPS 73:283-288. 2023.

Here, There, and Back Again ..... Again!

Andrew King

Stephen F. Austin (SFA) Gardens, Stephen F. Austin University, Nacogdoches, Texas 75962, USA, and King's Nursery, Tenaha, Texas, 75974, USA

Andrew.king@sfasu.edu

Keywords: academia, commercial industry experience, green industry, art and science of horticulture

Summary

Historically, individuals from the private sector of Horticulture/ Green Industry and those from the academic side have divergent backgrounds and experiences. Some academics have little or no commercial horticultural exposure. This paper highlights some of the challenges, opportunities and benefits of bridging the gap by employing people in academia with significant experience in private industry - "hybrid" professionals.

INTRODUCTION

I was raised by 3rd generation horticulturists on a retail nursery in deep east Texas. This upbringing shaped me in every conceivable way, but one thing that stands out from my childhood is the appreciation my parents and grandparents had for tangible results. This appreciation naturally led to a disdain for practices that were of questionable effectiveness. Fast forward 20 years from that little boy on the farm in Tenaha, Texas to a young man entering graduate school at Texas A&M University in College Station, Texas where I began to learn new ways of thinking about the challenges that faced the horticulture industry. I appreciated the newfound scope of work that I was able to take part in, as Texas A&M is the land grant University in Texas - and our stakeholders included the whole of the Texas green industry. However, I found myself missing the practicality with which my family had operated for generations. This juxtaposition became common fodder for my inner dialogue and many conversations with close friends and colleagues. The following is my effort to harmonize the world of academia and private industry and to display how beneficial it can be for individuals with interest in both realms - bridge this gap.

History Of King's Nursery

In 1915, J.B. King Sr. began growing strawberries in the sugar sands of east Texas (**Fig. 1).** With no other strawberry growers in the area, he developed a market for them by shipping the produce to Shreveport, Louisiana. Soon the idea became popular with many neighboring farmers, and J.B. realized that he needed to differentiate his farm.

Figure 1. The Kings' strawberry field in Tenaha, Texas in the early 1920s.

He began grafting and growing fruit trees and roses - and the King Nursery was born (**Fig. 2**). His slogan was: "Where the name of the firm indicates the quality of stock." In 1947, J.B. King Jr. was called into service at the nursery. He incorporated more ornamental plants into the nursery's inventory, until the early 1950s when it became a one-stop shop for retail customers. Even though the nursery has always depended on retail customers, J.B. Jr. had a wholesale sensibility. He was a tremendous propagator and preferred the solitude of plant production to the often-hectic nature of retail sales.

Figure 2. J.B. King, Sr. and Katie King standing in front of the King Nursery sign in the late 1930s.

In 1979, J.B. Jr.'s son, Aubrey King, began full-time work on the nursery. Like his father, he was an excellent plantsman; however, he was also highly skilled in the art of sales. Treating customers as students, he would spend hours with those who showed interest in horticulture. He placed his signature on the nursery by offering cutting-edge plant material, especially in the area of perennials and small trees.

In January of 2021, I took over as President and Operator of King's Nursery. It has been an incredible three years and most importantly my horticultural abilities have increased exponentially.

Obstacles To Practical Academic Horticultural Science

I was academically trained by some of the best horticulturists I know. Drs. David Creech **(Fig. 3)** and Mike Arnold are each fantastic in their own right and have conducted research that has influenced the plant production industry. I am grateful for these mentors. As I became more acquainted with the lay of the proverbial academic landscape - I realized that this was less common than I had assumed. A number of obstacles make it more difficult than it should be to conduct meaningful horticultural research that meets the needs of the industry that so many of us strive to serve.

First and most problematic: lack of funding is an issue. Genuinely relevant and needed research may not lead to a windfall for majority members of the industry and this often disincentivizes them from financially supporting these efforts.

Figure 3. Dr. David Creech of Stephen F. Austin University - outstanding in his field!

Historically, the Green Industry has underfunded applied research compared to other industries. The findings of needed research can lead to massive improvements in how the industry grows crops, what crops we can actually grow - and for the end-consumer it can mean success in their landscape. Long-term "trial and error" documented by research leading to successful landscapes results in more satisfied customers who will spend more money buying plants - increasing wholesalers' and retailers' bottom line. The problem is that this is a long-game model requiring patience. Many are unwilling to play the game. The academic is increasingly judged based on the amount of research dollars that are garnered by their program – and subsequent publication record.

Hence, many good scientists pass on needed research that is relevant - simply because it is underfunded and will not further their career. While not their fault, it is the reality of the academic system.

Second is the old axiom: "publish or perish." Unfortunately, unintended consequences are heavily weighted toward those scientists favoring short-term research that is quickly publishable versus long-term projects that are potentially industry-shaping; this skewed incentive also leads researchers to shift away from the goal of fulfilling industry needs, for which they were originally hired to do – to research that will allow them to remain employed. All of this may add up to research that the scientist is not passionate about - and of little relevance to the Green Industry.

Clearly, researchers who are members of IPPS-SR are scientists - motivated to help the Green Industry. Many of them fit the aforementioned description of a "hybrid" industry-experienced/academic professional, giving them greater credibility with Green Industry members. Still, there should be even more opportunities for industry-experienced individuals in academia than is currently available. These "hybrid" professionals bring unique perspectives/experiences that can lead to research that is both relevant – and answers future questions that no one is currently asking.

Academia Is Invaluable

I was initially drawn to the world of academia by the allure of teaching. The idea of being influential in the horticultural lives of students as Dr. Creech was in mine, was very appealing. In fact, to this day the professional accomplishments that I am most proud of revolve around students that I have been able to connect with, many of whom are currently becoming leaders in our industry (**Fig. 4**). I am grateful that I was able to teach them horticultural science.

Figure 4. The Kings (3rd - 5th from left) with former students from Texas A&M University.

Without at least some academic training, a horticulturist is generally left without understanding much of the "why" behind the "what" that they do. This is not to say that they cannot be an excellent grower or horticulturist, simply to say that colleges and universities often focus on training their students more in the science - than the art of horticulture. This scientific background is very useful in the development and understanding of a grower.

Again, the "hybrid" professional can bridge the gap between teaching the art and science of horticulture. One who has experienced planning, scheduling, propagating, growing, marketing, selling and delivering a crop on a consistent basis has a unique perspective for how much science versus art a student should learn.

Industry Experience Is Invaluable Too

Not every member of a horticulture department at a college or university will have experience working in the green industry, nor is that necessary. It is however exceedingly helpful to have people with experience working on your team. This is often accomplished at colleges and universities through industry cooperation, but it should be noted that the "hybrid" member of a department will likely be more available than those cooperators. These members of a team can assist with the direction of future research and give practical advice as to the needs of the industry.

In the past these "hybrid" professionals have been held at arm's length by the academic world, primarily out of concerns about conflict-of-interest. In certain cases, these are legitimate concerns. There have been and will continue to be individuals that take advantage of positions of trust or authority for personal gain, however the positives of greater academic involvement of industry members outweigh the negatives.

Formal Plant Trials

One example of truly useful work are formal plant trials. These trials allow industry, academia and the public alike to observe how select plant material will fare in a specified locale. For plant growers, this is valuable information particularly where new varieties and selections are concerned. Observing a new variety in trial enables a grower to get their initial observation of the plant - ***before*** trying to commercially grow it. This can prevent unnecessary financial losses for growers. At the same time, the trials can also lead to immediate demand for new varieties since the public can also observe the performance of the plants. This means less lag time for new varieties to become commercially successful with consumers – and growers.

In light of this, SFA Gardens has announced that in June of 2024 they will hold the inaugural Deep East Texas Annual & Perennial Plant Trials (DETAPPT). The goal will be to bring all of the previously-mentioned parties together to disseminate information about new annual and herbaceous perennial crops, to forge new relationships and cooperation amongst stakeholders, and to celebrate the rich horticulture industry that exists in east Texas.

CONCLUSION

Ultimately, the field of horticulture is stronger when there is a symbiosis of industry and academia - supporting, cooperating and working together. One way to better ensure that this is occurring is to enable more "hybrid" professionals with commercial experience in academia. In the future, these "hybrid" professionals should be sought out for their unique perspectives by academic institutions.

Combined Proceedings IPPS 73: 289-297. 2023.

General Observations of the Germination Requirements of New Zealand Native Flora

Philip Smith

Totara Glen Nurseries, 167 Staces Road, Palmerston North, Manawatu, New Zealand

totaraglennurseries@gmail.com

Keywords: native seed, seed propagation, dormancy

Summary

This paper is of interest to nurseries that collect and propagate native seed. Totara Glen Nurseries is primary a native plant supplier to New Zealand territorial authorities and infrastructure projects. Some 99% of seed is collected, processed and stored by us – and 95% of our plants are grown from seed. Challenges in dormancy and seed propagation systems - appropriate for seedling production of our select, New Zealand native plants and environs are discussed.

INTRODUCTION

This report covers general observations for germination requirements of New Zealand native plants. I include information from a seminal paper by Fountain and Outred (1991) on germination requirements of New Zealand native plants – which provides some excellent discussion points and observations. I also include my personal observations from sexually propagating native plants at Totara Glen Nurseries. Types of seed relative to seed germination, germination techniques – including the possibility of fire ecology in some New Zealand native seed – as well as increased infertility issues of native seed during the past 20-years are addressed.

There is a range of delayed seed germination phenomena in New Zealand, which is relevant to industry professionals who collect seed from the wild in other parts of the world.

Delayed germination is a critical factor in many New Zealand flora for the survival of many plant species. This could be because of the following environmental factors.

- Avoiding heavy frost periods. Low temperature and associated winter desiccation have a known effect in seedling survivability, both in the wild and in the nursery environment.
- Avoiding extensive drought periods (and ephemeral/ short-term environmental conditions), i.e. e.g. some of our native rushes and cultivars. A major abiotic factor in seedling loss after germination is summer moisture deficit periods, i.e. drought.
- Getting suitable light requirements for seed germination. It is often noted that once a large climatic plant species succumbs to either age or windfall – opening up large ground areas to sunlight - germination of some native species is dramatically increased.
- Setting up ecological seed banks via delayed germination.
- Combinations of the above.

Totora Glen Nurseries

Totara Glen Nurseries is primary a native plant supplier to New Zealand territorial authorities and infrastructure projects (**Fig. 1**).

Figure 1. Totara Glen Nurseries, Palmerston North, Manawatu, New Zealand

Some 99% of seed is collected, processed and stored by us – and 95% of our plants are grown from seed (**Fig. 2**). Most seed are sourced from the lower North Island, Taupo to Wellington, with some seed collected from the Auckland area and the South Island (**Fig. 3**).

Figure 2. Seed collection of native New Zealand plants and their subequent seedling production at Totara Glen Nurseries.

Figure 3. Palmerston North, Manawatu, New Zealand – North Island; see arrows.

In understanding germination processes, it is appropriate to discuss types of germination in respect to New Zealand flora.

Vivipary

Vivipary is the germination of seed while still attached to the parent plant. An example of this is *Pachystegia insignis*.

The propagules are pregerminated on mother plants - and naturally dispersed with an emergent root and shoot system (**Fig. 4**).

Figure 4. Vivipary of *Pachystegia insignis* with propagules pregerminated (arrow) while still attached to the mother plant – allowing dispersion with an emergent root and shoot system.

Recalcitrance

This word is defined as "obstinately defiant of authority or restraint" or "difficult to manage or operate and not responsive to treatment". Many native species exhibit recalcitrance. Recalcitrant seeds lose viability after drying, while orthodox seeds tolerate maturation drying.

Germination in recalcitrant seeds must proceed soon after maturity – or seeds must be stored under conditions that prevent drying (Davies, et al. 2018). Highly recalcitrant seed are intolerant of tissue moisture levels that drop below 10-12 percent. There are many examples in native New Zealand flora: Kahikatea, Pukatea, Kauri, Griselinia, beech, etc. (**Fig. 5**).

Figure 5. Some examples of native New Zealand species with recalcitrant seed include: Kahikatea, Pukatea, Kauri, and Griselinia,

Quiescent Seed

Quiescence is a period of inactivity. Most native seeds enter a quiescent stage during maturation associated with desiccation/ maturation drying. If seed is dried excessively, it may die or go into double dormancy (more than one dormancy requirement will need to be satisfied for seed to germinate). This is most likely a survival mechanism related to drought -and waiting for more optimal soil moisture conditions to assure seed germination and seedling survivability. Drought is the biggest factor in seedling death (survivability!) in the New Zealand natural forest ecosystem. Quiescence is distinct from dormancy where specific dormancy/priming treatments are required to initiate embryo growth. Examples are Manuka, Kanuka, Phormium, Nikau, forest margin species - *Pittosporum eugenioides, Pittosporum crassifolium*, and some Coprosma species.

Dormant Seed

Seeds of many species require an environmentally imposed stimulus to initiate germination. This could be chilling (stratification), light (duration, photoperiod) or even freezing. This is generally related to biochemical inhibitors (***internal dormancy***). Also, impervious seed coats which restrict either water or gas exchange cause ***external dormancy*** (Davies, et al., 2018).

Low Temperature Dormancy

A requirement of cold-period stratification to ensure germination is a characteristic of many New Zealand native seeds. My experience is that a stratification period not only ensures germination in some species, but it also promotes more uniform germination; this is critical for commercial medium- to large-scale native plant production systems - managing logistics of scheduling seedling production during the calendar year. In my experience, this reaction to temperature is very dependent on the ecological zone the seed was sourced from. There is a large difference in some species in germination time relative to the eco-source (provenance) of the seed. In general, the colder the eco-source area, the longer the germination period. Examples include *Pittosporum tenuifolium* and Podocarps (plum pine).

Light Induced Dormancy

The amount of light (irradiance) and the daylength (photoperiod) affects seedling emergence. My observations indicate that daylength, as well as temperature affect germination, i.e. Spinifex (coastal grass). Seeds sown in late autumn through late winter, often germinate in midspring. I suspect this is the case for many native seedlings, such as *Melicytus ramiflorus, Astelia* sp., *Pittosporum tenuifolium* and *Pittosporum ralphii*.

There are relationships of habitat to light requirements. Shaded habitats where light is enriched in green and far-red wavelengths, drive the light quality-sensing pigment phytochrome from an 'activating' state which is known to promote germination, i.e. recently open canopy situation or bush edge situations. Red light promotes seed germination, while blue light and a low red/far–red ratio condition inhibits seed germination.

Chemical Induced Inhibitors

At least 20 inhibitory compounds have been identified in seed, suggesting that these are also prevalent in some native seed. It is common for native plant propagators to carry out prolonged washing of seed (leaching), to overcome certain dormancies – and promote germination times, i.e., *Pittosporum eugenioides, Cordyline australis, Aristotelia serrata* and *Melicytus ramiflorus*. Past work has been done to show that the fleshy fruit layer of *Pseudopanax crassifolius* contains an inhibitory substance. Reasons for this could be to initially inhibit germination until sufficient rains and soil moisture occurs. Prolonged rain events will leach inhibitory compounds from the seed.

Rudimentary Embryos

Rudimentary embryos require further embryo development after the seed has been detached from the parent plant. Typically, there are double dormancy requirements where seed is sown to let the embryo further develop – and then a secondary dormancy condition must be met, i.e. cold-moist stratification for germination to proceed. Examples include conifers such as Miro, Kahikatea and Rimu (**Fig. 6**).

Figure 6. Rudimentary (underdeveloped) embryos require further embryo development after the seed has been detached from the plant, i.e. conifers such as Miro, Kahikatea and Rimu.

Seed Coat Imposition of Dormancy (External Dormancy)

Seed coat imposition is found in New Zealand native legumes, and as such germination is inhibited until seed coat disruption/rupturing/scarification has occurred.

Examples include *Sophora* sp. and native brooms. See seed trials of *Sophora microphylla* experiencing external dormancy - requiring scarification **(Table 1 and Fig.** 7**)**.

Table 1. Seed trials of *Sophora microphylla* experiencing external dormancy.

Sophora microphylla	Sth Kopuatai Peat Dome	Waikato	1992 ----	Sown 19th March	Germinated
Sophora microphylla	Trotters Gorge	South Island	1994 -----	Sown 19th March	Not Germinated Yet
Sophora microphylla	Gore	South Island	1996 -----	Sown 19th March	Germinated
Sophora microphylla	Hamner Springs	South Island	2000 ---	Sown 19th March	Germinated
Sophora microphylla	Haast	South Island	2001 ----	Sown 19th March	Not germinated yet

Figure 7. Seed trials of *Sophora microphylla* experiencing seed coat external dormancy.

Heat/Smoke Promotors of Germination

Do New Zealand Native Plants have an existing or hereditary fire ecology? Have we inherited this from seed coming over from Australia? Exposure to smoke has been shown to improve germination of Australian native species previously thought to be difficult or impossible to germinate (Dixon et al., 1995). I have noted that liquid smoke enhances germination of some New Zealand native species. In an IPPS publication, Bachman (2009) reported on enhanced seed germination with liquid smoke. Possible examples include Pomaderris, Spinifex, Sophora, Whau, and native orchids.

Ectomycorrhizal Fungi

These host plant associations are known particularly in Native Beech. Some 42 mycorrhizal genera are recorded in association with *Nothofagus* sp. (southern beech), *Leptospermum* (tea trees) and or Kunzea alone. We know they have an influence in seedling growth, so why not germination success as well? Mycorrhiza have been reported to enhance both nursery propagation and production systems (Davies, 2008).

Climate Change & Seed Viability

Is climate change affecting the viability of our Native Seeds? There have been large population declines in native insects. We know that pollination by native insects is an important part of native flora ecology. Reduction in insect pollinators could be indirectly affecting the viability of some native seeds. Many native nursery specialists in operation for 20 plus years, have experienced reducing viability of some native seeds. Examples are Spinifex (coastal grass) and Pittosporum species.

Record Keeping

Keeping accurate, detailed records of seed provenance, germination treatments, and observation of seed germination trials - is critical for the commercial seed propagation (**Fig. 8**). And modern technology allows information to be digitized (Davies et al., 2018).

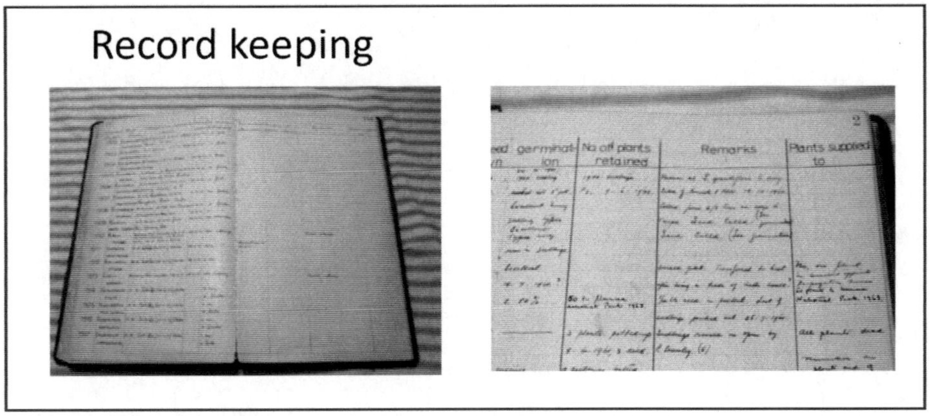

Figure 8. Accurate, detailed records and observation of seed germination trials is critical for the commercial seed propagation.

CONCLUSION

There is a huge variation in germination across New Zealand native flora. It is clear that once a strategy has been employed to engage germination (or break a dormancy) - there can be other environmental-genetic factors that affect select species germination systems, i.e. uniformity, production cycle period, etc. For smaller nurseries, fresh seed is best, but for larger nurseries where labor resources are closely managed - a clear strategy is needed to manage all the different variables of in play – from seed provenance to seed harvesting/processing to pre-germination to germination strategies. As previously stated, natural delayed germination is a critical factor in the survival of many plant species. Thus, better understanding how different seeds delay germination - helps develop more efficient germination strategies to efficiently unlock select dormancy requirements – to break dormancy – and ensure more uniform seed production systems.

LITERATURE CITED

Bachman, G.R. (2009). Enhancing germination with liquid smoke. Comb. Proc. Int. Plant Prop. Soc. *59*:557-561.

Davies, F.T. (2008). Opportunities from down under: how mycorrhizal fungi can benefit nursery propagation and production systems. Comb. Proc. Int. Plant Prop. Soc. *58*:539-548.
https://ipps.org/uploads/docs/58_128.pdf

Davies, F.T., Geneve, R.L., and Wilson, S.B. (2018). Hartmann and Kester's Plant Propagation: Principles and Practices. 9th ed. N.Y, N.Y. Pearson Education.

Dixon, K.W., Roche, S. & Pate, J.S. (1995). The promotive effect of smoke derived from burnt native vegetation on seed germination of Western Australian plants. *Oecologia 101*:185–192 (1995).
https://doi.org/10.1007/BF00317282

Fountain, D.W. and Outred, H.A. (1991). Germination requirements of New Zealand native plants: A Review. New Zealand J. Bot. *29*:311-316.
https://doi.org/10.1080/0028825X.1991.10416609

Combined Proceedings IPPS 73:298-307. 2023.

The Art and Science of Plant Propagation

Regina Coronado

Panoramic Farm, 3110 Tarlton Mill Rd., Marshville, NC 28103, USA

regina@panoramicfarm.com

Keywords: rooting hormones, auxins, cuttings, tissue culture, seed, propagation media, propagules

Summary

The goal of this paper is to motivate propagators to understand that plant propagation integrates both science and art in producing high quality liners. The underlining goal is maximizing efficiency and yields in propagation for commercial production success.

INTRODUCTION

Science can be defined as a rigorous, systematic endeavor that builds and organizes knowledge in the form of testable explanations and predictions about the world. Modern science is divided into three major branches: physical science, earth science and life science. Horticulture and plant propagation are part of the life sciences – which includes the study of living organisms and life processes as part of the natural sciences. Biology, chemistry, physics, mathematics (statistics) play important roles in understanding the *"why"/science/ principles* of plant propagation, while application of *technology/practices/art* is part of the *"how"* in commercial propagation.

IPPS Vol. 73 – 2023

All these sciences play an important role in propagation: how do (adventitious) roots grow, manipulating plant growth and development, understanding plant structures, water, oxygen/aeration, nutrient exchange, understanding ratios among elements such as nitrogen, phosphorus, potassium, calcium, magnesium, micro-elements – and determining fertilizer and pesticide rates. Utilizing science opens a whole new world of knowledge in propagation. It is important to understand plant biology and the different forms of plant reproduction - but it is also important to master their requirements with a high level of precision (art/application/technology) to maximize commercial propagation success.

In essence, plant propagation is the art and science of manipulating plant growth and development – utilizing select propagules and propagation techniques in commercially reproducing plants. Propagation is the process of creating new plants using different propagules: seeds, cuttings, tissue culture plantlets, grafts, layering, etc.

Challenges For the Plant Propagator

In developing a propagation systems approach, the propagator needs to determine sources and types of propagules: buying in unrooted cuttings (URC), seed, tissue culture produced microcuttings or plantlets, etc. (**Figs. 1 and 2**). Other questions include: propagation tray requirement, propagation media type, plant propagation/production requirements [electrical conductivity (EC), pH]; if propagating unrooted cuttings – what rooting hormones (auxin), application method and concentrations work best? (**Fig. 3**)

Figure 1. *Yucca filamentosa* 'Color Guard' tissue-cultured micro-propagules upon arrival from a tissue culture lab (left), and later, during nursery propagation in trays (right).

Figure 2. *Yucca filamentosa* 'Color Guard' after initially lining-out established propagules in containers (left), and in final container production (right).

Figure 3. Propagating *Abelia* × *grandiflora* 'Kaleidoscope' cuttings treated with auxin to enhance rooting under intermittent mist in cutting tray flats.

Propagating plants, regardless of the method of choice, is not a problem free system. It is up to the propagator to recognize potential problems, such as diseases, pests, rooting rate and acceptable top-growth (shoot) development. The propagator should be able to determine preventative and/or curative measures in order to produce a clean crop and avoid losses.

Selecting The Right Propagation Media

Regardless of plant species, the process starts with a cutting being stuck in suitable propagation media of choice - under the appropriate environmental conditions: temperature, light, mist, etc. The cutting should start rooting and a new plant will be produced with the same uniform characteristics of its parents.

Selecting good, affordable media for propagation can be challenging. Different aspects have to be considered, based on the plant requirements. There are many different choices of media, such as but not limited to: loose fill in open flats and plugs, autoplugs, green plugs, Elle pots/Elle plugs https://www.ellepot.com/, Preforma plugs https://jiffygroup.com/products/jiffy-preforma/,etc. Media can be peat based, fine bark or coir fibers. Media can be amended according to the needs of the plant. Some plugs can have paper holding the material together with loose or compressed media.

ROOTING HORMONES

Understanding rooting hormones such as indolebutyric acid (IBA), more water-soluble indolebutyric acid with potassium salt (K-IBA) and naphthalene acetic acid (NAA) are key factors for propagation. Determining the rate is highly important and is dependent on the plant species and variety (**Figs. 4, 5, and 6**).

ROOTING HORMONES

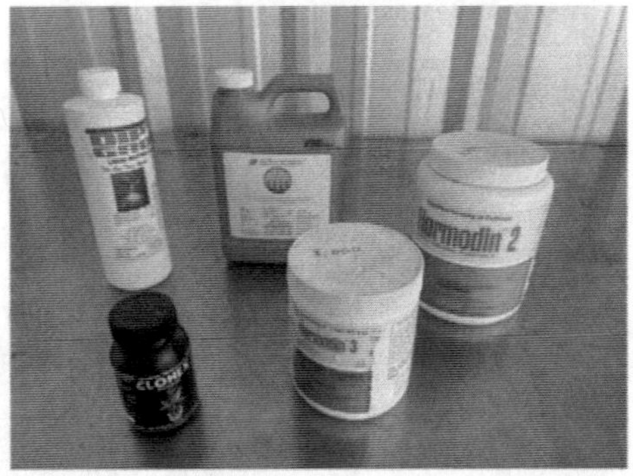

Figure 4. Rooting hormones: (top, left) Dip N' Grow (liquid combination of 1% IBA and 0.5 % NAA diluted in water to the desired concentration); (top, center) Advocate – 20% liquid IBA that is diluted to desired concentration and can be applied as a foliar application or basal dip; (bottom, left) Clonex – 0.31% (3100 ppm) IBA waterbased rooting gel; (bottom, center) Hormodin 3 – 0.8% IBA talc powder; (far-right) Hormodin 2 – 0.3% IBA talc powder.

HORMONE RATE TRIAL

Figure 5. Conducting a rooting study on *Hypericum kalmianum* 'Blue Velvet' using various auxin rates.

Figure 6. Rooting plug study of Camelia (10 weeks), Lysimachia (9 days) and Monarda (12 days) – showing species differences in rate of adventitious root formation and emergence.

Tank mixes can be used to help with stress, while dealing with unrooted cuttings. Always do a jar test when uncertain if materials are compatible. Tank mixing Advocate with Pageant and adjuvant can be beneficial. Advocate is an IBA compound (auxin) that promotes rooting, while

Pageant is a broad-spectrum fungicide, helps with stress and inhibits ethylene production in the cutting (**Fig. 7**). Not every variety reacts positively to use of liquid IBA or K-IBA. Some plants show toxicity and do not perform as expected. Other rooting hormone choices are gels and powders. Constant monitoring and attention to detail – and maintaining records are critical.

The production process has to be approached in different ways. Efficiency can be improved in multiple ways – and constant fine-tuning is required.

Understanding the physiology of an unrooted cutting, a miniature tissue culture plant or any other part of the plant used to propagate is critical - and so are all the interacting environmental factors: propagation mist (frequency and duration), light levels (percent shade), temperature (soil and air) and heat (source).

When all the possible variables have been put together, results will start showing. The most important and exciting moment for a propagator is to observe that first root (**Fig. 6**)!

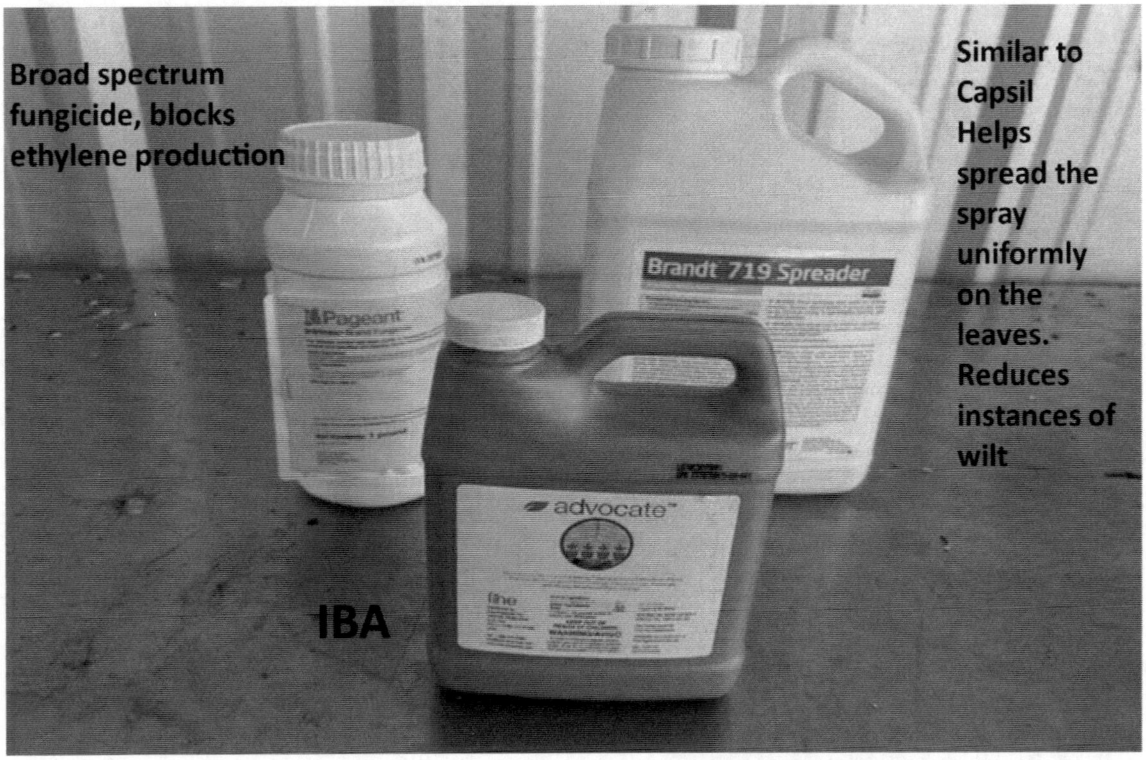

Figure 7. Fungicides and a wetting agent used in propagation with (bottom) Advocate (20% IBA) – to be diluted to desired concentration. (top, left) Pageant is a broadspectrum fungicide, that helps reduce stress and ethylene production in cuttings; (top, right) Brandt 719 is non-ionic surfactant, very similar to Capsil – that is mixed with auxins/chemicals – and helps reduce surface tension so the spray spreads more evenly with better penetration.

Merging Art and Science

Art is defined as something beautiful produced by a highly skilled person, who can put together all the pieces needed. Consistency and quality start from the moment the plant material is selected to be propagated. Top quality crops are not a coincidence or good luck.

Quality plant production is the result of dedication, attention to detail, research and understanding the plant's needs. Inconsistent crops to propagate are hard to work with - not only at the time of transplant but also while growing the finished crop. The main goal should be to increase efficiency and reduce percentage of waste at the production level - regardless if the liners are for in house production or outside customer sales (**Fig. 8**). Producing a finished crop with low quality liners, will produce a higher percent of waste than desired. Efficiency while planting will also be reduced since other tasks such as pruning, discarding and selecting will have to be added. There is a high demand for top quality in all the steps.

Figure 8. The merging of science/biology and application/technology/"the art" of producing quality rooted cuttings during cutting propagation (top, left); loading uniform rooted cutting liners for container production in the field (top, right), and (bottom) finishing off the rooted liners in containers ready for spring or fall sales.

Trialing – Experiments to Determine Optimal Media And Rooting Hormones

When trials are set up, every treatment must be labeled properly. Always include a control to verify the difference between treatments. *Simplify*: minimize the number of variables (treatments) in your propagation experiment to get more definitive results. Trials require constant monitoring, data analysis and if it shows positive results - it can go in two directions: either set up a bigger trial, or implement a new procedure.

The process for setting up a rooting trial is outlined in **Figure 9**.

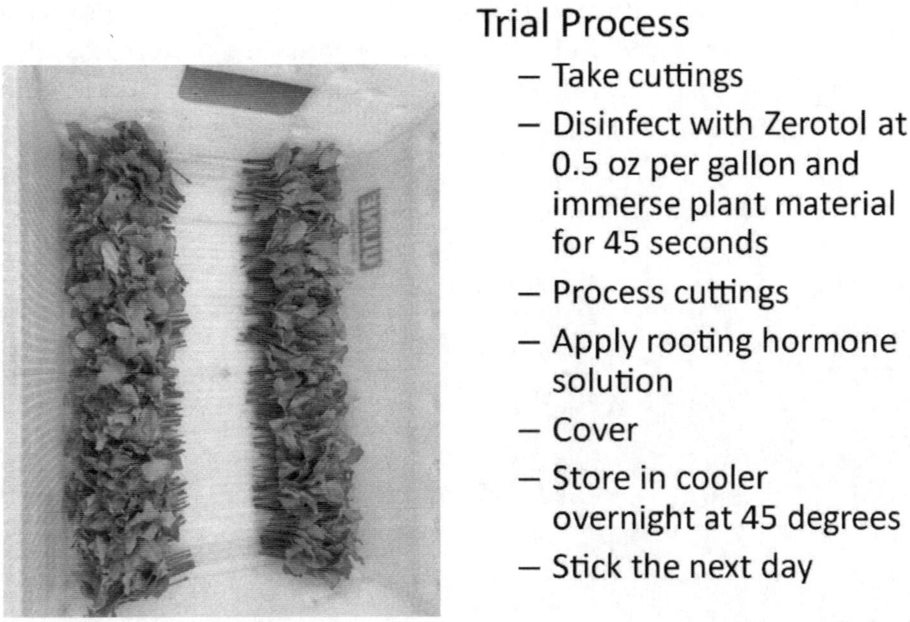

Figure 9. The trial process for setting up a rooting study, testing auxins of different sources and concentrations.

See examples of propagation media experiments (**Fig. 10**), and rooting trials of selected species using different auxin sources and concentrations (**Figs. 11 and 12**).

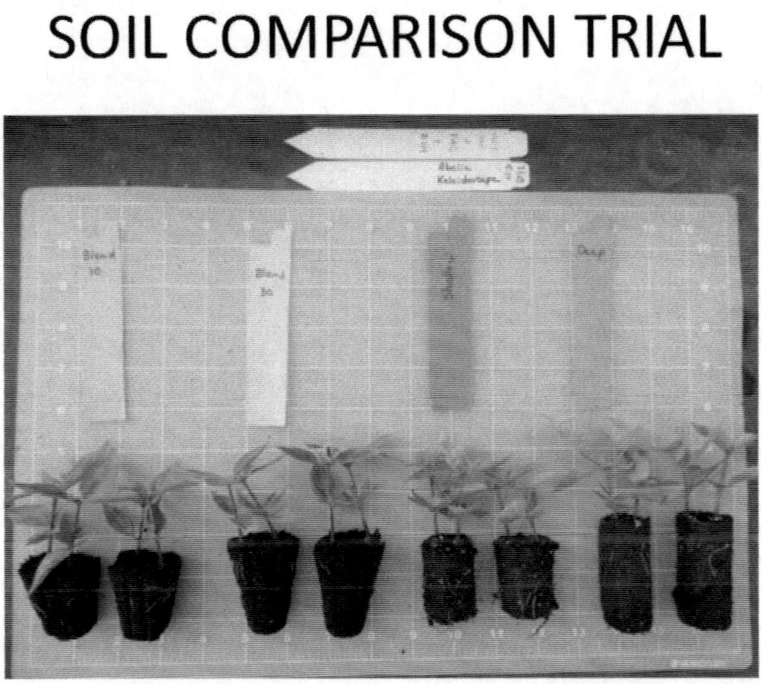

Figure 10. Propagation media experiments with *Abelia* × *grandiflora* 'Kaleidoscope'. (far-left)

Blend 10 is Preforma Blend 10 from Jiffy Products, coir fiber media, glued with a unique binding agent; (second, left) Blend 30 is Preforma Blend 30 from Jiffy Product, peat based media, glued with a unique binding agent; (second right) Elleplugs (pots), using a Sungro Propagation mix surrounded by biodegradable paper, consisting of 85% peat and 15% fine perlite with light starter charge; (far-right) Elleplugs (pots) with a greater plug depth.

Figure 11. (left) Shades of Pink™ Viburnum (*Viburnum tinus* 'Lisarose') initial rooting - double stuck in a 36 cell, Preforma Blend 10, treated with Advocate at 1,000 ppm IBA; (right).

Figure 12. (left) Rooting of *Buxus sempervirens* NewGen® Freedom *'SB300'* - 4 weeks after sticking, treated with Advocate at 1,500 ppm IBA; (right) – rooting of *Hypericum kalmianum* 'Blue Velvet' - 2 weeks after sticking, treated with Advocate at 1,500 ppm IBA.

Goals of a Propagator

Some important goals as a propagator are outlined in **Fig.13**. Useful textbooks on propagation and disease control are referenced in the literature cited. In summary, successful commercial plant propagation is maximized when the propagator understands both the science of propagation biology and applies it in the application and art of efficiently propagating plants.

GOALS AS A PROPAGATOR

- Continuing research and development
- Determine the best methods to propagate efficiently
- Set up high yield goals (90 % +)
- Pay great attention to detail
- Produce high quality plugs
- Train and motivate the crew
- Always strive for excellence and improvement

Figure 13. Some important goals of a plant propagator.

LITERATURE CITED

Davies, F.T., Geneve, R.L., and Wilson, S.B. (2018). Hartmann and Kester's Plant Propagation: Principles and Practices. 9th ed. N.Y, N.Y. Pearson Education.

Dirr, M, and Heuser, C.W. (2009). The Reference Manual of Woody Plant Propagation: From Seed to Tissue Culture. Timber Press, Oregon

Oliver, R. (ed) (2023) Agrios' Plant Pathology. 6th ed. Elsevier Pub., Amsterdam, Netherlands ISBN: 9780128224298

Combined Proceedings IPPS 73:308-324. 2023.

Starting a Nursery from Scratch

Doug Torn

Buds & Blooms Nursery, 7501 US-29, Browns Summit, NC 27214, USA

Doug@budsandbloomsnursery.com

Keywords: nursery layout, financing, markets, crop selection, production efficiency, personnel, cashflow

Summary

Doug Torn, founder and owner of Buds & Blooms, outlines the process of starting a nursery - locating in another state - without the luxury of inheriting land or having a rich uncle to bankroll the operation. Literally: starting from scratch, but doing so after gaining invaluable, professional nursery expertise – having a well-researched, thought-out business plan of what niche nursery products to produce, select markets to service, designing efficient/cost-effective production systems, growing methodically to not outpace sales, minimizing debt – and managing cash flow. The nursery has grown to 60 acres, with 25 employees - producing over 325,000 containerized plants. The four keys to his success are: 1) building a team of outstanding employees – hard-working with creative solutions, 2) managing dollars and cents, 3) efficiency, efficiency, efficiency, and 4) the love and joy of producing exceptional, quality plants – with satisfied customers.

INTRODUCTION

Starting a nursery was a goal of mine since my sophomore year of college. While working for various nurseries and growers during and after college, I would always search for possible better and more efficient ways to grow plants. When it came time to put my goals in motion, the very first question that had to be answered was, "What do I want to grow?". The choices are many: woody ornamentals, bedding plants, perennials, trees - to name a few. Your answer to this question will dictate so many of the decisions that follow. The plants you choose and the climate that is required to produce them will determine where you locate your nursery. You will also need to think about whether you want a container nursery or field production. Do you want to grow 1-gal, 3-gal or 15-gal containers? Who will be your main customer base? These are all important decisions that must be weighed before starting your nursery from scratch.

Dollars and Cents

Once you ave made a decision on what you will be growing and have started thinking about a location, the next big question you must ask yourself is, "How will I fund this endeavor?"

In other words - your dollars and cents. There are three main options for funding: banks, farm credit services, or investors. If you choose a bank, go with smaller banks where the banker can analyze your business and make decisions without going through multiple management levels for approval. Going the farm credit route, you may find they are easier to work with and understand agribusiness much better than your average bank. If you can find them, investors are another solid avenue and source of funding. But be warned, you may bite off more than you can chew with some of their interest rates.

So how much do you need to borrow? Well, that all depends on where you are thinking of starting your nursery and how many acres you want to purchase. Bear in mind, if you have chosen field grown production it will take considerably more land and you will need to factor in soil types before purchasing your land.

As a container grower, I started on 24 acres with an option on 10 more acres. I looked at land both visible and hidden from major highways and ultimately decided on a property with road visibility. Having visibility gave me immediate exposure, free advertising, and most of all, increased the value of the land over time. In my case, I borrowed $150K from an investor to purchase the land at $3,000/acre. In 1983, the local farmers said I was paying way too much for the property; but as of today, the land is now worth 6-10 times more than what I paid (or 5-6% return on investment). Had I chosen land with no highway visibility in the countryside, I would not have the same kind of valuation. After the major land purchase, the remaining $78K that I borrowed went towards grading, irrigation set up, payroll for one employee, and the purchase of our very first liners.

Determining Your Location

My decision to primarily grow ericaceous plants: Rhododendron, Azaleas, Pieris, Kalmia etc., meant I needed to be in somewhat of a cooler area - and definitely not near the coast due to its higher humidity and disease pressures.

After college while I was working at Mid-Western Nursery in Alabama, I subscribed to newspapers from various areas and began looking at ads for land and farms for sale. I called brokers from multiple states and narrowed my options down to northwest Georgia, northwest South Carolina, east Tennessee, or the western half of North Carolina.

We did not have all this fancy Google stuff back in those days. We were still operating rotary phones and typewriters (**Fig. 1**). Our first computer weighed over 30 pounds – which we would not utilize until the mid-late 1980's! So, what did I do? I pulled out the handy-dandy Statistical Abstract of the United States to do my research (**Fig 2).** I looked at everything from growth potential, road systems, logistics, proximity to metropolitan populations, and of course - climate. Then, for six months, I traveled on weekends and holidays to scope out my potential regions. I visited the future competition of the areas, networked through friends and family, and ultimately, I decided to locate on the Piedmont Triad in North Carolina.

Figure 1. Putting 1983 technology into perspective: (left top & bottom) an earlier desktop computer and IBM typewriter (arrow) – prior to word-processing; (right) the non-digital, non-cloud-connected - Ma Bell rotary telephone.

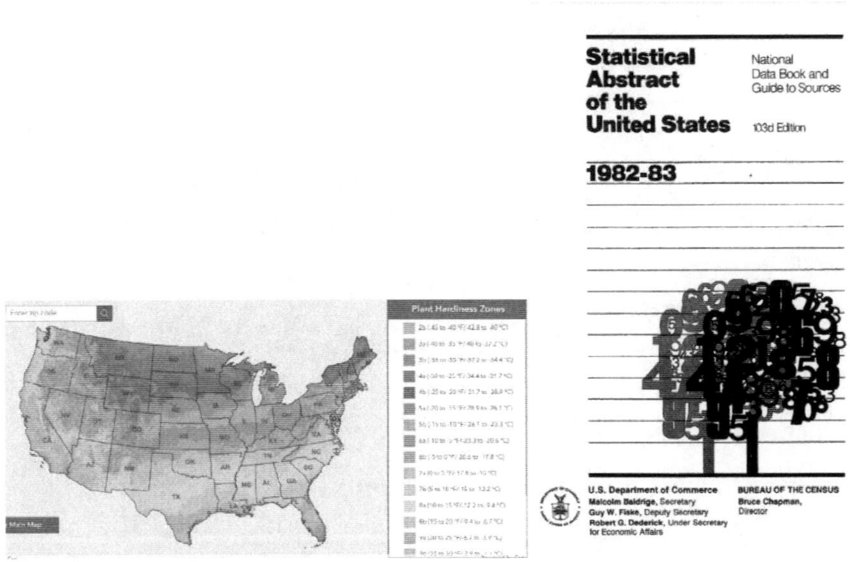

Figure 2. Determinining the nursery location. (left) USDA Plant Hardiness Zone Map, and (right) 1982-83 U.S. Dept. of Commerce Statistical Abstract Data Book & Guide to Sources.

Choosing the Piedmont Triad region meant that within 300 miles I could ship north to Washington D.C., south to Atlanta or west to Knoxville and within 500 miles I could access 44% of the nation's population. Back then, Greensboro was the 2nd largest city in the state. Now, it is third behind Charlotte and Raleigh and possibly on its way to 4th as the Wilmington area continues to grow.

I could have settled somewhere closer to the mountains where my crops surely would have thrived, but the growing season was much shorter. Additionally, had I chosen to go any further east where there are warmer temperatures and higher humidity, I would be dealing with a heck of a lot more disease pressure.

Last, but certainly not least, when choosing a location, you must consider your water source. How will you be irrigating your plants? Does the property already have a well or will you need to drill one? What about ponds and surface water? These are all factors that must be thoroughly evaluated before you purchase your land.

Land Preparation, Structures and Irrigation

Once your land is purchased you can start the grading. If possible, you should aim to grade at a 2-3% slope (**Fig. 3**). This will prevent water stagnation which could expose your plants to "wet feet". Any more than 2-3% you could end up experiencing washout. After grading you will need to decide the layout of your nursery. This is an extremely important step in your business planning and will be the foundation for running your nursery efficiently.

Figure 3. In the beginning: (left) Buds & Blooms Nursery, November 1983; (right) leveling and grading the former row cropped fields for containerized nursery production.

When planning our layout, we first had to determine the width of our beds and roadways. Initially, I chose a bed width of 50-ft, but have since then widened them to an average of 72-ft. Our roadways, which separate each of our blocks, are 14-ft wide. Each of these decisions was made with efficiency in mind. I chose these widths to reduce labor costs when potting plants, spacing plants, and pulling orders. It also allows great coverage from both sides when applying pesticides with our mist blower.

Congratulations, your land is prepped and you can finally start building some structures! But what kind of structure will you use? In the beginning we used shade structures rather than greenhouses. However, after a few very crucial winters, we swiftly made the switch to greenhouses. Today, the cold frames are pivotal to the protection of our plants. In the summer months, we cover netted shade cloth (between 30% - 50%) over the top and stretch it between greenhouses to protect plants from the heat. During the winter months we use poly to cover our houses, insulating our plants and protecting them from cold snaps and snowfall (**Fig. 4**).

Over the years, our regime for covering our houses with poly has changed significantly. Initially, we started out only using clear or 70% poly. Now, we use four different opacities: clear, 30%, 55%, and 70%. This not only allows us to push growth on our plants in the early spring, but it has also made a significant difference in the winter months. If you have ever spent a day manually knocking snow off greenhouses, you know that it is long and exhausting work. Adding the 30% and 55% opacities, we now see far less snow and ice sticking to those houses in comparison to the traditional 70% shaded poly.

Figure 4. Covered and uncovered overwintering production houses used today.

Your structures may have been built - but do not fill them with plants yet! How will you prevent disease pressure and weeds from spreading? Will you use gravel, ground cover, or both? Due to the sensitivity of our crops to Phytophthora - we use gravel and more recently have been putting ground cloth over gravel beds for additional weed control.

What about the irrigation? This may be one of the most important questions you ask yourself. Water is life. It is the key to everything. You will need to determine your source of irrigation, the type of sprinkler, what type of pipes to use, what irrigation system to use, and many other factors surrounding your water.

The main types of sprinklers typically used in nursery production are impact sprinklers, and wobblers or spinners. However, we chose something totally different. We first set up our irrigation using Rainbird shrub heads. However, it was a more costly endeavor, so we later began switching over to impact sprinklers to reduce installation costs, use less pipe, and dig fewer ditches. It was a big mistake! Impact sprinklers take 60-80% longer to wet plants than a shrub head, and when you are growing disease sensitive plants that should only be watered between 10:00 am and 2:00 pm, time is of the essence. Today, less than 20% of our nursery irrigates with impact sprinklers and we have continued to convert back to shrub heads **(Fig. 5)**. Efficiency is key!

Figure 5. (left) Rainbird shrub head irrigation sprinkler used for (right) container irrigation at the nursery.

Each day for the first 15 years or so, we manually opened valves by hand to water our zones. It was not until sometime in the late 1990's that we installed all electric 3-inch valves -which are controlled by two time clocks that are stationed in the office. This is the same system we use today.

Irrigation will be your saving grace some days and it can be your worst enemy. So, whenever it is your turn to make these decisions - the biggest piece of advice I can give you is to plan carefully and utilize professionals who are knowledgeable of these systems. Irrigation contractors, specifically those who work on golf courses, nurseries or even wineries, will be immensely helpful in creating the design and layout of your irrigation. Even if you chose to self-install like we did, they can guide you and help make many decisions regarding pipes, sizes, pressure, location, and so much more **(Fig. 6).** Happy irrigating!

Figure 6. (left) Setting up the irrigation system and (right) our first – and last! – shade structures.

POND DEVELOPMENT

If I have said it once, I have said it 1,000 times: water is the most essential element in growing container plants. Ponds, creeks, streams, or wells must be accessible on the property when looking at a site to locate a nursery. In our case, we started out with one pond (**Fig. 7**). Since then, we have built two more conventional ponds plus one small well-fed pond. Additionally, we have all our land graded so that we capture 100% of our run-off and can recycle our water. In doing so, we treat our water with chlorine using a Regal chlorinator. This helps clean the water of any pathogens and bacteria to prevent the spread of diseases on our already disease- prone plants.

Two of our ponds are located on a small creek that is fed by a 1,000-acre water shed. Thankfully, this means little rainfall is needed to fill them to capacity. In addition to our irrigation ponds, we have also added three silt ponds, now referred to as forebays.

When building ponds or forebays, or excavating silt from ponds, talk to your local Soil & Water Conservation District Service. They will help you engineer your damns and spillways. They can also provide cost-sharing funding. These are federal funds provided to the state for these intended purposes and are available to nurseries and farmers. By doing so, I have a 70% cost share on removing silt from one of our ponds.

Figure 7. (left) Building our first irrigation pond in the early 1990's and its current condition in 2023 as an established water source.

POTTING

Okay, now that you have your land graded, structures built, and irrigation set up you can finally get to the good stuff… the plants! But how are you going to pot without the proper equipment? Well, like many nurseries, when we first started, we were can filling and plugging plants. Luckily for us, light weight plastic pots were the standard means for potting, unlike some of our nurserymen predecessors who were using tin cans! But why would we spend all that time potting and plugging by hand?

It all comes back to your dollars and cents. Your costs can seem limitless when you are first starting out, so you have to very carefully and methodically consider where to invest your money.

One year later we built our first hand-potting machine that we hitched to the back of truck and would haul from block to block. Another 5-years later and still today, we have fully automated both potting and mixing by using a Javo Super potting machine and an HC Davis mixer (**Fig. 8**). These days, there are countless options at your disposal when choosing your potting equipment.

Continuing to keep efficiency in mind, we choose to pot our liners directly into the pot in which it will be sold, avoiding shifting. This means less time moving or handling the same plants and ultimately reduces our production time.

PROPAGATION

There are many factors to consider from a production and efficiency standpoint when propagating. Will you propagate all your plants? What size liners will you use? What kind of environment do your plants require? How long will your plants take to root? When do you need to finish the crop to make a profit?

In our case, we direct stick all cuttings into 3-inch, 4-inch, or 1-gal containers. For some crops, we will double or triple stick cuttings to speed up production time. This reduces losses in hard-to-root cultivars.

Figure 8. (top, left) container hand-filling in 1984, (top, right) a custom made portable potting machine (arrow), and (bottom) automated container potting machine used today.

Choosing to propagate rhododendron required even more planning. To propagate rhododendron, we use a Ray-Pac boiler and BioTherm Systems tubing on the ground for bottom heat (**Fig. 9**).

Figure 9. (left) BioTherm bottom heat, hot water tubing system for mist propagation, and (right) a Ray-Pac hot-water boiler.

This allows us to maintain 68-70°F soil temperature which helps initiate rooting. This is a must have for our crops - and it is important to consider any specialty needs that your crops may require in propagation.

We have toyed with producing our own tissue culture plants with a local basement lab but found there were too many variability issues from cultivar to cultivar (**Fig. 10**). These days, we purchase some tissue cultured plants from commercial labs.

Figure 10. (left) An earlier system of tissue-cultured rhododendrons produced at Buds and Blooms in collaboration with a local lab with plantlets being extracted from their shipping vessels; (right) tissue-cultured micropropagules are hand-planted with forceps into propagation trays to be acclimated, hardened-off, and grown-off as liners for future transplanting.

Figure 11. (left, right) Azaleas propagated under mist in propagation houses.

Ultimately, you will need to evaluate your upfront costs, maintenance costs, time, and margin before deciding which of your plants you will propagate and which you will purchase as liners (**Fig. 11**).

SHIPPING

Alright, now we are getting somewhere! Your plants are finished and ready to take to market. Now how are you going to get them to your customers? When it comes to shipping, we have always used shelved trucks to make sure our plants arrive looking as good as they look when they are loaded (**Fig 12).** In addition to shelving, there are two other commonly used methods to consider when shipping: lean stacking and racking (**Fig 13**).

Figure 12. (left) Handloading pulled nursery product from tracking trailers onto a delivery truck; (right) a shelving system to better protect plants in transit to retail nursery customers.

Lean stacking is efficient from both a time and space standpoint. It takes less time and labor not having to throw boards on a shelf and you can fit substantially more plants on a truck when they are bunched together rather than on a rack. However, with no stable structure, you run the risk of having plants arriving to your customer either damaged or flattened and therefore, less marketable.

Racking will get your plants to your customers looking good. It may also save you time and labor. However, racking requires up-front costs of purchasing either disposable wooden racks or metal racks. It also requires the use of a loader or forklift to load and unload trucks which is another added expense. With that in mind, this may not be the best option starting out when funds are limited.

Figure 13. (left) lean stacking and (right) racking containers with disposable wooden racks & pallets for moving with a forklift.

When we load our trucks, we pull our plants and place them on tracking trailers. From there, we load them straight into the truck. I am a firm believer in handling plants a minimal number of times. The more times you touch a plant, the more labor going into that plant and ultimately, the less profit margin you make on that plant (**Fig. 14**).

Figure 14. (left) Buds & Blooms 1st delivery truck, and (right) current trucking system.

Disaster Strikes

It is not a matter of if, but when you will experience a devastating disaster or problem. They can come in many forms, but ole' ruthless mother nature has been the source of most of our problems. And what has been her destruction weapon of choice? Snow: a four-letter word around here!

In our early years we only had shade structures which were wrapped with foam in the winter as a wind barrier. This was quite common in our area at the time. But twice we got caught with much more snow than predicted and had major issues with snow load damage.

The first time it broke a tremendous amount of our shade structures and ripped our shade cloth to threads. Remember all the time, money and energy spent putting those up in the first place? Now it is all crumbled-up on the ground. For the second snowstorm I tried thinking outside the box. I hired a helicopter pilot and had him fly over the shade structures in an attempt to blow the snow off. Unsuccessful! I ended up asking our employees and many friends to come in and cut the shade cloth along the isle below to reduce breakage of the plants and take the weight load off the structures. In the end it worked, and we only had to replace the shade cloth. All the plants and structures remained intact. Note to self: ***Sometimes you have to cut your losses, literally!***

Fast forward to 25 January 2000. After fully switching over from shade structures and building overwintering greenhouses, we still had one major snowstorm which dropped over 24 inches within 24 hours. Overall, in a two-day period, we had 30 inches of snow and ice. This time, 100 greenhouses were lost just like that. Including a newly built gutter connecting greenhouse (**Fig. 15**). We later replaced all of them with heavier gauge metal and more crossbow trusses. We also fortified the remaining 200 plus greenhouses with more crossbow trusses and some upright pipes along the center line of the houses.

The moral of the story? ***Learn from your mistakes, prepare for the worst, and pray for the best. Build back stronger!***

Figure 15. Disasters: (left) snowstorm with cut shade cloth, and (right) twisted house structure from the snow storm of January 2000.

Marketing

Now, after purchasing and prepping your land, setting up irrigation, constructing your greenhouses, and finishing your first potting, you finally have some plants to bring to market! But how will anybody know that? Marketing is how you speak to your customers.

Using our custom logo and tagline, "Bloom After Bloom, Year After Year", we hit the ground running by putting out ads in magazines. Those were the days when people still enjoyed reading a magazine with their morning cup of coffee. Now, you may not generate much buzz feed at all using ads.

In addition to our ads, I personally went door to door with plants to potential local customers and then worked my way out little by little. When you do not have word of mouth or a standing reputation regarding your quality, the customers have to be able to see the plants you are producing. They will not take a chance on you.

While door-to-door works, it takes a long time to cultivate a few potential customers and nothing beats the comradery of a good old-fashioned trade show.

Another way to generate conversation is sending out a company newsletter (**Figs. 16 and 17**). They can provide industry insights, company updates, and can spark conversation between customers the next time you see them. Our own *Blooming Journal* has since been retired, but if you have the time, newsletters can add a fun and personal touch to your nursery.

This generation is lucky. There are hundreds of ways to get the word out about your product. With just a few clicks of a button you can send pictures and information about your plants right to the customer's hand. Now, we primarily utilize our email, website, and social media to market to our customers https://budsandbloomsnursery.com/ However, every year we still take the time to stuff, stamp and snail mail our new catalog and spring order form for the upcoming year. What can we say… old habits die hard, and everybody loves a hard copy!

Figure 16. Marketing, advertising, and connecting with customers using (left) ads and (right) newsletters – *The Blooming Journal*.

Figure 17. (left, right) Creating our own brand and tagline: "Bloom After Bloom, Year After Year".

CONCLUSION

When I think about the many nuances that go into starting a nursery from scratch, there are four overarching elements that in my opinion have gotten me to where I am today.

The first and without a doubt the most important are my people. Buds & Blooms Nursery has flourished by hiring great personnel that have stuck with us for many years.

Over our 40 years of operation, I have been lucky to work with many incredibly hard-working individuals with strong minds and creative solutions **(Fig 18).** They are the foundation to any successful nursery.

The second is watching your dollars and cents. We have grown very methodically over many years making sure to minimize debt and only using a line of credit if necessary for cash flow purposes. ***Staying out of debt is of the utmost importance***! You never know when the next disaster or economic downturn will roll through and change everything.

Figure 18. Key nursery personnel at Buds & Blooms Nursery. Building a team of outstanding employees, hardworking with creative solutions – is critical to success.

The third element and a point I have mentioned many times, is having efficiency at the forefront of every decision. The decisions you make when setting up your nursery early on will have a major impact on you down the road. There is no amount of money that can make up for lost time, misused space, or poorly managed labor. It is crucial that you stop and take the time to envision the future and consider the potential consequences of your decisions. ***Efficiency is the key that can turn a single generational nursery into a multi-generational nursery*** **(Figs. 19 and 20).**

Figure 19. Buds & Blooms Nursery, 1988 – 5th year in operation.

Figure 20. Buds & Blooms Nursery, 2013, after 40-years of operation. The nursery has grown to 60 acres, with 25 employees - producing over 325,000 containerized plants.

While all these elements are important and play major roles in nursery development, none of them would matter without the love for growing plants. There is nothing that brings more joy or satisfaction than watching your nursery grow, producing exceptional quality plants, and receiving words of praise and gratitude from your customers. That is what it is all about and that is how you start a nursery from scratch!

A SPECIAL THANKS

I would not be where I am today without those who I work with and have supported me over the last 40 years. THANK YOU to Francisco Bribiesca, Rosa Rodriguez, Manuel Rodriguez and Belen Rosas for being outstanding leaders, production managers, and comrades. These four professionals collectively have 112 years of experience at Buds & Blooms Nursery.

Combined Proceedings IPPS 73:325-337. 2023.

Improving Air-Filled Porosity in Woody Propagation at the Hammond Research Station

Maureen Thiessen and Jeb S. Fields

Louisiana State University AgCenter, Hammond Research Station, 21549 Old Covington Highway, Hammond, Louisiana 70403, USA

mthiessen@agcenter.lsu.edu

Keywords: stratified substrate, rooting, Growcoon, woody, propagation, stem cuttings, wood fiber

Summary

Extreme heat during summer propagation of woody species often necessitates frequent mist applications to maintain proper leaf turgor. This frequent mist can result in excess moisture in propagation substrates, which tend to have low air space due to small container sizes and fine substrate particles. This study evaluated the use of substrate stratification, Growcoons, and wood fiber inclusion to increase air-filled porosity in propagation substrates and improve success and quality of rooted cuttings. Several woody species were selected to test each method and were rooted over the summer of 2023. Substrate stratification with 100% perlite did not yield significant benefit to rooting success or plug quality in most species, while Growcoons and wood fiber inclusion increased rooting success with little to no effect on root and plug quality.

INTRODUCTION

Achieving proper moisture content and aeration of rooting substrate is important for successful propagation of woody plants. The substrate must be lightweight with a sufficient water holding capacity, while also providing appropriate aeration to reduce fungal growth and gnats, and improve oxygen supply to the developing roots. The small container volumes and generally finer particles used for propagation, however, result in increased water holding capacity at the expense of air-filled porosity (Yafuso et al., 2019 and Milks et al., 1989). Common propagation substrates include varying proportions of peat, sand, perlite, vermiculite, and pine bark, typically targeting 25-40% air space (Dirr and Heuser, 2006).

Many woody species can be propagated in 100% perlite or pine bark, provided moisture levels are adequately maintained in the rooting substrate as well as on the leaf surface. When propagating woody species through the summer, maintaining proper moisture tends to be difficult in warm climates such as the U.S. Gulf South. In these climates, misting schedules that maintain sufficient leaf moisture can often result in excess substrate moisture, reducing propagation success. Increasing the air-filled porosity in the rooting substrate maintains adequate gas exchange under high mist schedules needed to prevent leaf drop.

There are various options available for propagators to increase air-filled porosity or airspace (AS). Commercial products exist, which can be lined in a propagation tray to maintain a boundary layer of air between the substrate and plastic cell, improving air flow. One such product, Growcoons, are a netted, cup-like seed and plug holder made of biodegradable material and designed to be transplanted with the plug (https://growcoon.com/en). These hold the plug roots together during transport and transplanting.

Substrate stratification is an emerging practice that involves the layering of substrate components of differing textures and physical properties on top of one another to achieve desired water and air holding properties. The practice is gaining momentum in container production where finer-textured, higher water-holding capacity substrates are layered on top of coarse, higher AS substrates. This improves aeration and oxygenation of the lower portion of the container (Criscione, et al., 2021; Fields and Criscione, 2023) providing plants access to the full container volume without saturated conditions. A similar phenomenon of substrate saturation in lower portions of the rooting cell trays often occurs, indicating potential benefit of stratification.

Wood fiber has also been gaining momentum as a substrate component. Previous research has shown that wood components and commercially available wood-fiber substrate products can significantly increase the AS of the substrate when blended (Jackson, 2018; Dickson et al., 2022; and Harris et al., 2020). Nitrogen immobilization concerns are less significant in propagation as nutrient demand of rooting plants is lower than those in larger containers, and the production time is often much shorter. Therefore, wood fibers lend themselves particularly useful to propagation systems.

Thus, the objective of this experiment was to evaluate the use of stratifying substrates, Growcoons, and wood fiber inclusion to alleviate excessive moisture in peat-based rooting substrate on rooting success and plug quality of several propagated woody shrub species.

MATERIALS AND METHODS

This experiment took place at the Louisiana State University Agricultural Center Hammond Research Station in Hammond, Louisiana. All four substrate treatments (**Table 1**) utilized 72-cell, star-shaped, deep propagation trays ((PL-72-STAR-DP-VH, T.O. Plastics, Clearwater, MN).

Table 1. Substrate treatments used in propagation study.

Treatment	Description
Standard Mix	2:1:1 bark fines: peat: perlite (non-stratified, loose-fill)
Stratified	Top layer – standard mix; bottom layer – 100% perlite
Growcoon	Filled with standard mix
Wood fiber blend	1:1 bark fines: southern yellow pine wood fiber

The standard (loose-fill, non-stratified) propagation substrate consisted of a 2:1:1 mix of pine bark fines (Phillips Bark Processing Co., Brookhaven, MS): peat (XL Commercial Peat Moss, Lambert Peat Moss, QC, Canada): perlite (Horticultural Grade, PVP Industries, Bloomfield, OH). Growcoons (Klasmann-Deilmann GmbH, Geeste, Germany) were placed in the propagation tray and filled with the standard propagation substrate. The stratified treatment consisted of filling the lower half of the cells of the propagation tray with 100% moistened perlite and the upper half of the cells with the standard propagation substrate (**Fig. 1**).

Figure 1. Stratified propagation trays contained 100% perlite in the bottom half and the standard mix in the top half.

The wood fiber treatment consisted of blending pine bark fines and an extruded southern yellow pine wood fiber at a 1:1 ratio. After filling, all individual cells were thoroughly moistened to drainage before sticking cuttings. Additionally, each substrate was evaluated for AS, container capacity (CC), and solid portion using a modified NCSU Porometer method (**Fig. 2**; Fonteno, Hardin, and Brewster, 1995).

Figure 2. Modified NCSU porometer method using individual 1-inch cells for determining air space, container capacity, and bulk density.

Propagation trays were filled with the substrate treatments and thoroughly moistened. Seven cells of each substrate treatment were individually cut from the trays and manually submerged in water to the level of substrate fill until the surface of the substrate glistened. A finger was placed over the drainage hole in the bottom of the cell and the cell was transferred to a funnel installed over a graduated cylinder and allowed to drain completely.

Precautions were used to ensure the cell remained vertical and was not squeezed. The volume of drainage was divided by the cell volume to calculate the AS of the substrate in the propagation cell. Cells were then weighed and the substrate dried in an oven at 105°C for 24 h for calculation of CC and solid portions. Container capacity, the total water holding capacity of a substrate in a given container, was calculated as the difference between the substrate's oven-dry weight and drained weight divided by the volume of the cell; solid portion was calculated as the volume remaining after AS and CC were subtracted from 100%.

Semi-hardwood stem cuttings of several woody shrub species were collected between May and July 2023 in the morning and stuck before noon the same day in four different rooting substrates, (**Tables 2, 3, and 4**).

Table 2. Species tested with substrate stratification

Species, cultivar	Date Stuck	Hormone used	Total Weeks
Callicarpa americana	6/8/2023	None	10.7
Callistemon 'Woodlander's Hardy'	6/20/2023	Hormodin 2, 3000ppm	9
Camellia 'Reverend Ida'	6/8/2023	Hormodin 2, 3000ppm	17.7
Ilex crenata	7/25/2023	Hormodin 2, 3000ppm	10.8
Ilex glabra	7/25/2023	Hormodin 3, 8000ppm	10.8
Ilex cornuta	7/26/2023	Hormodin 2, 3000ppm	10.8
Itea virginica 'Henry's Garnet'	7/24/2023	Hormodin 2, 3000ppm	11

Table 3. Species tested with Growcoons and Standard mix (2:1:1 Bark Fines: peat: perlite)

Species, cultivar	Date Stuck	Hormone used	Total Weeks
Callistemon 'Woodlander's Hardy'	6/20/2023	Hormodin 2, 3000ppm	9
Camellia 'Reverend Ida'	6/8/2023	Hormodin 2, 3000ppm	17.7
Itea virginica 'Henry's Garnet'	5/4/2023	Hormodin 2, 3000ppm	15.6
Rhododendron indicum 'Briliant'	6/8/2023	Hormodin 2, 3000ppm	15.6

Table 4. Species tested with wood fiber blended 1:1 with bark fines.

Species, cultivar	Date Stuck	Hormone used	Total Weeks
Callistemon 'Woodlander's Hardy'	6/20/2023	Hormodin 2, 3000ppm	9
Rhododendron indicum 'Fisher Pink'	6/20/2023	Hormodin 2, 3000ppm	12

Cutting material was taken from the new flush of growth as it began losing flexibility. Stem cuttings were approximately six cm long (**Fig. 3**). The lower half of the stems were stripped of their leaves, wounded, and dipped in a talc-based rooting hormone (Hormodin 2 & 3, OHP, Mainland, PA). Cuttings were then placed in an enclosed plastic-lined tent (**Fig. 4**) under intermittent mist at 4 s every 10 min. At four weeks, mist was adjusted to 6 s every 15 min; at 7 weeks, one side of the plastic sheeting was opened to allow air flow and the mist schedule was adjusted to 6 s every 20 min.

Figure 3. *Callistemon* 'Woodlander's Hardy' cuttings with lower leaves stripped.

Figure 4. Enclosed mist tent.

Cuttings were evaluated on one of two dates (21 August 2023 or 9 October 2023) depending on root establishment. A cutting was determined "rooted" when it could be gently pulled from its cell with 75% of the rooting substrate remaining intact.

The total number of rooted cuttings were counted for each species and treatment combination. Five random cuttings were taken from rooted cuttings, washed of all rooting substrate and/or Growcoons, and rated 1-5 on root quality (**Table 5**) and plug quality (**Table 6**).

Table 5. Root quality rating principles.

Rating Number: Root Quality	Description
1	Brown to black in color, or dried or slimy
2	Brown to tan in color, break easily
3	Tan to beige in color, healthy turgor
4	Beige to white in color, healthy turgor
5	White in color, healthy turgor and easily withstands washing

Plug quality was based on the visual appearance of the stems and foliage. For the standard, stratified, and wood blend treatments, roots were then individually teased apart and the average of the two longest roots measured. Root tissue was collected from stratified cuttings and placed in an oven at 70° C for 48 h and weighed.

All data were analyzed in JMP Pro 17 (SAS Institute, Cary, NC). Analysis of Variance (ANOVA) and means separation using Tukey's honestly significant difference (α=0.05) were used to evaluate substrate blend effect on air space. Within each substrate treatment and each species, root quality, plug quality, average root length, and root dry weight (N=5) were compared between the treatment substrate and the standard mix using a 1:1 t-test at α=0.05.

Table 6. Plug quality rating principles for the stems and foliage.

Rating Number: Plug Quality	Description
1	Dead and dried out
2	Stems alive and green, no visible foliage, roots low quality
3	Has some foliage, but foliage has signs of drying or disease; roots not filled out
4	Healthy foliage and root system
5	Significant and healthy foliage and root system

RESULTS AND DISCUSSION

Substrate Physical Properties. The mean AS (N = 7) for each experimental substrate was significantly different. The standard propagation mix had an AS of 18%, which corresponds to the lower end of the recommended range (18-40%; Dirr and Heuser, 2006). Stratification and wood fiber incorporation raised AS to 25% and 29%, respectively. The Growcoons increased AS the most, reaching 42%, just exceeding the recommended aforementioned range (**Fig. 5a**).

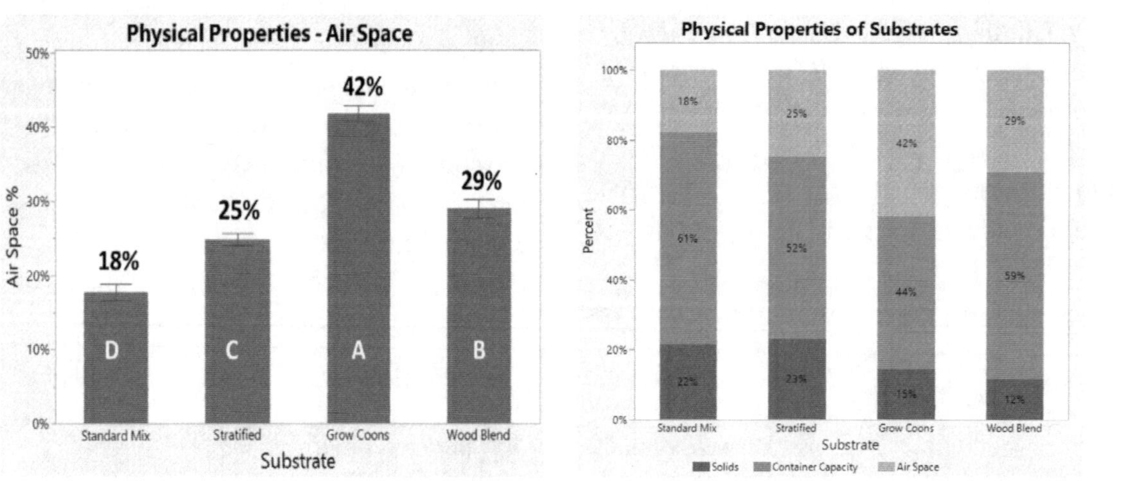

Figure 5a. Air-filled porosity in each substrate measured in 72-cell (1-inch) containers using a modified NCSU Porometer technique. **5b.** Air-filled porosity, container capacity, and solid portion in each substrate measured in 72-cell (1-inch) containers using a modified NCSU Porometer technique.

While each of these treatments increased the AS of the propagation substrate, it should be noted that the location of this added AS differs for each treatment – in the lower half of the cell for the stratified treatment, around the outer circumference of the cell for the Growcoon, and throughout the cell for the wood fiber blend. Container capacity followed a similar trend as AS; the standard propagation mix had the highest CC (61%). Stratification and addition of wood fiber reduced CC to 52% and 59% respectively, and the Growcoons had the lowest CC at 44%. The standard mix and stratified substrates had the highest proportion of solids, and Growcoons and wood blend had the lowest (**Fig. 5b**).

It should be noted that climactic extremes occurred at the Hammond Research Station during the summer of 2023 due to the heat dome that affected much of the U.S. midwestern and Gulf South states, resulting in higher temperatures and lower humidity than normal for our region.

Stratification Experiment. Rooting success in stratified substrate was 5-61% lower than in standard non-stratified substrate (**Table 7**), with the exception of *Ilex crenata*. Stratified substrate was associated with decreases in root quality, average root length, and root dry mass as well (**Table 7**) in the majority of species propagated. Roots in stratified substrates were less dense and showed more brown discoloration than the non-stratified counterparts (**Fig. 6**). Plug quality rarely differed between substrate treatments. It was also noticed that in *Camellia* and *Ilex cornuta*, little root exploration occurred in the bottom perlite layer (**Figs. 6C and 6D**). Regardless of root exploration, the lower portions of the plugs often disintegrated during the data collection process, an undesirable quality during transplanting. It is possible that excessive air space and insufficient water holding capacity were present in the lower half of the cells to provide a buffer against the especially hot and dry climate conditions experienced in 2023. Testing this method in subsequent years or stratification with a material having less drastic differences in water holding capacity and AS is needed.

Growcoon Experiment. Rooting success was similar in *Callistemon* and higher by 6% in all other species tested with Growcoons compared to loose-fill standard mix (Table 8). Root quality was reduced in all species with the Growcoons, but only statistically significant in *Rhododendron* and *Itea virginica*. Plug quality was statistically and visually similar between the Growcoon and loose fill standard mix in all species except *I. virginica*, where plugs had defoliated (**Fig. 7**). It was noted that the Growcoon treatments rooted faster and more densely than their standard mix counterparts, and it is believed that this higher root mass and further root development caused the plugs to have higher irrigation requirements, especially during the 2023 heat wave. Both treatments were irrigated similarly, and the Growcoon treatments were likely more susceptible to the effects of the 2023 heat wave, leading to increased root browning due to possible drying. Use of Growcoons should take this faster and denser root development into account in irrigation practices and production time.

Table 7. Rooting success, root quality, plug quality, average root length, and root dry mass of several woody species in standard (non-stratified) and stratified substrate.

Rooting Success	Non-Stratified	Stratified
Callicarpa americana	83%	67%
Callistemon 'Woodlander's Hardy'	61%	56%
Camellia 'Reverend Ida'	83%	22%
Ilex cornuta	83%	22%
Ilex crenata	50%	56%
Ilex glabra	50%	28%
Itea virginica 'Henry's Garnet'	44%	17%
Root Quality	**Non-Stratified**	**Stratified**
Callicarpa americana	4.2	3.0
Callistemon 'Woodlander's Hardy'	4.6	3.0*
Camellia 'Reverend Ida'	5.0	2.8*
Ilex cornuta	5.0	3.3*
Ilex crenata	4.4	5.0
Ilex glabra	3.2	2.8
Itea virginica 'Henry's Garnet'	3.0	3.0
Plug Quality	**Non-Stratified**	**Stratified**
Callicarpa americana	2.6	3.2
Callistemon 'Woodlander's Hardy'	4.8	4.8
Camellia 'Reverend Ida'	5.0	4.8
Ilex cornuta	5.0	5.0
Ilex crenata	5.0	4.6
Ilex glabra	5.0	5.0
Itea virginica 'Henry's Garnet'	3.6	2.0*
Average Root Length (cm)	**Non-Stratified**	**Stratified**
Callicarpa americana	7.45	8.45
Callistemon 'Woodlander's Hardy'	5.10	4.55
Camellia 'Reverend Ida'	5.20	3.31*
Ilex cornuta	5.00	1.88*
Ilex crenata	4.75	4.05
Ilex glabra	6.35	6.00
Itea virginica 'Henry's Garnet'	3.55	2.42
Root Dry Mass (g)	**Non-Stratified**	**Stratified**
Callicarpa americana	0.062	0.070
Callistemon 'Woodlander's Hardy'	0.036	0.010*
Camellia 'Reverend Ida'	0.070	0.043
Ilex cornuta	0.070	0.020*
Ilex crenata	0.032	0.026
Ilex glabra	0.034	0.028
Itea virginica 'Henry's Garnet'	0.014	0.013

Quality, root length, and mass data represent least-square means of five replicates. Means were compared using a 1:1 t-test at $\alpha = 0.05$, with significantly different means denoted with an asterisk (*).

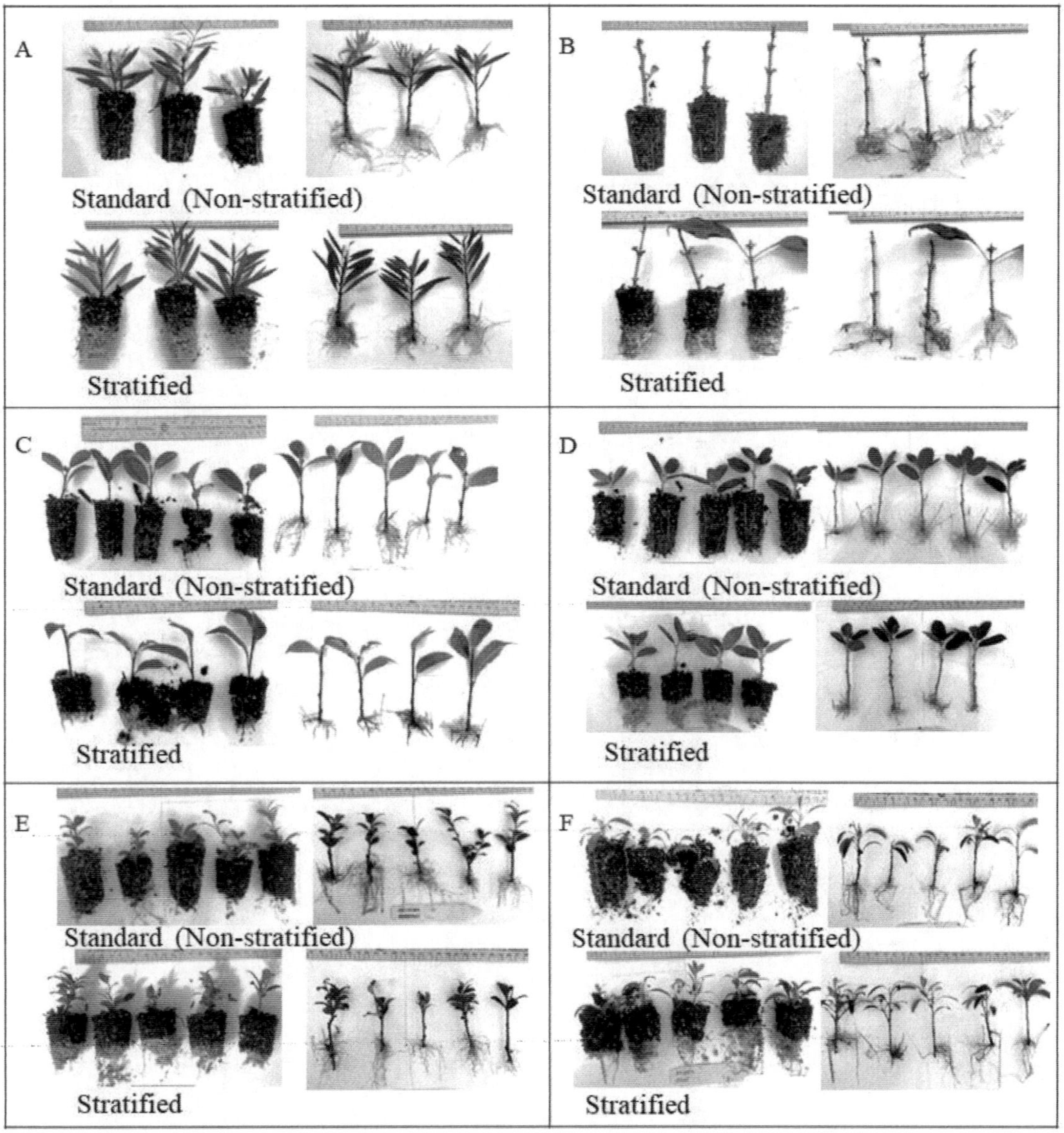

Figure 6. Plugs (left) and rooted cuttings washed of substrate (right) grown in standard (non-stratified) (top) and stratified (bottom) substrate in A) *Callicarpa americana*, B) *Callistemon* 'Woodlander's Hardy,' C) *Camellia* 'Reverend Ida,' D) *Ilex cornuta*, E) *Ilex crenata*, and F) *Ilex glabra*.

Wood Fiber Experiment. *Rhododendron indicum* cuttings died during the 2023 heat wave and this data was not included in this study. In *Callistemon*, rooting success was higher using the wood fiber blend substrate than the standard substrate by 11% (**Table 9**). Root quality and average root length were slightly lower and plug quality was slightly higher with wood fiber incorporation (**Fig. 8**); however, these figures were not statistically significant. The similar quality and rooting success in *Callistemon* propagated in wood fiber shows promise in using this material as an alternative to peat moss in propagation substrates.

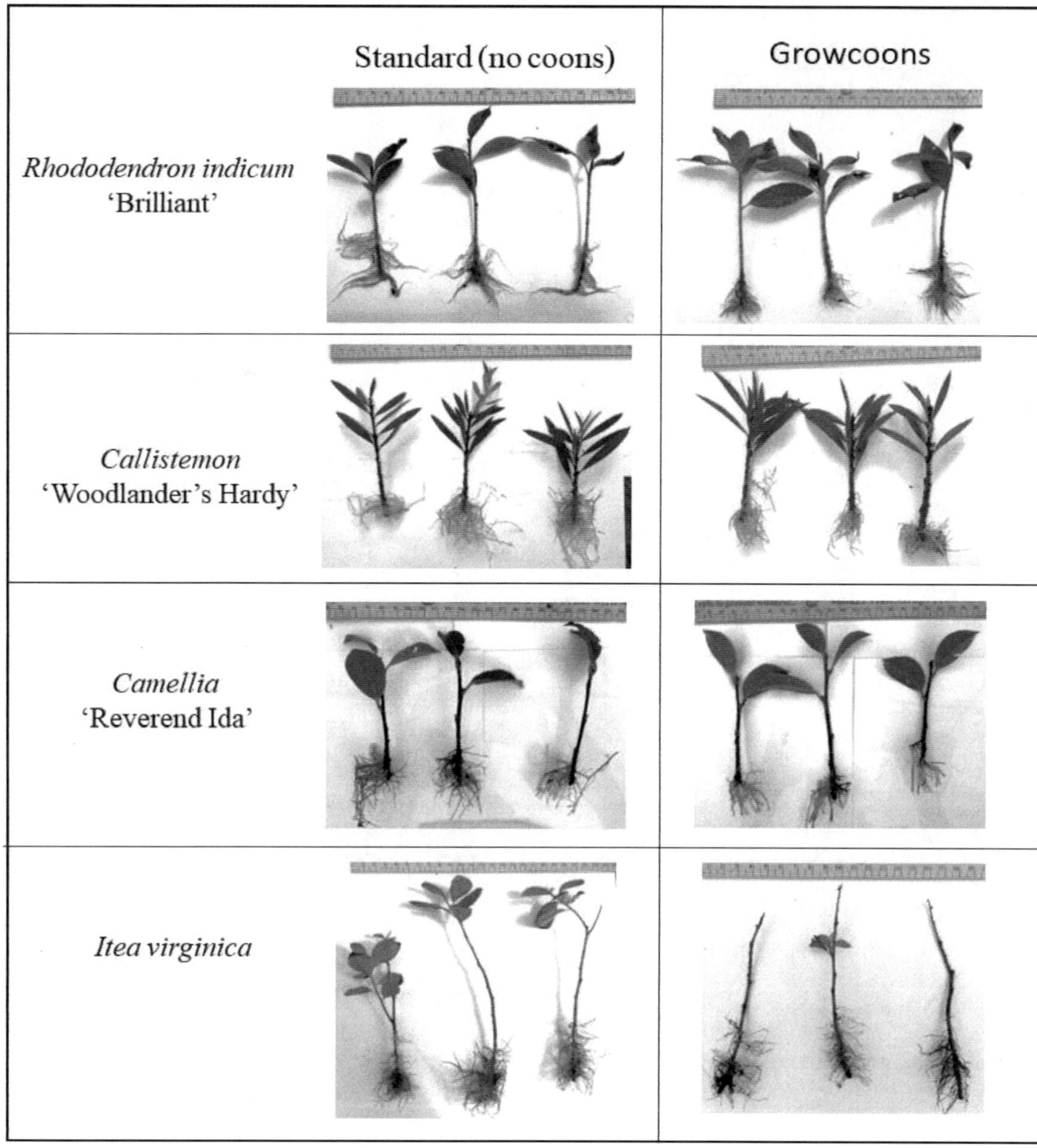

Figure 7. Rooted cuttings grown in loose-fill standard substrate (left) and in Growcoons filled with the standard substrate (right) with substrate and Growcoons removed.

Table 8. Rooting success, root quality, and plug quality of several woody species in standard (loose-fill) propagation mix and Growcoons.

Rooting Success	Standard	Growcoon
Camellia 'Reverend Ida'	44%	50%
Callistemon 'Woodlander's Hardy'	61%	61%
Itea virginica 'Henry's Garnet'	94%	100%
Rhododendron indicum 'Briliant'	44%	50%
Root Quality	**Standard**	**Growcoon**
Camellia 'Reverend Ida'	4.8	4.4
Callistemon 'Woodlander's Hardy'	4.8	4.4
Itea virginica 'Henry's Garnet'	3.2	2.6
Rhododendron indicum 'Briliant'	4.4	3.2*
Plug Quality	**Standard**	**Growcoon**
Camellia 'Reverend Ida'	5	5
Callistemon 'Woodlander's Hardy'	5	5
Itea virginica 'Henry's Garnet'	3.8	2.2*
Rhododendron indicum 'Briliant'	3.8	4.2

Quality data represent least-square means of five replicates. Means were compared using a 1:1 t-test at $\alpha = 0.05$, with significantly different means denoted with an asterisk (*).

Table 9. Rooting success, root quality, plug quality, and average root length in *Callistemon* 'Woodlander's Hardy' in standard propagation mix and 1:1 wood fiber and bark fines.

Callistemon 'Woodlander's Hardy'	Standard	Wood Blend
Rooting Success	61%	72%
Root Quality	4.7	4.4
Plug Quality	4.9	5.0
Average Root Length	5.4 cm	4.3 cm

Quality data and average root length represent least-square means of five replicates. Means were compared using a 1:1 t-test at $\alpha = 0.05$, with significantly different means denoted with an asterisk (*) (no significant differences found in this treatment).

Figure 8. Rooted *Callistemon* cuttings grown in loose-fill standard substrate (top) and in wood fiber blend (bottom) with the substrate intact (left) and removed (right).

CONCLUSION

The study's objective was to ameliorate the effects of excessive substrate moisture content often resulting from frequent misting associated with extreme heat in propagation environments. Substrate stratification, Growcoons, and incorporation of wood fiber all increased the substrate air space compared to a 2:1:1 bark fines: peat: perlite standard propagation mix. Stratification with 100% perlite did not offer any rooting benefits and should be further investigated with different materials. Growcoons improved rooting success and shortened rooting time in most species trialed, but quality results need further trials to hone management techniques for improved quality. A promising result is the improved rooting success and similar root and plug quality of *Callistemon* when rooted in a wood fiber and bark blend. These results indicate the suitability of wood fiber in reducing peat and perlite usage in propagation with no adverse effects on rooting percentage or resulting liner quality. Further trials with wood fiber in propagation of other woody species is therefore another favorable next step, as well as repeated trials assessing all these methods under more normal weather conditions.

ACKNOWLEDGEMENTS

The authors would like to recognize and thank the State of Louisiana Board of Regents Support Fund (AWS-005267) and the Klasmann-Deilmann Company, Germany for support of this research.

LITERATURE CITED

Criscione, K.S. Fields, J.S., and Owen, J.S. (2021). Exploring water movement through stratified substrates. Comb. Proc. Intl. Plant Prop. Soc. 71:116-124. https://sna.ipps.org/uploads/docs/6bstudent4_kcriscione2021.pdf.

Criscione, K.S. and Fields, J.S. (2023). Stratified substrates can reduce peat use and improve root productivity in container crop production. HortScience *58*(3):364-372. https://doi.org/10.21273/HORTSCI17019-22.

Dickson, R.W., Helms, K.M., Jackson, B.E., Machesney, L.M., and Lee, J.A. (2022). Evaluation of peat blended with pine wood components for effects on substrate physical properties, nitrogen immobilization, and growth of petunia (*Petunia x hybrida* Vilm.-Andr.). HortScience *57*:304-311.

Dirr, M.A. and Heuser, C.W. (2006). The Reference Manual of Woody Plant Propagation. Timber Press, Portland, OR.

Fonteno, W.C., Hardin, C.T., and Brewster, J.P. (1995). Procedures for determining physical properties of horticultural substrates using the NCSU Porometer. Horticultural Substrates Laboratory, North Carolina State University.

Harris, C.N., Dickson, R.W., Fisher, P.R., Jackson, B.E., and Poleatewich, A.M. (2020). Evaluating peat substrates amended with pine wood fiber for nitrogen immobilization and effects on plant performance with container-grown petunia. HortTechnology *30*:107–116. https://doi.org/10.21273/HORTTECH04526-19.

Jackson, B.E. (2018). Substrates on trial: wood fiber in the spotlight. 7 November 2022. <https://www.greenhousemag.com/article/substrates-on-trial-wood-fiber-in-the-spotlight/>.

Milks, R.R., Fonteno, W.C., and Larson, R.A. (1989). Hydrology of horticultural substrates: III. Predicting air and water content of limited volume plug cells J. Amer. Soc. Hort. Sci. *114*: 57-61. https://journals.ashs.org/jashs/view/journals/jashs/114/1/article-p57.xml?tab_body=pdf

Sanders, B. (2022). Alternatives to loose-fill media for improved plug handling. Com. Proc. Intl. Plant Prop. Soc. *72*:179-190. https://sna.ipps.org/uploads/docs/ipps_proceedings_vol_72_2022.pdf.

Yafuso, E.J., Fisher, P.R., Bohorquez, A.C., and Altland, J.E. (2019). Water and air relations in propagation substrates. HortScience *54*(11):2024-2030. https://journals.ashs.org/hortsci/view/journals/hortsci/54/11/article-p2024.xml?rskey=nwm51R&result=1#B31.

Combined Proceedings IPPS 73:338-348. 2023.

IPPS European Exchange 2022

Erika Ramos

J Berry Nursery, 201 Private Road 5180, Grand Saline, Texas 75140, USA

erika@jberrynursery.com

Keywords: Early-Career IPPS Southern and European Regions' Exchange Program

Summary

The International Plant Propagators Society-Southern Region (IPPS-SR) and IPPS-European Region support a joint Early-Career IPPS Exchange Program – reciprocally hosting and exposing new professionals to the Green Industry in their region – and attending their annual conference. Erika Ramos of the IPPS-SR was the 2022 professional awardee – and reports on her experience touring and learning about the European Nursery Industry.

INTRODUCTION

In 2022, I was lucky enough to have been selected as the IPPS-SR International Representative to the Early-Career IPPS-SR-European Exchange Program. It has been one of the most rewarding experiences of my life! It gave me the opportunity to meet some wonderful people and visit some amazing places. Horticulture was not a primary focus of mine while in college, but ironically, my very first job out of college was at J Berry Nursery. Organizations like IPPS and the opportunities it provides has helped my career - and continue helping me learn and expand on my knowledge within the industry. It was neat being able to go on this trip because although horticulture is very similar across the world, the methods plants are grown and the ways they are applied in other landscapes are very different from region to region. So being able to experience this firsthand is something I will never forget.

The IPPS Europe Exchange program lasts 10-14 days and attendees can expect to be hosted by different members of IPPS Europe. While staying at each member's house, you will be shown nurseries, garden centers, or gardens in their area. In 2022, IPPS Europe took place in Bad Zwischenahn, Germany.

My journey began in London, on 5 October 2022 where I was picked up by my first host, Tim Lawrance. Tim is also the Chair of the IPPS International Board. We took a short drive just south of London to his home in Chichester, where I met his wife, Annette. After resting for a bit, we took a walk to Bosham where we saw some beautiful scenery while the tide came in for the night (**Fig. 1**).

After our walk, we stopped for a short rest at the cutest English pub, which just so happened to be my very first. We finished the first day by having a delicious homemade dinner prepared by Annette. On my second day with Tim, we visited a couple of nurseries- Hills Nursery and Walberton Nursery.

Figure 1. A dock on the Chichester Channel.

ENGLAND

Hills Nursery. Hills is a 7-acre family owned nursery that was started in 1920. Now on their 4th generation, Hills has a very good understanding of the UK market, and it shows. Their facilities were impeccably clean and organized. They use the rail system in their greenhouses to move whole sections at a time, making them very efficient. Hills is also the first nursery I had ever seen with flood irrigation, so I found it very interesting (**Fig. 2**). After Hills we drove over to Walberton which was not too far.

Walberton Nursery. Walberton Nursery was established in 1973 and is one of four nurseries that compose The Fairplants Group. Tim, having just retired from working there a few years ago, was the one showing me around as he had a very good understanding of the processes and location.

Figure 2. Floor irrigation and belt system and Hills Nursery.

Two things stood out for me at Walberton, the first was the wind barriers that had been placed across different sections of the nursery (**Fig. 3**). I had never seen that before and I found it pretty cool. Secondly, I saw that many greenhouses had a green shade cloth over the top of the poly, as opposed to the black cloth I am used to seeing (**Fig. 4**). After inquiring about using green instead of black, I was intrigued to find out it was to simulate the shade of a tree.

Figure 3. Wind barrier at Walberton.

After a long morning of driving across Chichester and lots of walking we returned to the house to have a light lunch and afternoon tea. Tim showed me his back yard which was delightful and very comfortable with its assortment of flowering plants, trees and even some veggies. After spending some time outside, I retreated to my room and took a nap for a couple of hours. I was really feeling jetlagged at this point. After the refreshing nap, we walked to downtown Chichester for dinner, where we had some authentic Indian food with some friends of Tims.

Chatsworth House. The next day it was time for me to leave. Tim dropped me off at the local train station and I was very nervous about this since I had never traveled by train before. It turned out great: the countryside was beautiful, peaceful and the ride seemed all too short. My next host was Vicky Endersby in Sheffield where I spent two days. That afternoon we took a walk and visited Sheffield's Winter Gardens which were beautiful and provided much contrast to the outside greenery. After visiting the gardens, we had a delicious dinner at The Botanist.

Figure 4. Green shade on some of the greenhouses at Walberton to mimic natural tree shade.

On my second day with Vicky, we drove to Chatsworth House and spent the majority of the day there. Upon arriving, Vicky surprised me with reservations for afternoon tea at Chatsworth. After experiencing a proper four-course English afternoon tea, I'd have to say I wish it was something I could do on a daily basis. It was so much fun (**Fig. 5a**)!

Figure 5. a.) Erika Ramos at formal British afternoon tea. **b.)** Gardens on the trail around Chatsworth House. **c.)** Yew maze in the center of the trail.

Next, we walked the Chatsworth gardens, which were so beautiful, tidy, and wild at the same time (**Fig. 5b**). We had a little fun and completed the yew maze located on the grounds (**Fig. 5c**), and finally finished off in the greenhouses which housed their tropical plants. Tired as we were that evening, Vicky asked me if I would like to go to a light show that night, and I could not say no. Matlock Bath Illuminations was a lot of fun - the illuminated and decorated boats along with the fireworks where a good way to end a very busy day.

Osberton Nursery. The next day was time to go. Before dropping me off at the train station, we stopped at Osberton Nursery, where Vicky works. Their set-up was one I was familiar with, and similar to what I work with. I did like how they set their pots in the ground which is not something I had seen before, but would become a common site in the tours to come (**Fig.6**).

Figure 6. Pots being set in the ground at Osberton.

The train ride to Richards, was the longest yet. I headed back south to Sevenoaks where Richard, my last and final host lives. He and his family welcomed me into their home. After resting for a bit and having tea, we went for a walk on a trail near their house. Walking though the greenery was beautiful and I have never seen so many mushrooms on a trail before in my life!

ROADTRIP TO IPPS

On Oct 10, my fifth day of the trip we were up and off a couple of hours before dawn on our two day road trip to Germany. Two colleges of Richards accompanied us, so we had a full car. Our trip would take us under the English Channel to France then to Belgium, Netherlands, and finally Germany.

Solitair. The first stop of our roadtrip was at Solitair in Belgium. Solitair specializes in large tree production. Their property is huge, with row after row of beautiful and large decorative trees. Every tree was manicured to perfection and we could clearly see its full potential (**Fig. 7**). Many of the trees are decades old and are just waiting for the right person to come and hand select their favorites to add to their own landscapes. Hands-down, Solitair was my favorite nursery.

M. van den Oever. Our second stop was at M. van den Oever, in the Netherlands. They are a container nursery that produces deciduous trees, fruit trees, and a variety of perennials. They had a very efficient shipping facility that was running when we passed through. Their incorporation of belts, watering system, and wrapping machine made their process flow very efficiently. After M.van den Oever, it was time to call it a day. We had delicious dinner and spent the night at a hotel in Hengelo, Netherlands.

Figure 7. Some of the trees at Solitair.

Heinie. On the 6th day we had an early start again as we were headed to Heinie in Germany. They are a container and in-ground nursery with a wide variety of trees and perennials. A few things stood out for me: the first was that throughout their nursery they used retainers that the pot would sit in to prevent them from falling over due to wind. Each one of these wind containers were on the edges of each growing pad (**Fig. 8**). This is a common issue in nurseries back home and it looked like this method could prove to be an easy fix. The second thing that stood out to me was that many of the growing pads had permanent spraying booms installed. This allows their spraying to be very efficient.

Figure 8. Retainers used to keep edge-row plants from falling over at Heinie.

Renke zur Mühlen. Our next stop was a Renke zur Mühlen, in Bad Zwischenahn, Germany as a pre-conference tour. Here I met many other members of IPPS. Their facility was very neat and tidy - but what was really interesting to me was their propagation house. They had a no-mist propagation house, which I am not very familiar with (**Fig. 9**). After the conference I realized that it is a common method in the area.

Next, we took a short ride to the hotel where the conference was going to take place. HansenS Haus am Meer was a very cute hotel on the water of Zwischenahner Meer in Bad Zwischenahn. After settling into our rooms, we had dinner at a local restaurant where I met the rest of the IPPS committee and the '6-packers'. The 6-packer program is designed to help anyone new to the horticulture industry. They help throughout the conference and at the end one of them gets chosen to be the following years exchange representative to IPPS-SR.

Figure 9. No-mist propagation at Renke zur Mühlen.

The conference took place over three days. The first two days where presentations in the morning with tours of local nurseries, gardens, and other facilities in the afternoon. The third day was presentations in the morning with a departure time after lunch.

Zu Jeddeloh Pflanzen. The first nursery stop of the conference was at Zu Jeddeloh. They are a container nursery with some very neat and organized facilities. Their plants were impeccable. I saw the permanent booms on individual growing beds here again, and this made me think that it might be a common practice. The thing that impressed me the most though was their shipping facility. Everything is about efficiency and the rolling tables and belts make the process flawless (**Fig. 10**).

Figure 10. Shipping area at Zu Jeddeloh Pflanzen.

Bruns Pflsnzen. Our last stop of the day was at Bruns. Bruns is a large specialty tree nursery, and like Solitair - they had the most amazing trees. We saw trees destined for Disney and Mercedes-Benz, and we even got to see some special orders (**Fig. 11**). After seeing some stunning trees we headed to the Bruns Rhododendron park where the family has developed a huge park primarily composed of rhododendrons that are decades old. In the center, there is a beautiful event center overlooking the pond in the middle of the park. We had dinner here while a few people gave presentations.

Figure 11. Specialty trees at Bruns.

Jens Meyer Jungpflanzen. Our first stop of the second conference day was at Jens Meyer Jungpflanzen. Jens Meyer is a young plants nursery that does the majority of their own production.

Their propagation houses are also no mist houses, but they do have some portable mist sprinklers that can be moved around when needed (**Fig. 12**). They had a very impressive automatic planting machine that would plant liners and set them on a platform to be taken out to the field.

Klasmann K Deilmann. The final stop of the conference was at Klasmann K Deilmann; a media mixing facility. They did not allow any cameras in their facilities, but they are very clean and organized. It is obvious that media sterilization is a priority for them. We saw how different media is mixed, bagged, and palletized. It was fascinating to see how the process that we normally do not think of occurs. They talked a bit about the issues they are having, specifically with the peat shortages happening throughout Europe.

After the tours we went back to the hotel to freshen-up and prepare for the conference dinner and auction. The dinner was delicious, and the auction was a complete success! I even got to go home with a couple of souvenirs. The next morning was the last day of presentations, but we had to leave a little early since we were going to make the full drive to Richard's house in one day. Additionally, we were going to try and make two more nursery stops on the way!

Figure 12. Propagation house at Jens Meyer.

Rötjes Young Pants. We made it to Rötjes in the Netherlands by late afternoon. We were lucky that they let us take a tour even though they were about to close. They showed us their propagation stations, mist houses, and storage areas. Rötjes has the most pristine propagation houses I have ever seen. The facilities consisted of acres of glass greenhouses. There were liners in different stages of rooting with some covered like the no-mist nurseries, and some of it was under a portable boom mist system that would roll over the newly stuck liners as needed.

Euro Tree. Down the road form Rötjes was Eruo Tree, where Richard had a friend that was happy to give us a quick tour before the day ended. Euro Tree is primarily a container tree nursery with a wide range of varieties. We saw many of the same type of specialty trees that we saw in other nurseries. When we left Euro Tree, it was dark and we had to stop at a nearby restaurant to have a quick bite to eat. Once we were done, we drove though until we made it back to Richards house a little after midnight.

Figure 13. Assortment of plants at Provender Nurseries.

Provender Nurseries. The day after getting back, we had a late breakfast and then Richard took me to his workplace at Provender Nurseries. Provender primarily sells to landscaping companies who are part of their system. They have a huge selection of plants for all types of landscapes. Everything looked so good and ready to go (**Fig. 13**).

After getting back, Richard and his family took me to a local Mote. Ightham Mote is a 14th-century manor house, that is beautifully preserved and the gardens around the home are just stunning (**Fig. 14**). We toured the home and saw some great English architecture. It was a wonderful way to end a great conference.

The next day I took a train ride back to London but was not ready to go home quite yet. In London, I spent an additional three days of my own time exploring and just being a tourist. I went to the palaces, the National Gallery, the London Eye - Europe's tallest cantilevered observation wheel, ate some very good food, and even attended a showing of The Phantom of the Opera!

Figure 14. Ightam Mote.

CONCLUSION

After reading this, you may think that this trip was exhausting with all the places we visited. And you would not be wrong! To be honest, I truly don't know how we fit so many stops in such a short amount of time, but I am forever grateful that we did. I am extremely thankful for my generous host families for welcoming me into their homes. The time they took to be personal tour guides in showing me around many nurseries will be something I will never forget. Finally, I would like to say a special thank you to IPPS-SR for providing me with the opportunity of a lifetime. This is something I never dreamed I would be able to partake in and I highly encourage others to apply to have the same opportunity I was afforded.

Combined Proceedings IPPS 73:349-363. 2023.

The Adventures, Challenges and Rewards of Propagating Rare and Unusual Perennials

Aaron Selby

Juniper Level Botanic Garden and Plant Delights Nursery, 9241 Sauls Road Raleigh, NC 27603, USA

Aaron@plantdelights.com

Keywords: gardens, nothospecies, propagation techniques, offsets, scooping

Summary

Over the past twenty years, I have participated in the production of countless perennials at Juniper Level Botanic Gardens Juniper Level Botanic Garden and Plant Delights Nursery Plant Delights Nursery,. This is a 28-acre ex situ conservation and research institution that also includes over 10-acres of gardens. During my time at Plant Delights Nursery and Juniper Level Botanical Gardens, an estimated 60,000 new garden accessions have been introduced and annually the collection increases by an average of 2,000 plants.

IPPS Vol. 73 – 2023

INTRODUCTION

With over 27,000 different active records in the garden, we are always on an "Adventure" in the world of propagation. Often this leads us into many propagation "Challenges," but also presents "Rewards" in many forms. Our specialty collection consists of 44 genera, 1,073 species, 162 nothospecies (a hybrid that is formed by the direct hybridization of two species, not other hybrids), 13,274 different clones and a cold hardy fern collection that boast roughly 400 species and 783 clones. It is often difficult for propagators to be aggressive with such rarities. But sometimes you just have to go for it (**Figs. 1, 2 and 3**).

Figure 1. (top, left) Sunrise over Juniper Level Botanic Garden (JLBG) Trillium Trial beds, (top, right) *Yucca filamentosa* flower spike node, (bottom, left) , *Agave x nickelima* (*asperimma x nickelsiae*) cultivar, and (bottom, right) Begonia 'Cotes de Castillon'.

Figure 2. (top, left) *Trillium discolor*, (top, right) *Agave ovatifolia*, (bottom, left) *Hosta* 'Cathedral Windows', and (bottom, right) *Lycoris sprengeri* 'Pink Floyd'.

Figure 3. (top, left) Scooping Agave meristem with a power drill, (top, left) rhizome cutting of *Alocasia* 'Architexture', (bottom, left) leaf cutting of Amorphophallus x kachjac (*konjac x kachinensis*), and (bottom, right) large *Agave ovatifolia* 'Flipper' prepped for scooping (arrow).

With so many clones to play with, we are often dipping into uncharted waters. Many times, we may be the original source for things and many times the benefactor of such rarities. Some of the adventures dip into mutilating extremely rare plants such as *Musella lasiocarpa* var. rubribracteata (**Fig. 4**). Before we potentially killed this rare specimen, I crosscut the straight species *lasiocarpa*. I was confident that this would work rather well and most likely in a short amount of time. As expected, the plant produced offsets and divisions within a few weeks. Since we noticed this behavior, we hoped the results would be mirrored in our new variety *rubribracteata*. When I was given this plant to produce, it was only one of two specimens in the U.S. We went ahead and used a cross cutting technique to destroy the apical meristem and forced multiple offsets to break. With this technique you cut through the basal plate making the apical meristem the bullseye of the cut. To our surprise, this variety was rather slow to offset.

Figure 4. (top, left) *Musella lasiocarpa* var. rubrobracteata (in-situ), (top, right) *Musella lasiocarpa* [producing offsets with new, smaller shoots (arrow) - after cross-cutting the meristem], (bottom, left) *Musella lasiocarpa* var. rubrobracteata [producing offsets (arrow) after cross-cutting meristem], and (bottom, right) *Musella lasiocarpa* var. rubrobracteata [macro of offsets, (arrows)] bud development after cross-cutting meristem.

Other times we may be collecting rare and unusual plants from a wild population, and instead of consuming a whole plant we may be able to get a leaf cutting or small portion of the plant to reproduce, study and share. We have done this with one of our rare riparian natives *Helonias bullata*. We found that you can asexually propagate *Helonias* via leaf cuttings and when those bulk up after a couple of years, perform a crosscut technique through the basal plate **(Fig. 5).** What this does is to break apical dominance by reducing levels of the phytohormone, auxin, from the shoot apex - thus allowing a higher cytokinin/auxin ratio – stimulating dormant lateral buds - creating plantlets. We then go through those offsetting plants and divide them into individual crowns.

Figure 5. Offsetting with new shoots developing from a cross-cut (scooping) of *Helonias bullata*.

Cross-cutting and or scooping are techniques we use to isolate random leaf patterning in plants such as Agave (**Figs. 6 and 7).** We often find streaked mutations in plants that we then isolate through this technique. At the base of each leaf is a clone of the leaf it is attached to. We also use this technique to get a variegated plant to offset and clone its phenotype. We have done this with hundreds of agave specimen, but one of the first was to take Agave 'Craziness' - and stabilize it to become Agave 'Bareback Rider'. We will also isolate new sports of mutating plants like *Agave filifera* 'Golden Sword' and force clonal offsets on non-offsetting species like *Agave ovatifolia* 'Flipper'.

Figure 6. (Agave sport isolation). (top, left) Agave 'Craziness' (*cupreata x asperrima*) (unstable streaked seedling), (top, right) Agave 'Bareback Rider (isolated and stabilized sport of 'Craziness'), (bottom, left) *Agave filifera* 'Golden Sword' (1/2 gold mutation), and (bottom, right) *Agave filifera* 'Golden Sword' (offsetting gold sport induced by scooping).

Figure 7. (top, left) *Agave ovatifolia* 'Flipper' core (only specimen at time of scooping), (top, right) *Agave ovatifolia* 'Flipper' (freshly scooped, arrow), (bottom, left) *Agave ovatifolia* 'Flipper' freshly stuck 3/4-in. offset, and (bottom, right) *Agave ovatifolia* 'Flipper' [freshly rooted (arrow) from the base of 3/4-in offset).

Some of our adventures includes plants such as *Bambusa multiplex* (**Fig. 8).** This is a more difficult-to-root species – but after a mild winter, and nice humid summer, the plant develops plantlets along the cane that we propagate from cuttings.

No rooting hormone/auxin is needed: just place it in perlite under mist. The cuttings typically start rooting within 20 days with a strike rate of nearly 100%. Once rooted, we simply pot these specimens up, bulk their size and then resume pot divisions once they mature.

Figure 8. (top, left) *Bambusa multiplex* 'Green Giant' (prior to cutting), (top, right*)* *Bambusa multiplex* 'Green Giant' (initial stage of adventitious roots on branch nodes), (bottom, left) *Bambusa multiplex* 'Green Giant' (cuttings stuck in perlite and under mist), and (bottom, right) *Bambusa multiplex* 'Green Giant' [6-day strike (rooting) and image of adventitious roots (arrow) at 12 days).

We are always getting new and interesting *Alocasia* and banana species in (**Fig. 9**). These are often rare, variegated and marginally cold hardy forms. When we need more, we simply use to rhizome cuttings. We will cut a rhizome (specialized shoot) that has developed into a "sub-trunk" with multiple eyes. We cut rhizomes somewhere between the meristem and the first set of roots. We then treat the cut portion with auxin (Hormodin 3) and stick it like a normal cutting into a pre-watered container. Sparingly water these plants for the next 2-3 weeks before they start to root into the pot.

Most adult *Alocasia* tubers will produce multiple offsets as well. Many times we take slowly offsetting forms of banana plants the same way. We bareroot the plant and examine the main rhizome for vegetative eyes. Once we confirm the eyes are developed enough, and there is enough rhizome to establish a new plant with, we cut the rhizome. We have the best results doing this in our heated propagation house with bottom heat from October to May.

Figure 9. (top, left) *Alocasia* 'Architexture' (bare rooted rhizome prepped and examined for cutting), (top, right) *Alocasia* 'Architexture' (rhizome cutting with developed eyes both above and below cut), (bottom, left) *Alocasia* 'Architexture' (Hormodin 3 applied to fresh cutting), and bottom, right) *Alocasia* 'Architexture' (freshly stuck cutting placed in Pacific Organic potting soil).

Over the years we have adventured into doing leaf cuttings on aroids, but particularly the genus *Amorphophallus* (**Fig. 10**). Many of the species in this genus are difficult to root, and others not. However, some of the nothospecies that have two difficult species are now rooting with ease. We have tried to root leaf cuttings of the species *A. konjac*, along with many of its cultivars and forms. After sticking hundreds of different cuttings, we found that it simply does not root. But when this species was crossed with another enigmatic species *A. kachinensis*, a highly textural plant was formed. We took a leaf cutting and it has now produced multiple tubers along the petiole.

Figure 10. (top, left) *Amorphophallus x kachjac (konjac x kachinensis)* [in-situ at Juniper Level Botanic Garden (JLBG)], (top, right) *Amorphophallus x kachjac (konjac x kachinensis)* (leaf/petiole cutting), (bottom, left) *Amorphophallus x kachjac (konjac x kachinensis)* (rooted leaf/petiole cutting), and (bottom, right) *Amorphophallus x kachjac (konjac x kachinensis)* [tuber development (arrow) along cutting).

Along "adventures" come the "challenges." Many of the plants we want to introduce simply do not root with any good percentages, if even at all. Many of these challenges come out of our breeding program for *Baptisia* (**Fig. 11**). Many cuttings root with exceptional strike rates and then there are those that do not. Unfortunately, sought after clones like *Baptisia perfoliata* 'Flying Saucers' are super difficult to root from the garden. With the addition of a large cooler, we force plants into dormancy in August by keeping them at 32° F for 90 days, remove and then place them on bottom heat in a heated greenhouse that is kept at a high of 75°F and a low of 58°F. We force the plants and after flowers are pinched, we take tip cuttings. Since this change, we now have an average strike (rooting) rate of 75% and the rate of perennialization has increased by 90%.

Figure 11. (left) *Baptisia perfoliata* 'Flying Saucers' (in-situ at JLBG), and (right) failed *Baptisia* cuttings during propagation.

We find propagation challenges with other taxa like *Castanopsis cuspidata* 'Nakafu' or even *Ilex chinensis* 'Cherry Ice' (**Fig. 12**). We have stuck every cutting imaginable, used every hormone known, and even tried during multiple seasons. At best you end up with a couple here or there. We will often take the rooted clones and use them as mother plants for future cuttings.

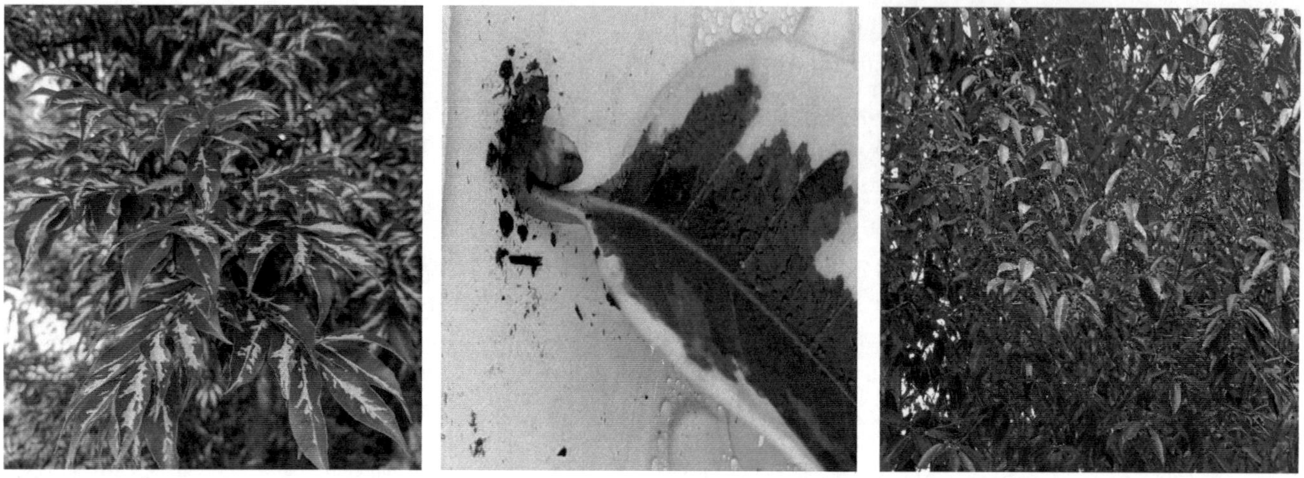

Figure 12. (top, left) *Castanopsis cuspidata* 'Nakafu' (in-situ at JLBG), (top, right) *Castanopsis cuspidata* 'Nakafu' (experimental bud cutting with Dyna-Gro IBA/NAA gel), and (bottom) *Ilex chinensis* 'Cherry Ice', which is difficult to clone.

Often the challenge may be to produce difficult and slowly offsetting plants like *Musa* 'Ae Ae'. (**Fig. 13**). We received our original clone in 2000 and our nursery mother plant in 2007. Since then, we have only been able to produce a few dozen plants. The demonstrated plant was a supreme specimen with a market value of several hundred dollars. When we find ourselves handling these rare and expensive clones, we aggressively process them in the manner I mentioned above for elephant ears and bananas. This rhizome cutting is placed on bottom heat through the cool months and kept at a temperature range of 75°F to 58°F and through the summer months as a normal cutting. The success rate drops through the heat of the summer – so I encourage aggressive techniques to be done during the cool season. We have also trialled crosscut techniques with the use of the plant growth regulator/cytokinin 6-BA. "Configure" to encourage offsetting. Overall, the process is slow and this plant will be expensive.

CONCLUSIONS

What I find to be the most satisfying is that there are so many different "Rewards." One of the most unique rewards I have ever been granted was when one of our 2009 leaf cuttings of *Amorphophallus titanum*, opened its bloom on its 9th birthday. This was one of three tuber offsets and the first of two to bloom (**Figs. 14 and 15**). Since then, we have flowered another *titanum* in the summer of 2023.

Other huge satisfactions are learning from peers, hobbyists, and colleagues that have become such an intricate part of my career. I have learned techniques like repotting cut *Sarracenia rhizome* that look good-as-dead - and resprouting them under mist. It may be the difficult plant to root, or the breakthrough technique you stumble across - but more often than not, it is just simply showing up to such a beautiful world class collection of plants to play with.

Figure 13. (top, left) *Musa x Paradisicaca* 'Ae Ae' variegated banana leaf, (top,right), (top, right) *Musa x Paradisicaca* 'Ae Ae' (bare rooted rhizome prepped and examined for cutting), (bottom, left) *Musa x Paradisicaca* 'Ae Ae' [verifying developed eyes (arrow) before cutting the rhizome], and (bottom, right) *Musa x Paradisicaca* 'Ae Ae' (non-rooted rhizome cutting above developed eyes).

Figure 14. (top, left) *Amorphophallus titanum* 'Peter Grande', (top, right) *Sarracenia* rhizome division w/ no growth, (bottom, left) foliage re-flush of *Sarracenia* rhizome divisions placed in a communal pot), and (bottom, right) *Amorphophallus sp*.

Figure 15. (top, left) *Narcissus sp.* (in-situ at JLBG), (top, right) *Opuntia santa-rita* 'Baby Rita' (in-situ at JLBG), (bottom, left) *Cylindropuntia x multigeniculata* (in-situ at JLBG), and (bottom, right) *Baptisia x* OP (un-named open pollinated trial plant).

Combined Proceedings IPPS 73:364-375. 2023.

Cornus, Benthamidia, Dendrobenthamia, and *Swida*: Oh My – Making Taxonomy Less Taxing

Qiu-Yun Jenny Xiang

Department of Plant and Microbial Biology, Gardner Hall 2115, North Carolina State University, Raleigh, NC 27695-7612, USA

Jenny_xiang@ncsu.edu

Keywords: classification, *Cornus*, dogwood, PhyloCode, phylogeny, taxonomy

Summary

Different opinions among taxonomists lead to differences in classification schemes and result in frequent name changes and confusion in communication. New evidence from phylogenetic studies using genetic/genomic data have also led to the need for reclassification of many groups, leading to new names being given to many previously familiar plants. Frequent name changes not only lead to misunderstandings in communications but also cause problems in data storage and information retrieval. We need new plant classification systems that are resilient to name changes resulting from the splitting of large genera or families, lumping of small genera, or due to personal opinions on plant characteristics instead of new names being frequently proposed, accepted, and then rejected. Examining the highly controversial taxonomy and classification history of dogwoods demonstrates evident limitations of the traditional Linnaean System that organizes taxa hierarchically from the kingdom to the species level and gives each species a unique two-word Latin name. The limitations present the need for a classification system that is rank-free and aims to make taxonomy and names more stable. PhyloCode is an alternative to the traditional Linnaean system that names taxa/clades without assigning ranks and can be used concurrently with the Linnaean system. The Fundamental ideas about PhyloCode will be introduced and a PhyloCode-based classification of dogwoods is proposed.

INTRODUCTION

Taxonomic Chaos in the Dogwood Genus *Cornus* L. and Major Limitations of the Traditional Linnaean Classification System

The dogwood genus *Cornus* was first published by Linnaeus (1753) who included five species *C. florida* L., *C. mas* L., *C. canadensis* L., *C. suecica* L., and *C. sanguinea* L. (**Fig. 1**). These plants are fundamentally similar in vegetative and reproductive morphology, such as in their simple, opposite leaves with entire margin, arched lateral veins, and appressed two-armed hairs, small 4-merous flowers with parts that are free, a hypanthium fused with a 2-carpellate ovary each with a single pendulous ovule, and a fleshy drupaceous fruit. Later, additional species sharing these common features were discovered and added to the genus, including two alternate-leaved species, *Cornus alternifolia* L. f. and *C. controversa* Hemsl. The genus now consists of approximately 55-60 species (Xiang and Boufford, 2005; Murrell and Poindexter, 2016; Xiang et al, 2006).

Figure 1. Species included by K. Linnaeus in *Cornus* in Sp. Pl. 1: 117, 1753. A-E: *C. florida* L., *C. canadensis* L., *C. suecica* L., *C. mas* L., and *C. sanguinea* L.

Despite the similarities, the original five species of *Cornus* differ dramatically in some detailed morphology of the inflorescence and fruit (**Fig. 2**), as detailed below. *Cornus florida*, the flowering dogwood tree, bears four large, petaloid bracts subtending a capitulum/head inflorescence. Species like *C. florida* are often referred to as the Big-Bracted Dogwoods (BB dogwoods). The BB dogwoods now have 9 species and nine subspecies species disjunctly distributed in eastern Asia and North America, extending to Mexico and Central America. The three American species bear simple, red fruits in clusters. *Cornus florida* occurs in the eastern U.S. with subspecies disjunct in Mexico, *C. disciflora*, whose bracts fall off before expansion, extends from Mexico to Costa Rica, while *C. nuttallii*, whose bracts vary from 4 to 6, is restricted to mountains of the Pacific Northwest. The eastern Asian BB dogwoods include the kousa dogwood, *C. kousa* H. Bürger ex Hance and the likes, which make compound red fruits. They occur in most parts of China, except in the northwest, and adjacent countries to the northeast and southwest (Du et al., 2023a, b). *Cornus canadensis* and *C. suecica*, the dwarf cornels or bunchberries, are rhizomatous perennial herbs that produce minute, condensed, dichasial inflorescences subtended by four, enlarged petaloid bracts and simple red fruits in clusters. They are often referred to as the Dwarf Dogwoods (DW group). They now include four disjunct species in circumboreal regions and in the high mountains of Myanmar (Burma) (Wahlsteen et al., 2020).

Figure 2. Examples of inflorescence and fruit variation in *Cornus* L. 1. Determinate head with petaloid bracts; 2. Determinate umbel with non-petaloid bracts; 3. Corymbose compound cymes without apparent bracts; 4. minute compound dichasia with petaloid bracts; 7. compound/multiple fruit. Remainders: simple fruits in clusters.

In contrast, *C. mas*, the European cornelian cherry, has flowers in umbels and distinct red elongate fruits in clusters, subtended by four small, non-petaloid bracts. The cornelian cherries (CC group) now include six medicinally valuable species in eastern Asia (*C. officinalis* Sieb. & Zucc., *C. chinensis* Wangerin, *C. eydeana* Q. Y, Xiang & Y. M. Shui, Europe (*C. mas*), western North America (*C. sessilis* Torr. ex Durand), and Africa (*C. volkensii* Harms) (Xiang et al., 2003; refs). *Cornus sanguinea*, the blood twig dogwood, represents a group of shrubs and trees that bear elongated compound, corymbose or paniculate cymes that have rudimentary and early deciduous bracts and simple, blue, white, or black fruits **(Figs. 1, 2)**. This group, often referred to as the Blue- or White-fruited Dogwoods (BW Group), is the most diverse, containing the remaining species of the genus. Interestingly, the evident differences in inflorescences and fruits plus additional variation within the BB, CC, and BW groups were emphasized variously among subsequent taxonomists. As a result, the Linnaean concept of *Cornus* has been split into multiple genera by some (e.g., Hutchison, 1942; Pojakova, 1950) whereas others recognized the morphological subgroups as subgenera or sections within *Cornus* (e.g., Wangerin, 1990; Xiang, 1987; Ferguson, 1966). Some better-known genera that have been segregated from *Cornus* include *Swida* Opiz (all BW dogwoods), *Bothrocaryum* (Koehne) Pojark. (the alternate-leaved BW dogwoods), *Thelycrania* (Dumort.) Fourr. (all opposite-leaved BW dogwoods), *Afrocrania* Hutch. (African Cornelian Cherry), and *Macrocarpium* Nakai (CC group; synonym of *Cornus* s. s.), *Chamaepericlymenum* Hill (the DW dogwoods), *Benthamidia* Spach. (all BB dogwoods), *Cynoxylon* Raf. (all BB dogwoods or American BB dogwoods), *Discocrania* (Mexican 'BB' dogwood) and *Dendrobenthamia* (Asian BB dogwoods). Each author differed in the circumscriptions and compositions of genera or infrageneric subgroups (subgenera or sections) in one way or another (see Hutchinson, 1942; Hara, 1948; Pojarkova, 1950; Ferguson, 1966; discussion in Eyde, 1987, 1988; Xiang et al., 1993, 1996).

Therefore, a species of dogwood often has more than one name and depending on personal preferences, it can be labelled or annotated with different names, which has resulted in confusion and obstacles in communication and information retrieval from herbaria and databases. For instance, in the herbaria of Smithsonian Institutions and Harvard University, specimens of the giant pagoda dogwood are filed under *Cornus controversa*, while in China and European countries, some herbaria may file the species under *C. controversa*, while others may file it under *Swida controversa*, still others may file it under *Bothrocaryum controversum*, or under all these names, based on annotations on the specimens. One may not be able to find all specimens of the species in an herbarium if he/she looks only for specimens under one name. Similarly, one may not find all information for the species in a database if only one name is used in searching. The flowering dogwood tree has been called *Cornus florida* L, *Benthamidia florida* (L.) Spach., or *Cynoxylon floridum* (L.) Raf. ex B.D. Jackson, and the kousa dogwood has been called *Cornus kousa* Hance, *Benthamia japonica* Sieb. & Zucc., *Benthamidia japonica* (Sieb. & Zucc.) Hara, *Cynoxylon japonica* (Sieb. & Zucc.) Nakai, or *Dendrobenthamia japonica* (Sieb. & Zucc.) Hutch. at different times in different places. In America, the flowering dogwood tree has long been called *C. florida* until recently (see Weakley et al., 2022). Due to the change of classification to recognize the four major clades of *Cornus* revealed in phylogenetic studies (Xiang et al., 2006; 2011; Fu et al., 2019; Thomas et al., 2021; Du et al., 2023) as four distinct genera by Weakley et al. (2022), the name of the flowering dogwood tree was changed to *Benthamidia florida* (L.) Spach.).

Clearly, the taxonomic controversy and species name variations in *Cornus* are results of differences in personal opinions and compliance to the rules of a rank-based nomenclature. Taxonomic ranks (Division, Class, Order, Family, Genus, Species and the ranks below them; names of higher ranks above the genus level need to have prescribed endings) are fundamentally arbitrary and assigned subjectively by authors. In a well-resolved phylogeny, two authors can derive contrasting classification schemes giving different ranks at a given node **(Fig. 3)**, resulting in changes of names due to the prescribed name endings of ranks or difference in assigning the genus rank.

Taxa of the same rank are often thought to be equivalent and comparable in some ways, but not in many ways (e.g., ages, diversity level, or ecological breadth). For example, the ages of flowering plant families and orders currently recognized varies widely (Stevens. 2001 onwards; Kumar et al, 2022; Santiago et al., 2020). Naively treating taxa at the same rank as equivalents can lead to flawed science or wrong actions in biodiversity conservation. Clearly, the need to maintain the hierarchy of the ranks leads to instability of names (names being changed without good reasons).

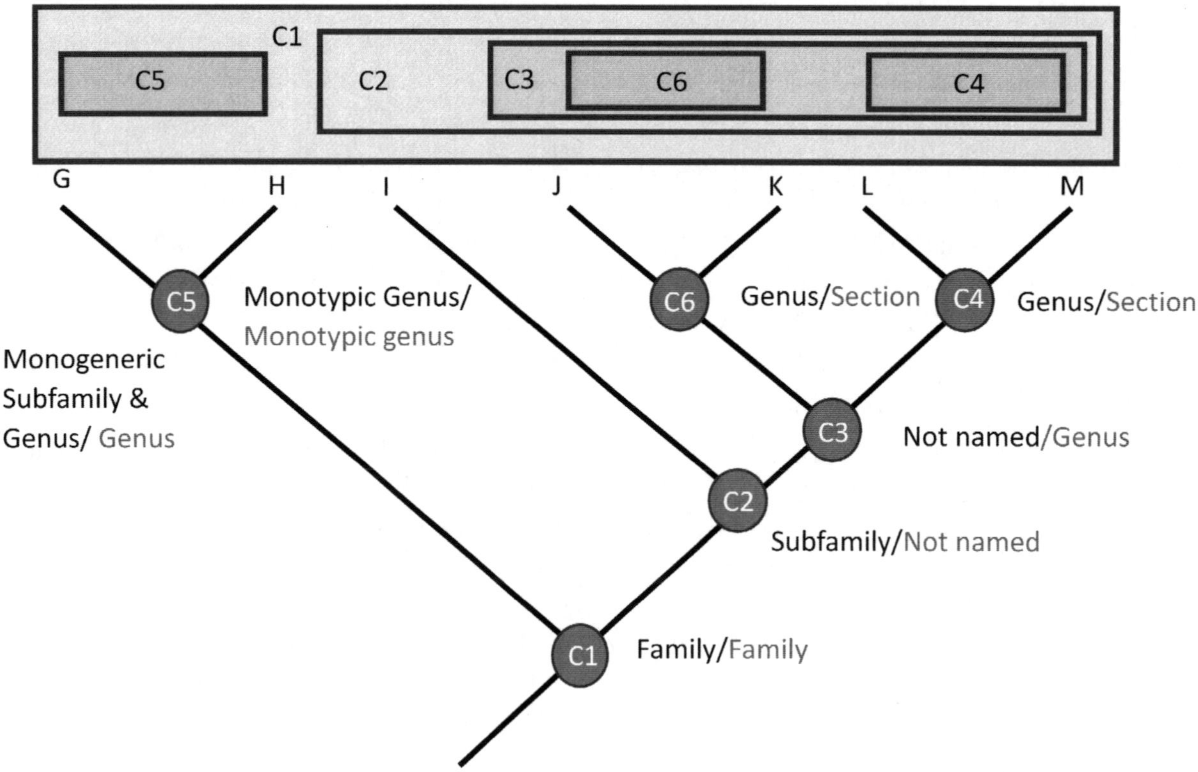

Figure 3. A hypothetical phylogeny showing clades C1 through C6 which are hierarchically nested within one another. Clade C1 consists of Clade C2 and C5; Clade C2 consists of species I and Clade C3 while Clade C5 consists of species G and H. Clade C3 consists of Clade C6 and Clade C4, each consists of two species, J and K in clade C6 and L and M in Clade C4. The classification with reference to this phylogeny using PhyloCode will give a name to each of these clades without assigning a rank. In traditional rank-based classification, authors may differ in the rank assignments of clades (black vs red illustrating one of the many ways of possible different classification schemes), leading to name changes.

In addition to these issues, the traditional Linnean nomenclature is limited by ranks to suffice in classifying the tree of flowering plants, not mentioning the tree of life that has millions of branches. Given these limitations and disadvantages, a rank-free nomenclature would be desirable to resolve the rank-associated problems.

PhyloCode for a Rank Free Classification and Stabilizing Names

An alternative method to the traditional Linnaean System is the PhyloCode (http://phylonames.org/code/), which eliminates the rank associated problems (de Queiroz and Cantino, 2020). The PhyloCode is a set of principles, rules, and recommendations governing phylogenetic nomenclature and a system for naming taxa by explicit reference to phylogeny (the evolutionary history of organisms drawn as a branching line pattern or called phylogenetic tree to show ancestor-descendent relationships, e.g., **Fig. 3**). In contrast to the Linnaean system, PhyloCode attaches names to clades (or branches of the tree) without assigning ranks, such as taxa C1, C2, and C3 in Figure 3, each representing a progressively less inclusive clade (i.e., a monophyletic group or an ancestor and all its descendants). It is a system of nomenclature developed to explicitly name taxa by reference to phylogeny, using 'phylogenetic definitions' to delineate the clade with 'specifiers.' The phylogenetic definitions of a named clade can be node-based, apomorphy (derived features)-based, or branch- or stem-based (**Fig. 4**). For example, in **Figure 4**, X is for the least inclusive clade containing specifiers B and C. It is node based, while Y is for the most inclusive clade exhibiting the red character synapomorphic (derived and shared) with that in B and/or C, which is apomorphy based, and Z is for the most inclusive clade containing C but NOT A, which is branch based.

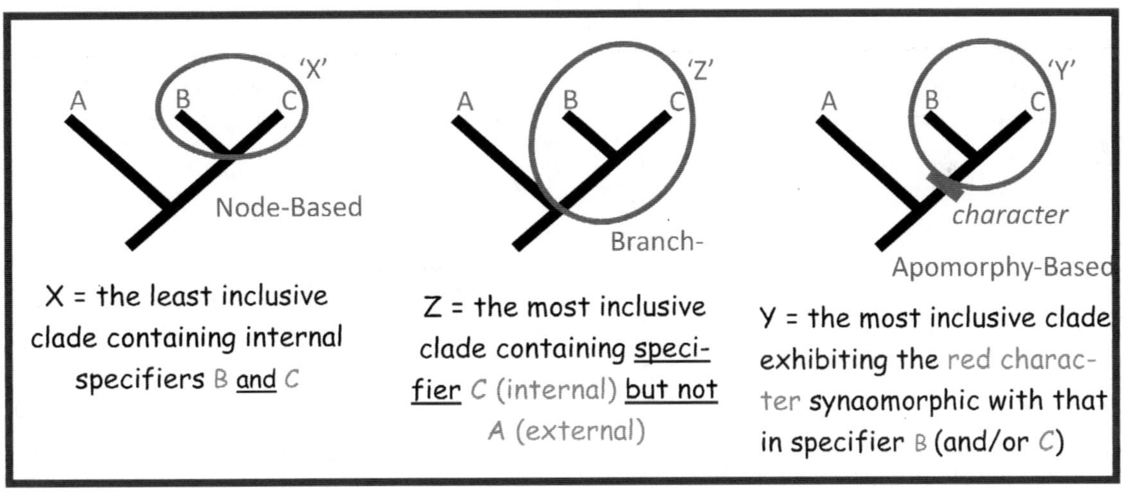

Figure 4. Three forms of phylogenetic definitions and specifiers. Taxa 'X', 'Y', and 'Z' are clades defined using the node-based, branch-based, and apomorphy-based definitions, respectively. Apomorphy: derived feature; Synapomorphy: shared, derived features.

These different definitions allow precision and enable differentiating a crown clade from a stem clade that contain the exact same species-membership. Application of the definitions with care can avoid future name changes if specifiers are carefully chosen in the phylogenetic definitions. In this system, future name changes may occur, but they will be due to changes in our understanding of phylogeny or relationships, rather than arbitrary decisions on taxonomic rank. Here are three examples illustrating the naming of angiosperm clades based on different phylogenetic definitions with PhyloCode in reference to the phylogeny (APG IV, 2016). The *Asteridae* A. Takhtajian 1967 [R. G. Olmstead & W. S. Judd 2020] clade is 'least inclusive clade containing *Lamium purpureum* Linnaeus 1753 (Lamiidae/Lamiales), *Cornus mas* Linnaeus 1753 (Cornales), *Aster amellus* Linnaeus 1753 (Campanulidae/Asterales), and *Arbutus unedo* Linnaeus 1753 (Ericales)'. It is a node-based definition or a minimum/smallest-crown clade definition. The *Superrosidae* D.E. Soltis, S. Smith & N. Cellinese 2011 [W. S. Judd, D. E. Soltis & P. S. Soltis 2020] clade is the 'maximum clade containing *Rosa cinnamomea* L. 1753 (Rosidae/Rosales) but not *Aster amellus* L. 1753 (Asteridae/Asterales)'. Here one internal and one external specifier were used in the definition. It is a branch-based definition, or a maximum/largest-crown clade definition. The *Tricolpatae* P. D. Cantino, J. A. Doyle, S. W. Graham, W. S. Judd and R. G. Olmstead 2007 [Donoghue, M.J., J.A. Doyle, and P.D. Cantino 2020] clade is defined as 'The most inclusive clade exhibiting tricolpate (or derivative) pollen grains synapomorphic with those found in *Platanus occidentalis* Linnaeus 1753 (Eudicotyledoneae). A tricolpate pollen grain is one having three elongate, furrow-like apertures (colpi) located at and oriented perpendicular to the equator.' (RegNum, https://phyloregnum.org/?). It is an apomorphy-based definition with one internal specifier and an apomorphy.

The PhyloCode focuses squarely on reflecting phylogenetic relationships and eliminates the reliance of taxonomic ranks. It is designed to allow concurrent uses with the rank-based code to provide an alternative system for governing the application of both existing and newly proposed names. The code currently only governs the names of clades; species names are still governed by traditional codes. However, in PhyloCode, 'the first part of the species binomen is not interpreted as a genus name but simply as the name of a taxon that includes that species' (Chapter X, Article 21.2, PhyloCode http://phylonames.org/code/articles/21/). There have been proposals for a completely rank-free PhyloCode without the species rank, which is also considered arbitrary. The proposal suggested using SNaRC (Smallest Named Registered Clade) in the place of species. In such a system, all taxon names are uninominal (Gellinese et al., 2012; Mishler and Wilkins 2018; Mishler, 2022) with the smallest named clade treated like other levels and given a formal (uninominal) name registered in a database. However, it is still controversial within the PhyloCode community whether the rank of species should be removed. Mishler (2022) argued that a complete rank-free system better serves today's research in ecology, evolution and systematics as well as conservation management.

The ideas of PhyloCode initially developed in several key papers in the 1990s (de Queiroz and Gauthier, 1990, 1992, 1994; For more literature, see: http://phylonames.org/literature/). It has gone through some hot debates (for Critiques, see http://phylonames.org/literature/#critiques, and Replies to Critiques, see http://phylonames.org/literature/#replies). PhyloCode (http://phylonames.org/code/articles/20/; de Queiroz and Cantino 2020) is a product of 30 years of thought by Kevin de Queiroz and Philip Cantino. The publication of the *PhyloCode* was accompanied by the volume *Phylonyms* (de Queiroz et al., 2020), an implementation of PhyloCode thatdocuments the real-world uses of PhyloCode. An online registration database

'*RegNum*' for names created using the rules of the PhyloCode, including those in '*Phylonyms*,' has been created. Taxonomists do not have to name all clades on the phylogeny of their study organisms, but only name clades that are well supported by evidence. Clades with uncertainty can be assessed by future phylogenetic studies and can be named when new evidence is available.

In summary, the following quotes speak well of the need of adopting PhyloCode in taxonomy in the genomic era when phylogeny can be robustly determined using genomic data.

"The traditional codes of nomenclature were first developed long before there was any knowledge of evolution and phylogeny. In this context, unfortunately, emphasis was placed on taxonomic ranks. The PhyloCode was developed specifically to connect nomenclature to evolution and phylogeny, and it works better (e.g., eliminating name changes based on arbitrary rank changes) in the current era where biologists of all types are focused on evolution and phylogenetic relationships. This is especially important as systematists have more important work to do as biodiversity is being lost. We should not be wasting our time on changing names based on outdated nomenclatural procedures that are tied to the wrong metric — that is, they are tied to ranks instead of to phylogenetic relationships." (Michael Donoghue, Yale University).

"Why do we keep trying to put what we know about evolution in a system that wasn't built to reflect it?" (Pamela Soltis, University of Florida).

Dogwood Phylogeny and PhyloCode-Based Classification

Several phylogenetic studies have been conducted in the past to elucidate species relationships and dating the divergence of clades, each varied in the scale of taxon and data sampling (Xiang et al., 1996, 1998, 2006, 2011; Fan et al., 2003; Fu et al, 2019; Thomas et al., 2020; Du et al., 2023). The most recent comprehensive phylogenetic study of the *Cornus* used three genomic datasets with the most complete species sampling and derived a robust phylogenetic tree of the dogwoods with estimates of the age of clades (Du et al., 2023). In the intention of stabilizing the naming of groups of dogwoods, names were proposed for clades with strong support using preexisting names without assigning a rank, following PhyloCode (**Fig. 5**). Minimum clades of the BB, CC, DW, and BW groups were named *Benthamidia, Macrocarpium, Arctocrania* and *Swida*, respectively. Within *Benthamidia*, the American clade and the Asian clade were named *Cynoxylon* and *Syncarpea*, respectively; within the *Swida* clade, the three subclades previously treated as genera were named *Yinquania, Mesomora* and *Kraniopsis*; all are preexisting names. Formal registration of these names and the phylogenetic definitions of the clades is in preparation (Du et al., in revision). In a practical way, all species of dogwoods can keep their names under *Cornus*. In nurseries and botanical gardens, the species can be labelled with reference to their respective clades withing *Cornus*. For example, one can label *Cornus florida* as *Cornus florida* L. (*Benthamidia/Cynoxylon*; Cornaceae) and *Cornus controversa* Hemsl. (*Swida/Mesomora*, Cornaceae) to indicate the clade to which they belong.

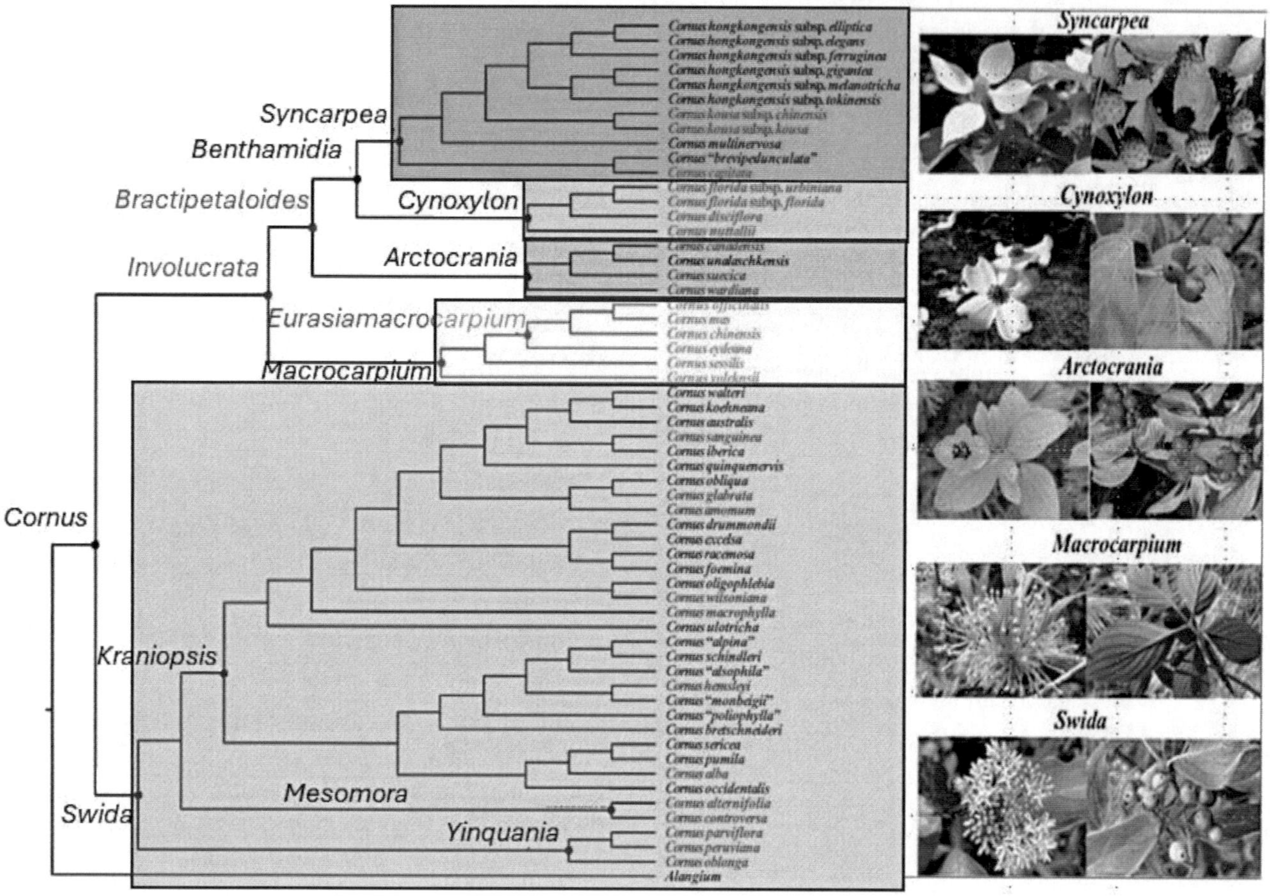

Figure 5. Phylogeny of *Cornus* L. based on three sets of genomic data from Du et al., (2023). PhyloCode-based hierarchical classification of the *Cornus* clade in reference to the phylogeny is shown. Clades marked with small dots are named using preexisting names except those in blue. Shaded clades are those shown with images at the right of the tree. Figure is modified from Du et al., 2023 in American Journal of Botany **https://doi.org/10.1002/ajb2.16116**

ACKNOWLEDGEMENTS

I am in debt to Michael Donoghue at Yale University and Brent Mishler at the University of California, Berkeley for sharing slides (for the oral presentation) and ideas on the PhyloCode.

I am grateful to David E. Boufford for proofreading the article and English edits.

LITERATURE CITED

APG IV. (2016). An update of the Angiosperm Phylogeny Group classification for the orders and families of flowering plants: APG IV. Bot. J. Linn. *181*:1–20, https://doi.org/10.1111/boj.12385

Cellinese, N., Baum, D.A., and Mishler, B.D. (2012). Species and phylogenetic nomenclature. Syst. Biol. *61*: 885-891.

De Queiroz, K., and Gauthier, J. (1992). Phylogenetic taxonomy. Annu. Rev. Ecol. Syst. *23*: 449–480.

De Queiroz, K., and Gauthier, J. (1994). Toward a phylogenetic system of biological nomenclature. Trends Ecol. Evol. *9*: 27–31.

De Queiroz, K., and Cantino, P. (2020). International Code of Phylogenetic Nomenclature (PhyloCode). CRC Press, Taylor and Francis Group, LLC, Boca Ralton, FL, USA.

De Queiroz, K., Cantino, P., and Gauthier, J. (2020). Phylonym- A companion to the PhyloCode. CRC Press, Taylor and Francis Group, LLC, Boca Ralton, FL, USA.

Du, Z., Soltis, D.E., Soltis, P. S, and Xiang, Q.-Y. (J.). *in Prep*. Addition to "An updated phylogeny, biogeography, and PhyloCode-based classification of Cornaceae based on three sets of genomic data" Bulletin of Phylogenetic Nomenclature.

Du, Z.Y., Xiang, Q.-Y.(J.), Cheng, J., Zhou, W.B., Wang, Q.-F., Soltis, D.E. , and Soltis, P.S.. (2023). An updated phylogeny, biogeography, and PhyloCode-based classification of *Cornaceae* based on three sets of genomic data. Amer. J. Bot. https://doi.org/10.1002/ajb2.16116

Eyde, R. H. (1987). The case for keeping *Cornus* in the broad Linnaean sense. System. Bot. *12*: 505-518.

Eyde, R. H. (1988). Comprehending *Cornus*: Puzzles and progress in the systematics of the dogwoods. *Botan. Rev. 54*: 233-351.

Fan, C., and Xiang, Q. Y. (2003). Phylogenetic analyses of Cornales based on 26S rRNA and combined 26S rDNA-*mat*K-*rbc*L sequence data. Amer. J. Bot. *90*: 1357-1372.

Ferguson, I. K. (1966). Notes on the nomenclature of *Cornus*. J. Arnold Arbor.*47*: 100-105.

Fu, C. N., Mo, Q., Yang, J. B., Ge, X. J., Li, D. Z., Xiang, Q. J., and Gao, L. M. (2019). Plastid phylogenomics and biogeographic analysis support a trans-Tethyan origin and rapid early radiation of Cornales in the Mid-Cretaceous. Molec. Phylogen. Evol. *140*: 106601.

Hutchinson, J. (1942). Neglected generic characters in the family Cornaceae. Ann. Bot. *6*: 83-93.

Kumar, S., Suleski, M., Craig, J.E., Kasprowicz, A.E., Sanderford, M., Li, M., Stecher, G., and Hedges, S.B. (2022). TimeTree 5: An Expanded Resource for Species Divergence Times. Mol. Biol. Evolut. DOI: 10.1093/molbev/msac174.

Mishler, B.D. and Wilkins, J.S. (2018). The hunting of the SNaRC: a snarky solution to the species problem. Philos. Theor. Pract. Biol. *10*: 1-18

Mishler, B.D. (2022). Ecology, evolution, and systematics in a post-species world. In Wilkins, J., Zachos, F., and Pavlinov, I. (eds.) Species Problems and Beyond: Contemporary Issues in Philosophy and Practice. CRC Press, Boca Raton, FL.

Murrell, Z.E., Poindexter, D.B. (2016). *Cornaceae.* In Flora of North America Editorial Committee [eds.], Flora of North America north of Mexico, Vol. 12. Magnoliophyta: Vitaceae to Garryaceae, 443-457. New York, NY, USA: Oxford University Press.

Pojarkova, A. (1950). De systemate generis Linneani *Cornus* L. *Botaniceskie Materialy Gerbarija Botaniceskogo Institutiimeni V. L. Komarova Akademii Nauk SSSR 12*: 164-180.

Santiago, R.-B., Sauquet, H., and Magallon, S. (2020). The delayed and geographically heterogeneous diversification of flowering plant families Nature Ecol. Evol. *4*:1232–1238. DOI - 10.1038/s41559-020-1241-3.

Stevens, P. F. (2001 onwards). Angiosperm Phylogeny Website. Version 14, July 2017 [and more or less continuously updated since]." will do.
http://www.mobot.org/MOBOT/research/APweb/

Thomas, S. K., Liu, X., Du, Z. Y., Dong, Y., Cummings, A., Pokorny, L., Xiang, Q. Y..and Leebens-Mack, J. H. (2021). Comprehending Cornales: phylogenetic reconstruction of the order using the Angiosperms353 probe set. Amer. J. Bot. *108*: 1-10.

Weakley, and the Southeastern Flora Team. (2022). Flora of the Southeastern United States. University of North Carolina Herbarium, North Carolina Botanical Garden, Chapel Hill, NC, USA.

Wahlsteen, E., Zhou, W., Xiang, Q., and Rushforth, K. (2020). Rediscovery of the lost little dogwood *Cornus wardiana* (Cornaceae)—Its phylogenetic and morphological distinction and implication in the origin of the Arctic-Sino-Himalayan disjunction. J. System. Evol. *59*: 405-416.

Wangerin, W. (1910). Cornaceae. *In* A. Engler [eds.], Das pflanzenreich, series IV, heft 41. W. Engelmann, Leipzig.

Xiang, Q.-Y. (1987a). A neglected character of *Cornus* L. s. l. with special reference to a new subgenus-*Sinocornus* Q. Y. Xiang. J. System. Evol. *25*: 125-131.

Xiang, Q.-Y. (1987b). System and synopsis of *Cornus* subgen. *Syncarpea* (Nakai) QY Xiang (Cornaceae). Bull. Botan. Res. *7*: 33-52.

Xiang, Q.-Y., Soltis, D. E., Morgan D. R., and Soltis, P. S. (1993). Phylogenetic relationships of *Cornus* L. sensu lato and putative relatives inferred from rbcL sequence data. Ann. Missour Botan Garden *80*: 723-734.

Xiang, Q.-Y., Brunsfeld, S. J., Soltis, D.E. and Soltis, P. S. (1996). Phylogenetic relationships in *Cornus* based on chloroplast DNA restriction sites: implications for biogeography and character evolution. Syst. Bot.*21*: 515-534.

Xiang, Q. Y., Soltis, D.E. and Soltis, P. S. (1998). Phylogenetic relationships of Cornaceae and close relatives inferred from *mat*K and *rbc*L sequences. Amer. J. Bot. *85*: 285-297.

Xiang, Q.-Y., Thomas, D. T., Zhang, W., Manchester, S.R., and Z. Murrell. (2006). Species level phylogeny of the genus *Cornus* (Cornaceae) based on molecular and morphological evidence—implications for taxonomy and Tertiary intercontinental migration. Taxon *55*: 9-30.

Xiang, Q.-Y. (J.), Thomas, D.T, and Xiang, Q.P. (2011). Resolving and dating the phylogeny of Cornales- effects of taxon sampling, data partitions, and fossil calibrations. Mol. Phylogen. Evol. *59*: 123-138.

Combined Proceedings IPPS 73:376-382. 2023.

Managing Water Quality On-Farm

Sarah A. White

Clemson University, Department of Plant and Environmental Sciences, E-143 Poole Agricultural Center, Clemson, South Carolina, 29634-03100, USA

swhite4@clemson.edu

Keywords: Clean Water[3], water storage, water availability, water-use strategy, remediation, recycling

Summary

Water availability and quality contribute to crop health and impact the profitability of crop production. Water security is likely the next big challenge for ornamental producers worldwide. Enhancing on-farm water storage (e.g., reservoirs) capacity may be one of the savviest investments currently that can be made. Grower use of poorer-quality water will likely increase in the coming years. Managing water quality on-farm can be achieved by strategically integrating chemical or biologically-based treatment technologies within production areas or water-management systems.

INTRODUCTION - WATER CHALLENGES AND CONCERNS

Water Availability. Water sources used for irrigation include municipal (potable or reclaimed), ground, surface, and captured stormwater. Geographic and economic variables regulate water availability. In the eastern US, riparian water rights govern reasonable water uses, while in the western US, prior-appropriation, water rights give rights to a certain volume of water to the first person/business who used it. Thus, depending upon operation location, the capacity to gain access to surface waters for irrigation may be simple or difficult. Any one

IPPS Vol. 73 – 2023

of these sources can serve as either primary source or a backup source.

It is critical to understand how much water is used for irrigation per day at peak application times to truly be able to plan for scenarios when water availability might be limited due to drought. Figure out how much water you apply per day. Multiply that volume by the number of days you want your water to last if an alternate backup water source is not available. Then develop a plan to invest in on-farm infrastructure (over time, all at once) to have enough water available to meet your production goals.

Water Quality. Water quality influences plant growth rates and quality. Meeting plant production goals requires understanding your water quality and managing it so plant growth outcomes are attained. Do you have pathogens in your water that infest crops and cause inventory shrinkage? Consider sanitizing your water (e.g., chlorine, peroxides, etc.) to kill the pathogen before it can infest your crops. Do you have issues with clogged emitters? Consider adding disc/media filters that can remove debris from water that contributes to clogging. Are your plants growing slower than you think they should? Check your water pH and alkalinity. Sometimes, water can be too pure – and nutrients are less available than they should be because water pH is too low. In this instance, you may need to add additional buffer to the substrate (e.g., lime) or irrigation water (e.g., flowable lime) to moderate pH to enhance nutrient availability.

WATER QUALITY MANAGEMENT SOLUTIONS

Water quality management is not about one single solution that solves all your problems. It's about a concerted, strategic effort to combine treatment options to attain water quality outcomes specific to your operation.

Reservoirs. Reservoirs are typically considered water storage structures. Reservoirs are also called tailwater recovery ponds or containment ponds (Sahoo et al., 2021). But in recent years, we have come to realize that reservoirs add benefits beyond simple water storage (Yazdi et al., 2021). Reservoirs should be designed to intercept operational water before it leaves your operation (**Fig. 1)** but can also be used to store "clean" water.

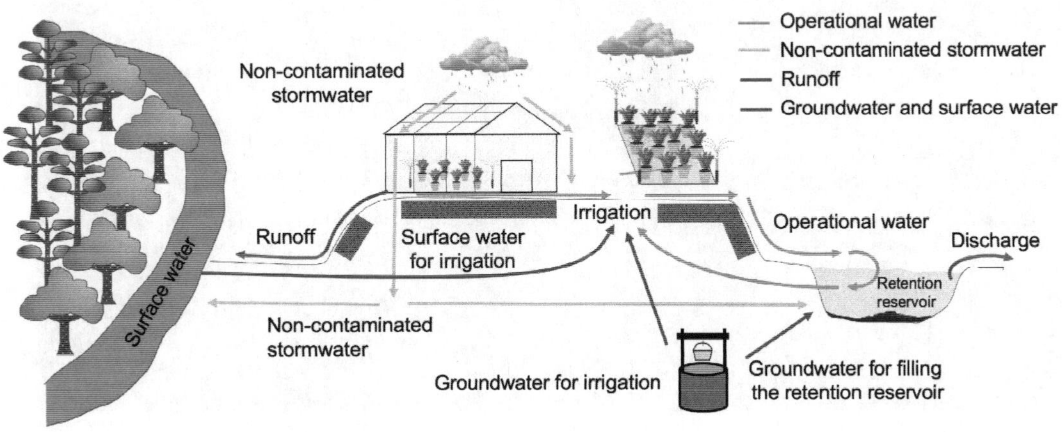

Figure 1. Types of on-farm water.

Operational water is runoff from stormwater and captured irrigation return flow. Ornamental producers should consider using a reservoir to (1) conserve water via capturing operational water, irrigation return flow, and stormwater, (2) reduce reliance on surface and ground water via enabling water recycling and reuse, and (3) clean additional contaminants from water prior to its discharge into receiving waters.

Yazdi et al. (2021) proposed changes to reservoir design to enhance water security on-farm. Sahoo et al. (2021) translated those recommendations into actionable items for use on-farm (**Fig. 2**).

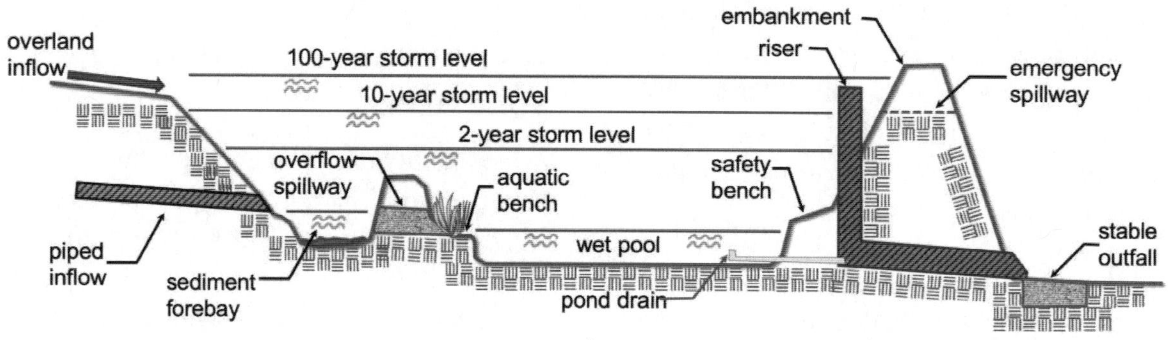

Figure 2. New design recommendation to consider for irrigation reservoirs (Sahoo et al., 2021)

When designing a reservoir, consider ways to manage sediment and detain it before it enters the reservoir (e.g., a sediment forebay, **Fig. 3**) to help maintain reservoir storage capacity and to have a means to remove sediment from the reservoir without dredging the entire reservoir.

Figure 3. Sediment forebay designed for easy clean out.

Biological Treatment Solutions. Researchers collaborating on the Specialty Crops Research Initiative project – Clean, WateR^3 (https://cleanwater3.org/) evaluated various treatment technologies that harness plant and microbial-based biological systems to clean water. All of these systems operate under the principle that the longer the amount of time water (and contaminants) are held in one place, the more contaminants can be removed from the water. Each of the following treatment technologies have a minimum treatment capacity that is enhanced as the contact time between the water+contaminant and the treatment technologies lengthens. These technologies work best when placed strategically on-farm to capture operational water and when used in concert with each other. As each technology may work better for specific contaminants, grouping/pairing technologies (i.e., treatment train) enhances treatment efficacy (**Fig. 4**).

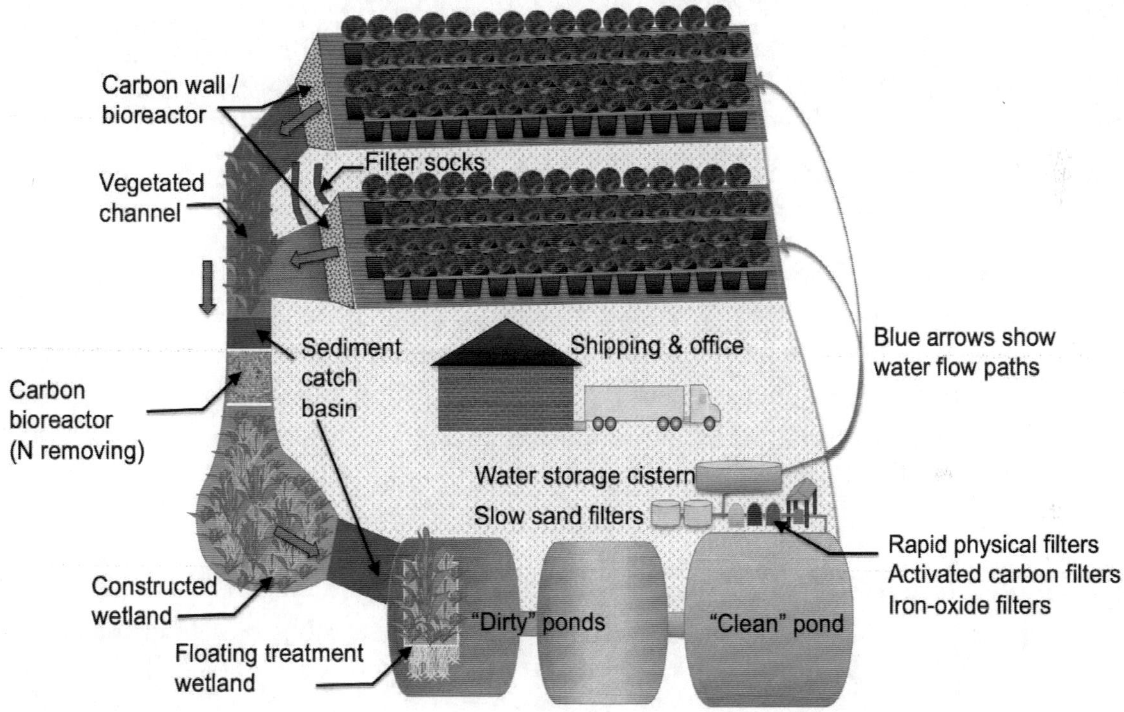

Figure 4. Integration of treatment technologies and on-farm practices to manage water quality and availability.

Vegetated Waterways and Filter Strips. Vegetated waterways, ditches, or channels serve to help reduce contaminant entry into reservoirs by prefiltering contaminants from water prior to their entry into the reservoir (Majsztrik et al., 2017). The plants in these vegetated waterways help to diminish flow velocity and disperse the flow so sediment can settle from the water (**Fig. 5**). The plants within the buffer serve to absorb nutrients. The microbial communities within their rhizosphere also increase the degradation or transformation of pesticide and nutrient (e.g., nitrate) contaminants.

Figure 5. Vegetated buffer and channel to help minimize movement of sediment or other agrichemical contaminants in operational waters into reservoirs or off-farm.

Floating Treatment Wetlands. Floating treatment wetlands are a style of constructed wetland that has been modified for use (i.e., made buoyant) within existing infrastructure (e.g., reservoirs or ponds) (White, 2013). Floating treatment wetlands comprise a scaffold or buoyant structure that helps keep the plant crowns above the water surface and plants whose root systems extend within the water column (**Fig. 6**).

Figure 6. Floating treatment wetland deployed in a nursery reservoir in Florida.

As long as the plant crowns remain above the surface of the water, a wide range of plants (even ornamental plants) can be grown within these wetlands. If used as a secondary production area on-farm, Garcia Chance et al. (2022) estimated a return on investment can be attained within as little as 2.2 years. Bell et al. (2018) reported that the plants in floating treatment wetlands helped diminish the viability of zoospores transported in simulated irrigation runoff and that none of the plants screened were infected by *Phytophthora*. However, subsequent work has noted that three species (*Carex stricta, Panicum virgatum,* and *Typha latifolia*) became infested with *Phytophthora*. These three species should be avoided if growers are worried about the circulation of *Phytophthora* propagules in irrigation water.

CONCLUSIONS

Water management on-farm is not a once-and-done plan, it is a constantly evolving strategy that considers crop mix, economic goals, water security needs, and the treatment technologies needed to meet your goals. It is the integration of these factors that makes for successful on-farm water management. As the weather becomes more variable and the prevalence of flash (micro)-droughts and intense storms come more frequently and at atypical times of the year when we may be less prepared to manage them. It becomes all the more important to plan ahead and invest in the infrastructure and treatment technologies needed to help your operation maintain its capacity to produce plants profitably.

ACKNOWLEDGEMENTS

The summary of research presented herein was a collaborative effort with colleagues Jim Owen, Chris Wilson, Tom Fernandez, Debabrata Sahoo, Mohammad Yazdi, Damon Abdi, and Bill Strosnider.

LITERATURE CITED

Bell, N.L., Garcia, L.M., Chance, and White, S.A., (2018). Clean water3: Evaluation of three treatment technologies to remove contaminants from recycled production runoff. Acta Hort. *1191*:199-206. 10.17660/ActaHortic.2018.1191.28.

Garcia Chance, L.M., Hall, C.R., and White, S.A., (2022). Viability assessment for the use of floating treatment wetlands as alternative production and remediation systems for nursery and greenhouse operations. J. Environ. Manag. *305*. 10.1016/j.jenvman.2021.114398.

Majsztrik, J.C., Fernandez, R.T., Fisher, P.R., Hitchcock, D.R., Lea-Cox, J., Owen, J.S., Jr., Oki, L.R., and White, S.A., (2017). Water use and treatment in container-grown specialty crop production: A review. Water, Air, Soil Pollut. *228*(4), 151. 10.1007/s11270-017-3272-1.

Sahoo, D., Yazdi, M.N., James S. Owen, J., and White, S.A., (2021). The basics of irrigation reservoirs for agriculture. Land-Grant Press by Clemson Extension, Clemson (SC), LGP 1121. https://lgpress.clemson.edu/publication/the-basics-of-irrigation-reservoirs-for-agriculture/

White, S.A. (2013). Wetland technologies for nursery and greenhouse compliance with nutrient regulations. HortScience *48*(9):1103-1108. https://doi.org/10.21273/HORTSCI.48.9.1103 .

Yazdi, M.N., Owen, J.S., Lyon, S.W., and White, S.A. (2021). Specialty crop retention reservoir performance and design considerations to secure quality water and mitigate non-point source runoff. J. Cleaner Prod. *321*. 10.1016/j.jclepro.2021.128925.

Combined Proceedings IPPS 73:383-389. 2023.

J Berry Nursery Marketing Influence

James Berry

J. Berry Nursery, 201 Private Road 5180, Grand Saline, Texas 75140, USA Email:

jim@jberrynursery.com

Keywords: branding, licensed growers, brand partners, trademarks, customer relationships

Summary

J Berry Nursery stands as a testament to the power of innovation and dedication. Jim's love for plants and his commitment to delivering quality have been the guiding principles behind this horticultural endeavor. We are going to unveil the comprehensive marketing strategy that has propelled J Berry Nursery to its current heights of success. Jim Berry, the visionary leader and co-founder of J. Berry Nursery, shares insights into their dynamic sales and marketing strategy. With a passion for our exceptional plant products, Jim is on a mission to cultivate not just growth but flourishing partnerships with our valued customers. At the heart of our strategy is innovation and market disruption. J Berry nursery is known for collaborating with a collective group of growers to serve the needs of big-box stores, ensuring that our plants stand out in a crowded market. But it does not stop there. Jim's meticulous planning and keen market insights are instrumental in shaping the modern landscape with solid science. Real-time data analysis is the backbone of our approach, guiding decisions to optimize sales growth and profitability.

IPPS Vol. 73 – 2023

INTRODUCTION

J Berry Nursery is a 17-year-old nursery located in Grand Saline, Texas. The J Berry team constantly seeks to improve science-based genetics and visuals of our plants, setting them apart in terms of function and appeal. Innovation is a way of life; our strategy is not just about meeting market expectations; it's about exceeding them. It's about fostering partnerships that grow and flourish, creating a win-win scenario for everyone involved in the journey. We are not just growing plants – we are growing success. J Berry Nursery has influenced growers, retailers, and end users. The domestic and foreign market are charged through licensing and making of patents, plant breeding to rights and trademarks. This effort, spearheaded by Jim Berry and his team's unwavering dedication, serves as a testament to the company's commitment to growth and innovation. Over the years, J Berry Nursery has made significant strides in marketing that have had a lasting impact on our industry.

One of the most notable moments was our introduction of the Black Diamond® brand in 2012. The occasion was the Texas Nursery & Landscape Association (TNLA) annual show, with the theme "Hidden Treasures." At this event, we unveiled around 300 Black Diamond® Crape Myrtle plants that had never been seen before, placing them strategically in approximately 75 different areas. The response was nothing short of astonishing, as a true market disruption unfolded. By the end of the show, the frenzy for these remarkable treasures had grown to the point where security had to be called in due to civil disobedience disputes among attendees wanting the Black Diamond® Crape Myrtle plants. The allure of the Black Diamond® brand had captivated everyone at the event, making it a watershed moment in our marketing journey.

Around 2018, we embarked on another marketing milestone when we participated in a Home Depot event, bringing with us the exciting Hollywood® Hibiscus in a competitive "Shark Tank" sales promotion theme. Our presentation ignited a fervour among Home Depot merchants from across the nation, who were eager to embrace this new product and branding. This initial spark has continued to burn brightly, establishing a lasting relationship and enthusiasm for our products.

Our third pivotal move last year 2022 involved breaking new ground by entering the grocery store market, particularly through our success in the Go Fresh Go Local Kroger event (**Fig. 1**).

There was a small business stimulation competition with 1600 applications – selected down to 15 invitees during the 1st round, further selected to 5 groups making the final round. J Berry Nursey was ultimately the winner! This strategic expansion into the grocery sector opened a wealth of new merchandising opportunities and market prospects. It marked a significant shift in our approach, broadening our reach and making our plants accessible to a whole new audience. These three monumental moments in our marketing history reflect how each move have not only disrupted the market but also reshaped the landscape of gardening and horticulture, propelling us forward in our mission to make gardening a source of joy and inspiration for all.

Figure 1. Projecting the J. Berry Nursery brand.

At the very heart of J Berry Nursery's achievements lie its mission and values. The company's mission, deeply rooted in delivering excellence, is complemented by values such as integrity, leadership, resilience, and innovation (**Fig. 2**). These values form the core of every operation and interaction. Behind every successful sales strategy is a dedicated team of sales managers. At J Berry Nursery, sales managers play a pivotal role in bridging the gap between the company and its cherished customers. Their insights, guidance, and dedication ensure that J Berry Nursery consistently meets customer expectations. J Berry Nursery's approach to potential customers goes beyond traditional sales tactics. The company seeks to understand each customer's unique needs, aspirations, and preferences. This personalized approach lays the foundation for trust and sets the stage for customers to a satisfying partnership for their individual business goals.

Figure 2. Core values of J. Berry Nursery: leaders, innovative, integrity, and resilient.

The art of maintaining and managing customer relationships is vital to J Berry Nursery's success. Through proactive communication, personalized service, and a steadfast commitment to addressing customer concerns, the company ensures that customers not only return but become lifelong advocates. The company believes that each customer interaction is an opportunity to create an enduring relationship. This emphasis on personalized service has been pivotal in establishing strong customer loyalty and fostering lasting connections.

J Berry Nursery's vision for the plant market extends far beyond the present. The company is a pioneering force in innovative horticulture, consistently introducing new and captivating plant varieties. This vision serves as the driving force behind the company's unwavering commitment to research and development. Expanding market share is a strategic imperative for J Berry Nursery. The marketing efforts are rooted in a commitment to excellence in every aspect of our products, from the plants themselves to the packaging and consumer communication. We recognize the profound importance of presenting our plant varieties with exceptional packaging and tags that are not only clear and concise but also reflective of our strong branding.

We are growing brands with a whole lot of personality! We create brands that are like the life of the plant party – they are not just green; they are the coolest kids in the garden! This unique and spirited approach perfectly aligns with our sales strategy, making our products unforgettable and irresistible to our customers. Instead of basic product promotion, we focus on creating well-curated collections and solution-based programs. By doing so, we offer customers a gardening experience that goes beyond individual plants. We believe in making purchasing and gardening enjoyable and exciting, allowing our customers to explore and embrace a world of possibilities through our thoughtfully designed collections and programs. At J Berry Nursery, we are dedicated to not just delivering plants but crafting an entire gardening adventure that both novice and seasoned enthusiasts can relish.

The art of naming our plant varieties at J Berry Nursery is a process that we do not take lightly. Each name is carefully considered to ensure that it perfectly encapsulates the essence of the plant, making it catchy and memorable. Moreover, the names harmoniously align with the specific program or collection they belong to, creating a cohesive and engaging experience for our customers. At J Berry Nursery, we believe that a name not only distinguishes a plant but also tells its unique story, and we strive to make that narrative as vibrant and captivating as the plants themselves. To achieve this, our naming process is a collaborative effort, with the J Berry team participating in discussions and voting to select names that truly resonate with the essence of each plant and the overall vision of our programs and collections.

The company actively explores new markets and regions while fortifying its position in existing ones. Collaborations and partnerships with key players in the industry are integral to achieving this expansion. In a rapidly evolving market, the key to competitiveness lies in innovation. J Berry Nursery is unflinching in its dedication to research and development. The company regularly introduces new plant varieties and enhances existing ones. J Berry Nursery deploys a diverse array of sales channels to effectively connect with customers. These encompass: J Berry Nursery's online presence provides customers with the flexibility to explore and purchase plants from the comfort of their homes. (**Fig. 3**) The user-friendly website and integration with Shopify make the online shopping experience seamless.

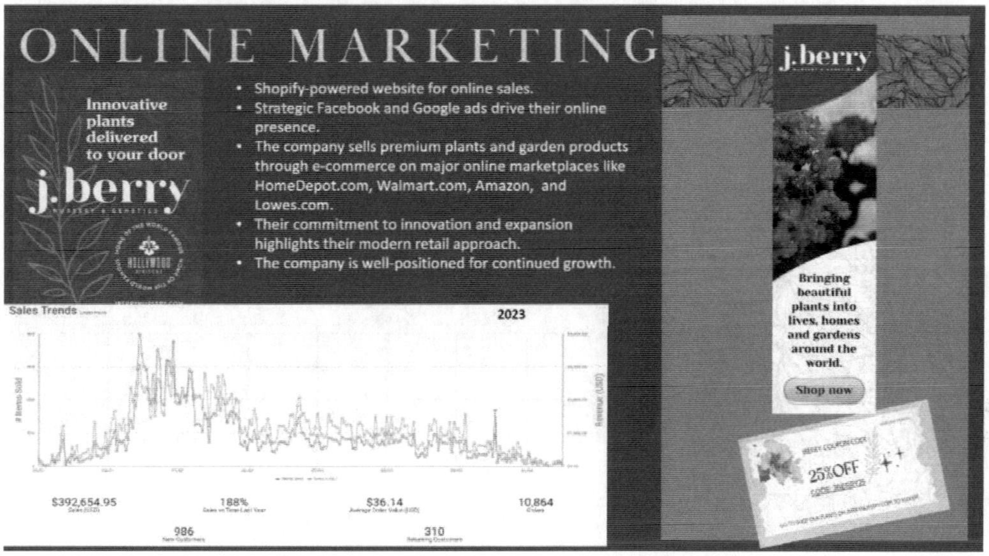

Figure 3. Online marketing at J. Berry Nursery.

Partnerships with Major Online Retailers. Collaborative agreements with online retail giants such as Amazon, Home Depot Online, Lowe's Online, and Walmart Online significantly broaden the company's reach and make J Berry Nursery's products accessible to a vast online audience.

Utilization of FedEx for Online Orders: Reliable and efficient shipping services through FedEx ensure that online orders are delivered to customers in impeccable condition and within expected timeframes. This reliability enhances customer satisfaction and trust in J Berry Nursery's online sales channel.

J Berry Nursery's marketing endeavors are marked by dynamism and diversity. The company harnesses the power of social media platforms to engage customers and promote its products. These platforms serve dual purposes as both marketing channels and avenues for sales, creating a harmonious customer journey. The integration of Shopify into J Berry Nursery's online presence has been a transformative step. It empowers direct sales through social media platforms and websites, streamlining the purchasing process and enhancing the customer experience. This integration aligns seamlessly with the company's commitment to making plant purchases effortless and enjoyable. A one-size-fits-all approach does not suffice in the dynamic landscape of horticultural sales. Customized strategies are designed to maximize impact and relevance for each channel. J Berry Nursery is renowned for its innovative promotional activities and strategic partnerships.

Collaboration with other breeders and strategic partnerships with companies like Plant Heaven, The Royalty Administration International, Proven Winners, First Editions, Endless Summer, and Star Roses and Plants. (**Figs. 4** and **5**).

Figure 4. Licensed growers for J. Berry Nursery.

By working closely with these strategic partners, we combine our strengths, harnessing their expertise and brand influence to enhance our marketing strategy. It is a powerful alliance where we not only showcase the best but also actively support and amplify these esteemed brands in return, helping the garden industry thrive together. We have forged more strategic alliances with growers and suppliers who play pivotal roles in the plant supply chain. These partnerships are instrumental in ensuring a seamless flow of products from our nursery to customers, ultimately enhancing the accessibility of our unique plant varieties Our brand partners, such as Plant Heaven and The Royalty Administration International, Proven Winners, First Editions, Endless Summer, and Star Roses and Plants are not only valuable collaborators but also possess their own robust marketing initiatives. We work closely with these partners to synergize our marketing efforts and consolidate our collective strengths, ultimately leading to a higher degree of success in bringing our innovative plant varieties to market. J Berry Nursery thrives on this extensive network and mutually beneficial relationships with breeders, growers, and companies, all of which are integral to our mission of providing exceptional plant varieties to the market.

Figure 5. Brand partners of J. Berry Nursery.

CONCLUSION

It is essential to look forward. J Berry Nursery's future is one marked by continued growth, innovation, and customer-centricity. The company aims to expand its reach globally, bringing its exceptional plant varieties to customers worldwide.

Research and development will remain at the forefront of J Berry Nursery's activities. The company will continue to invest in developing new plant genetics and enhancing existing ones. This commitment to innovation will ensure that customers always have access to cutting-edge plant varieties.

Combined Proceedings IPPS 73:390-398. 2023.

Native Seed Germination at the Rae Selling Berry Seed Bank

Gabriel Campbell-Martínez, Roxy Olsson, April Hersey, Stephanie Meikle, and Kris Freitag

Rae Selling Berry Seed Bank, Portland State University, 1719 SW 10th Ave, Rm B1-81, Portland, OR 97201

gec2@pdx.edu

Keywords: native seeds, germination trial, seed bank, *ex situ* conservation

Summary

A study at the Rae Selling Berry Seed Bank in Portland, Oregon, conducted 187 seed-germination trials on 154 species native to the Pacific Northwest. Seeds, stored for 0.1 to 34 years, underwent various treatments, including cold stratification and gibberellic acid soaking. Results showed 51% germination across trials, with notable improvements for specific taxa. Light conditions influenced germination outcomes, and gibberellic acid overcame light or dark requirements for certain species. Long-term storage trials (14 to 38 years) revealed that 52% of taxa maintained viability, indicating orthodoxy of many native plant species for seed bank storage. This research provides valuable insights into the germination and storage potential of Pacific Northwest plant seeds, contributing to biodiversity-conservation efforts.

INTRODUCTION

The Rae Selling Berry Seed Bank and Plant Conservation program is a non-profit institution affiliated with Portland State University. It is dedicated to conservation and restoration of native plants in the Pacific Northwest of North America. Major activities include management of a seed bank which includes research concerning seed germination and seedling establishment often in rare or threatened plants.

MATERIALS AND METHODS

A total of 187 preliminary germination trials were conducted on 154 taxa native to the Pacific Northwest at the Rae Selling Berry Seed Bank located at the Portland State University campus in Portland, OR (Table 1). Trials were conducted throughout 2022 on seeds stored for 0.1 to 34 years in a cold, dry room (5° C and 25% relative humidity). Seeds were sown without pretreatments (C) or prior to sowing seeds were imbibed and cold-stratified for 4 months at 5°C (CS) or soaked in 1,000 ppm gibberellic acid for 24 hours (GA). Seeds were either germinated under laboratory conditions (25/15 °C day/night; 12-hour photoperiod) or in a climate-controlled greenhouse under light (L) or dark (D) conditions (see Table 1 footnotes). Seeds in Fabaceae were scarified prior to sowing by vigorously rubbing seeds between 100-grit sandpaper until approximately 5% of the seeds were shattered to overcome physical dormancy common in the family.

Seeds were observed daily, and water was added when necessary. Germination was recorded approximately 4 weeks after sowing. Seeds were considered germinated when the radicle emerged from the seed coat >1mm for laboratory trials and when cotyledons were visible above the potting mix for greenhouse trials. Germination percentages were estimated visually and placed into one of the five following classes: 0 (0%), 1 (0.1 to 5%), 2 (6 to 25%), 3 (26 to 50%), 4 (51 to 75%), 5 (76 to 100%).

RESULTS and DISCUSSION

Germination occurred in 51% of the trials and a total of 89 taxa were successfully propagated (Table 1). The seed pretreatments CS or GA improved germination for *Iris missouriensis, Navarretia squarrosa, Plantago maritima, Penstemon fruticosus, Rhamnus purshiana, Ribes cereum,* and *Rosa woodsia*, indicating seeds of these species may have physiological dormancy and require pretreatments prior to sowing. There was >50% germination for seeds without pretreatments (C) and CS or GA did not improve germination for *Juncus breweri, Navarretia intertexta, Philadelphus lewisii, Physocarpus capitatus, Prunella vulgaris* and *Vaccinium deliciosum*, indicating seeds of these species may be non-dormant or have non-deep physiological dormancy which can be overcome with short times in dry storage.

Light (L) improved germination for seeds of *Juncus breweri* and *Physocarpus capitatus*, reduced germination for *Plantago maritima*, or had no effect compared to dark for seeds of *Gilia capitata, Lonicera utahensis, Lupinus argenteus,* and *Lupinus littoralis.* The application of GA overcame light requirements for germination of *Prunella vulgaris* and dark requirements for *Lathyrus japonicus.*

Of the 90 trials conducted on seeds stored long-term (14 to 38 years), 52% of taxa had germination, indicating that many species of Pacific Northwest native plants are orthodox and can be stored long-term using standard *ex situ* seed-banking techniques. Notable taxa which maintained high viability (>50% germination) during long-term storage include *Agrostis exarata, Beckmannia syzigachne, Deshcampsia elongata, Geum macrophyllum, Gilia capitata, Plectritis congesta*, and *Vicia nigricans* var. *gigantea*. Many *Carex* and *Juncus* taxa consistently retained viabilities >25% in long-term storage, indicating that these genera generally have potential for long-term storage in seed banks.

Table 1. Summary of germination trials conducted during 2022 at The Rae Selling Berry Seed Bank in Portland, OR. Species in bold had >0% germination while species in regular text had 0% germination after approximately 4 weeks.

Species[1]	Yr.[2]	Date	#[3]	Cond.[4]	Seed pretreatments[5] and germination data[6]
Abronia latifolia	U	7/22/2022	~50	L	C=0, GA*=0
Acer circinatum	21	5/6/2022	9	L	C=0, CS=0
Achlys triphylla	22	5/6/2022	~25	L	C=0, CS=0
Acmispon americanus	21	7/21/2022	~100	L	C=4, GA*=4
Acmispon americanus	21	5/6/2022	~250	L	C=0, CS=0
Actaea rubra	0.5	4/16/2022	~5	GH	GA+D=1, GA+L=0
Adenocaulon bicolor	21	5/3/2022	~25	L	C=0, GA=0
Agoseris grandiflora	U	7/25/2022	~100	L	C=2, GA*=2
Agrostis exarata	21	5/6/2022	~250	L	C=4, CS=4
Alisma triviale	22	4/29/2022	~25	L	C=0, CS=0
Alisma triviale	22	4/16/2022	~25	L	C=0
Allium acuminatum	23	6/25/2022	~100	L	C=0, CS=0
Amelanchier alnifolia	21	5/6/2022	~10	L	CS=0
Anaphalis margaritacea	0.5	2/11/2022	U	GH	C+D=U, C+L=U, GA+D=U, GA+L=0
Arbutus menziesii	0.5	2/25/2022	5	GH	C+D=0, C+L=0, GA+D=2,
Arctostaphylos columbiana	0.5	2/11/2022	~5	GH	C+D=0, C+L=0, GA+D=0, GA+L=0
Baccharis pilularis	0.3	2/11/2022	U	GH	C+D=0, C+L=0, GA+D=0, GA+L=0
Balsamorhiza sagittata	0.5	2/11/2022	8	GH	C+D=0, C+L=0, GA+D=0, GA+L=0
Beckmannia syzigachne	31	6/25/2022	~250	L	C=4
Berberis aquifolium	21	5/6/2022	~100	L	C=0, CS=0
Berberis aquifolium	0.5	2/11/2022	3	GH	C+D=0, C+L=0, GA+D=0, GA+L=0
Berberis nervosa	0.5	2/11/2022	3	GH	C+D=0, C+L=0, GA+D=0, GA+L=0
Bidens frondosa	U	7/21/2022	~100	L	C=1, GA*=1
Bistorta bistortoides	0.5	2/11/2022	7	GH	C+D=0, C+L=0, GA+D=0, GA+L=0
Bromus sitchensis	21	5/6/2022	~100	L	C=0, CS=0
Bromus sitchensis var.	U	7/21/2022	~100	L	C=1, GA*=1
Bromus vulgaris	22	5/6/2022	~50	L	C=0, CS=1
Calandrinia menziesii	2	7/21/2022	~50	L	C=1, GA*=1
Calyptridium umbellatum	U	7/25/2022	~100	L	C=3, GA*=3
Camassia leichtlinii	21	5/3/2022	~25	L	C=0, GA=0

Camassia quamash	0.5	2/11/2022	7	GH	C+D=0, C+L=0, GA+D=0, GA+L=0
Carex aperta	21	5/6/2022	~250	L	C=0, CS=1
Carex leptopoda	21	5/11/2022	~500	L	C+D=1, C+L=2
Carex macrocephala	U	7/22/2022	~100	L	C=0, GA*=0
Carex microptera	U	7/21/2022	~200	L	C=3, GA*=4
Carex obnupta	0.5	6/3/2022	~25	L	C=0, GA=1
Carex scoparia	21	5/6/2022	~250	L	C=2, CS=2
Carex unilateralis	21	5/5/2022	~250	L	C=3, CS=3
Castilleja hispida	U	7/22/2022	~50	L	C=0, GA*=0
Ceanothus velutinus	0.5	2/11/2022	1	GH	GA+L=0
Chamaenerion angustifolium	0.5	2/11/2022	U	GH	C+D=U, C+L=U, GA+D=U, GA+L=U
Claytonia parviflora	2	7/21/2022	~200	L	C=1, GA*=1
Clintonia uniflora	0.5	2/11/2022	5	GH	C+D=0, C+L=0, GA+D=0, GA+L=0
Collomia grandiflora	21	4/29/2022	~100	L	C=1, CS=1
Collomia grandiflora	21	4/16/2022	~50	GH	C+D=0, C+L=0
Collomia heterophylla	21	5/6/2022	~25	L	C=1, CS=1
Coreopsis tinctoria	21	5/5/2022	~250	GH	C+D=2, C+L=3
Coreopsis tinctoria	21	5/11/2022	~250	GH	C+D=3, C+L=3
Cornus sericea	21	5/6/2022	~100	L	C=0, CS=0
Cornus sericea	21	5/6/2022	~25	L	C=0, CS=0
Cornus sericea	0.5	2/11/2022	4	GH	C+D=2, C+L=2, GA+D=2, GA+L=2
Corydalis aquae-gelidae	21	4/16/2022	~25	GH	C+D=0, C+L=0
Corydalis caseana	14	4/16/2022	~25	GH	C+D=0, C+L=0
Dasiphora fruticosa	0.5	2/11/2022	~5	GH	C+D=3, C+L=0, GA+D=0, GA+L=4
Delphinium trolliifolium	U	7/21/2022	~100	L	C=0, GA*=0
Deschampsia danthoni-	U	7/21/2022	~100	L	C=4, GA*=4
Deschampsia elongata	21	5/6/2022	~100	L	C=4, CS=4
Dicentra formosa	0.5	2/11/2022	5	GH	C+D=0, C+L=0
Downingia elegans	23	5/6/2022	~500	L	C=1, CS=1
Eleocharis sp.	30	6/25/2022	~500	L	C=1, CS=1
Elymus glaucus	21	5/6/2022	~100	L	C=2, CS=2
Elymus glaucus	21	4/16/2022	~25	GH	C+D=1, C+L=0
Epilobium densiflorum	21	6/25/2022	~250	L	C=0, CS=1
Ericameria nauseosa	0.5	2/11/2022	~250	GH	C+D=0, C+L=0, GA+D=0, GA+L=0
Erigeron peregrinus	U	7/22/2022	~100	L	C=5, GA*=5
Eriophorum chamissonis	U	7/22/2022	~50	L	C=0, GA*=0

Eriophorum viridicarinatum	0.5	2/11/2022	5	GH	C+D=0, C+L=0, GA+D=0, GA+L=0
Eriophyllum lanatum	21	5/5/2022	~250	L	C=0, CS=2
Eriophyllum lanatum	22	5/6/2022	~25	L	C=0, CS=0
Eryngium castrense	31	6/25/2022	~25	L	C=1, CS=1
Eryngium petiolatum	2	7/21/2022	~100	L	C=1, GA*=1
Erythronium sp.	0.5	5/15/2022	5	GH	C+D=0, C+L=0
Festuca occidentalis	24	5/6/2022	~100	L	C=0, CS=0
*Festuca ovina**	21	5/6/2022	~100	L	C=0, CS=1
Frangula purshiana	21	5/6/2022	~25	L	C=0, CS=0
Garrya elliptica	0.5	2/11/2022	4	GH	C+D=0, C+L=0, GA+D=0, GA+L=0
Geum macrophyllum	21	4/16/2022	~25	L	C=3, GA=5
Gilia capitata	21	5/5/2022	~250	L	C=1, CS=2
Gilia capitata	21	5/11/2022	~100	GH	C+D=4, C+L=4
Glyceria elata	22	5/6/2022	~100	L	C=1, CS=0
Glyceria elata	27	5/6/2022	~250	L	C=0, CS=0
Grindelia integrifolia	31	6/23/2022	~25	GH	C=3, GA=3
Grindelia integrifolia	31	6/25/2022	~250	L	C=0, CS=0
Holodiscus discolor	0.5	2/11/2022	U	GH	C+D=0, C+L=0, GA+D=0, GA+L=0
Holodiscus discolor	21	4/16/2022	~25	GH	C+D=0, C+L=0
Hosackia pinnata	U	7/22/2022	~50	L	C=2, GA*=0
Hymenoxys cooperi	21	4/29/2022	~100	L	C=0, CS=0
Hymenoxys cooperi	21	4/16/2022	~100	GH	C+D=0, C+L=0
Iris missouriensis	0.1	6/25/2022	~25	L	C=0, CS=3
Iris tenax	22	5/6/2022	~25	L	C=0, CS=0
Juncus breweri	0.5	2/11/2022	~12	GH	C+D=2, C+L=5, GA+D=0, GA+L=5
Juncus bufonius	U	7/21/2022	~500	L	C=2, GA*=2
Juncus oxymeris	23	5/6/2022	~500	L	C=3, CS=3
Juncus tenuis	23	5/6/2022	~500	L	C=3, CS=3
Juncus torreyi	23	5/6/2022	~500	L	C=3, CS=3
Lathyrus japonicus	0.5	2/25/2022	3	GH	C+D=6, C+L=0, GA+D=6, GA+L=6
Lathyrus littoralis	U	7/22/2022	~50	L	C=3, GA*=3
Lomatium bradshawii	38	5/2/2022	~100	L	C=0, CS=0, GA=0
Lomatium macrocarpum	U	7/25/2022	~25	L	C=0, GA*=0
Lomatium triternatum	2	7/21/2022	~100	L	C=0, GA*=0
Lomatium utriculatum	U	7/22/2022	~100	L	C=0, GA*=0
Lonicera involucrata	0.5	2/11/2022	5	GH	C+D=0, C+L=0, GA+D=1, GA+L=0
Lonicera utahensis	0.5	2/11/2022	2	GH	GA+D=3, GA+L=3

Lupinus albicaulis	21	5/6/2022	~100	L	C=3
Lupinus albicaulis	21	5/11/2022	~250	L	C=3, GA=3
Lupinus albicaulis	21	5/2/2022	~100	L	C=3
Lupinus argenteus	21	4/16/2022	~25	GH	C+D=3, C+L=0
Lupinus argenteus	21	5/4/2022	~100	L	C=3
Lupinus littoralis	0.5	2/25/2022	~5	GH	C+D=2, C+L=3, GA+D=2, GA+L=3
Luzula parviflora	U	7/22/2022	~50	L	C=0, GA*=0
Luzula parviflora	21	5/3/2022	~25	L	C=0, GA=0
Madia elegans	2	7/21/2022	~50	L	C=0, GA*=0
Maianthemum stellatum	0.5	2/11/2022	5	GH	C+D=0, C+L=0, GA+D=0, GA+L=0
Melica subulata	3	7/21/2022	~100	L	C=0, GA*=0
Menyanthes trifoliata	0.5	2/11/2022	2	GH	GA+D=6, GA+L=0
Microseris laciniata	31	6/25/2022	~250	L	C=1, CS=0
Montia chamissoi	21	2/11/2022	U	GH	GA+D=U, GA+L=U
Myrica californica	0.5	2/11/2022	3	GH	GA+D=0, GA+L=0
Navarretia intertexta	2	7/21/2022	~100	L	C=2, GA*=0
Navarretia squarrosa	2	7/21/2022	~100	L	C=1, GA*=4
Oemleria cerasiformis	21	5/6/2022	~25	L	C=0, CS=0
Oemleria cerasiformis	21	5/6/2022	~25	L	C=0, CS=0
Oenanthe sarmentosa	21	5/6/2022	~200	L	C=1, CS=0
Oenothera wolfii	33	4/29/2022	~100	L	C=0, CS=0
Oplopanax horridus	0.5	2/11/2022	3	GH	GA+D=0, GA+L=0
Oreostemma alpigenum	U	7/22/2022	~100	L	C=2, GA*=2
Osmorhiza berteroi	22	5/6/2022	~100	L	C=0, CS=0
Osmorhiza berteroi	22	5/6/2022	~100	L	C=0, CS=1
Pedicularis sp.	0.5	2/11/2022	5	GH	C+D=0, C+L=0, GA+D=0, GA+L=3
Penstemon fruticosus	0.5	2/11/2022	~25	GH	C+D=0, C+L=0, GA+D=3, GA+L=6
Phacelia nemoralis	0.5	4/16/2022	~25	GH	GA+D=0, GA+L=0
Phacelia sp.	0.5	2/11/2022	~25	GH	GA+D=1, GA+L=0
Philadelphus lewisii	0.5	5/5/2022	200	L	C+D=2, C+L=3, CS+D=4, CS+L=4
Philadelphus lewisii	0.5	2/11/2022	~25	GH	C+D=1, C+L=5, GA+D=4, GA+L=1
Physocarpus capitatus	0.5	2/11/2022	10	GH	C+D=0, C+L=6, GA+D=1, GA+L=3
Plagiobothrys figuratus	U	7/21/2022	~250	L	C=2, GA*=2
Plagiobothrys figuratus	31	6/25/2022	~250	L	C=0, CS=1
Plantago maritima	0.5	2/11/2022	~25	GH	C+D=0, C+L=0, GA+D=6, GA+L=3

Plectritis congesta	22	5/6/2022	~200	L	C=4, CS=5
*Potentilla glandulosa**	21	5/6/2022	~100	L	C=1, CS=2
Potentilla gracilis	31	6/25/2022	~500	L	C=1, CS=2
Poteridium occidentale	U	7/21/2022	~50	L	C=5, GA*=5
Prunella vulgaris	31	6/25/2022	~250	L	C=0, CS=0
Prunella vulgaris	0.5	2/11/2022	~7	GH	C+D=0, C+L=5, GA+D=6, GA+L=6
Prunus emarginata	21	5/3/2022	~25	L	C=0, GA=0
Ranunculus occidentalis	31	6/25/2022	~250	L	C=1, CS=1
Ranunculus orthorhynchus	U	7/21/2022	~50	L	C=1, GA*=1
Rhamnus purshiana	0.5	2/11/2022	2	GH	C+D=0, C+L=0, GA+D=6, GA+L=0
Ribes cereum	0.5	2/25/2022	4	GH	C=0, GA=5
Ribes cereum	0.5	4/16/2022	~5	L	C=0, GA=0
Ribes cereum	0.5	2/11/2022	~5	GH	GA+D=0, GA+L=0
Ribes sanguineum	21	5/6/2022	~250	L	C=0, CS=1
Ribes sp.	0.5	2/11/2022	~5	GH	C+D=0, C+L=0, GA+D=0, GA+L=0
Ribes sp.	0.5	2/11/2022	~5	GH	C+D=0, C+L=0, GA+D=0, GA+L=0
Ribes sp.	0.5	2/11/2022	~5	GH	C+D=0, C+L=0, GA+D=0, GA+L=0
Rosa gymnocarpa	0.5	2/11/2022	5	GH	C+D=0, C+L=0, GA+D=0, GA+L=0
Rosa woodsi	0.1	6/25/2022	~100	L	C=0, CS=2
Rubus leucodermis	0.5	2/11/2022	5	GH	C+D=0, C+L=0, GA+D=0, GA+L=0
Sambucus mexicana ssp. *cerulea*	0.5	2/11/2022	5	GH	C+D=0, C+L=0, GA+D=0, GA+L=0
Sambucus mexicana ssp. *cerulea*	21	5/6/2022	~25	L	C=0, CS=1
Sambucus racemosa	21	5/6/2022	~100	L	C=0, CS=0
Sambucus racemosa	21	5/3/2022	~25	L	C=0, GA=0
Sambucus racemosa	0.5	2/11/2022	~25	GH	C+D=0, C+L=0, GA+D=0, GA+L=0

Sanicula bipinnatifida	U	7/22/2022	~50	L	C=0, GA*=0
Sarcobatus vermiculatus	0.5	2/11/2022	2	GH	GA+D=0, GA+L=0
Schoenoplectus americanus	21	5/6/2022	~250	L	C=1, CS=2
Silene sp.	0.5	2/11/2022	5	GH	C+D=0, C+L=0, GA+D=0, GA+L=0
Solidago lepida	21	5/6/2022	~25	L	C=0, CS=0
Sorbus sitchensis	0.5	2/11/2022	5	GH	C+D=0, C+L=0, GA+D=0, GA+L=0
Sparganium emersum	21	5/6/2022	~250	L	C=0, CS=0
Symphoricarpos albus	21	5/2/2022	~100	L	C=0, CS=0, GA=0
Symphoricarpos albus	0.5	2/11/2022	5	GH	C+D=0, C+L=0, GA+D=0, GA+L=0
Symphyotrichum chilense	0.5	2/11/2022	~25	GH	C+D=0, C+L=0, GA+D=0, GA+L=0
Symphyotrichum hallii	31	6/25/2022	~100	L	C=0, CS=0
Tanacetum bipinnatum	0.5	2/11/2022	~5	GH	C+D=0, C+L=0, GA+D=2, GA+L=0
Thalictrum polycarpum	U	7/21/2022	~100	L	C=0, GA*=5
Thermopsis montana	0.5	2/25/2022	~5	GH	C+D=4, C+L=3, GA+D=6, GA+L=2
Tolmiea menziesii	22	5/6/2022	~25	L	C=0, CS=1
Tolmiea menziesii	22	5/3/2022	~100	L	C=0, GA=0
Toxicoscordion vene-	31	6/25/2022	~250	L	C=0, CS=0
Toxicoscordion vene-	31	6/25/2022	~250	L	C=0, CS=0
Triantha occidentalis	U	7/25/2022	~200	L	C=0, GA*=0
Vaccinium deliciosum	0.5	2/11/2022	~25	GH	GA+D=3, GA+L=3
Vancouveria hexandra	22	5/6/2022	~25	L	C=0, CS=0
Vicia nigricans var. *gigantea*	21	5/6/2022	~25	L	C=4
Wyethia angustifolia	U	7/21/2022	~100	L	C=1, GA*=3
Xerophyllum tenax	0.5	2/11/2022	~5	GH	C+D=0, C+L=0, GA+D=0, GA+L=0

[1]Taxonomy follows OregonFlora.org except for species followed by an "*", which are listed given the original taxon that has since been split into multiple taxa.
[2]Years in cold, dry storage (5°C and 25% relative humidity). Taxa with "U" were purchased from local native plant nurseries and their storage conditions and time in storage are unknown. [3]The number of seeds sown per treatment. The "~" indicates an estimated number. [4]Germination conditions: **L** (laboratory; seeds sown in transparent plastic germination boxes

[11×11×14 cm] and placed within a growth chamber [Model 136LLVL, Percival Scientific, Perry, IA, U.S.A.] set at 25/15 °C daily high/low temperatures with a 12-hour daily photoperiod that corresponds to high temperatures) or **G** (greenhouse; seeds sown in 48-cell liner trays [6.0×6.6×4.3 cm] containing a commercial bagged potting mix [Greenhouse Mix, Northwest Inc., Canby, OR, U.S.A; 45% mulch, 25% peat moss, 20% pumice, 5% vermiculite, 5% perlite, and a wetting agent]).

[5]Seed pretreatments: **C** (control; sown without pretreatments), **CS** (cold stratified; seeds imbibed and placed in 5°C for 4 months), **GA** (Gibberellic Acid; seeds soaked in a 1,000 ppm Gibberellic Acid solution [Research Products International, Mt Prospect, IL, USA] or when "GA*" seeds were immersed in 1,000 ppm GA [ProGibb, Valent BioSciences] within a germination box and were thereafter hydrated with distilled water), **D** (dark; seed lightly covered with potting mix), or **L** (light; seed sown on top of the potting mix). Treatments with a "+" indicate two treatments (GA + D = seeds soaked in Gibberellic Acid and placed in dark conditions).

[6]Seeds were considered germinated when the radicle emerged from the seed coat >1mm for laboratory trials and when cotyledons were visible above the potting mix for greenhouse trials. Germination percentages were estimated visually and placed into one of the five following classes where 0=0%, 1=0.1 to 5%, 2=6 to 25%, 3=26 to 50%, 4=51 to 75%, 5=76 to 100%.

Combined Proceedings IPPS 73:399-403. 2023.

How to Reduce Complexity for Better Processes

Della Fetzer

Rebel Cultures, 44 E 8th Street, Holland, MI 49423

della@rebelcultures.com

Keywords: process complexity, process improvement, current state, future state

Summary

In this exploration of process complexity in production and research, I underscore the pivotal role of effective processes in achieving our most critical goals. I delve into the common pitfalls of process improvement, attributing failures to the inadvertent introduction of complications through unnecessary features, drawing parallels from the software industry. I advocate for intentional efforts to simplify processes, proposing a systematic approach involving understanding the current state, collaboratively simplifying the future state, and rigorously validating changes to reduce complexity. I emphasize the importance of recognizing and optimizing for innate human abilities, such as adaptation and contribution. Ultimately, I highlight the need for resilience, collaboration, and proactive enthusiasm for positive changes to foster effective process improvement and contribute to a self-sustaining cycle of progress.

IPPS Vol. 73 – 2023

INTRODUCTION

Process complexity in production and research processes accumulates gradually and can prevent the achievement of our most important goals when not intentionally removed **(Fig. 1)**.

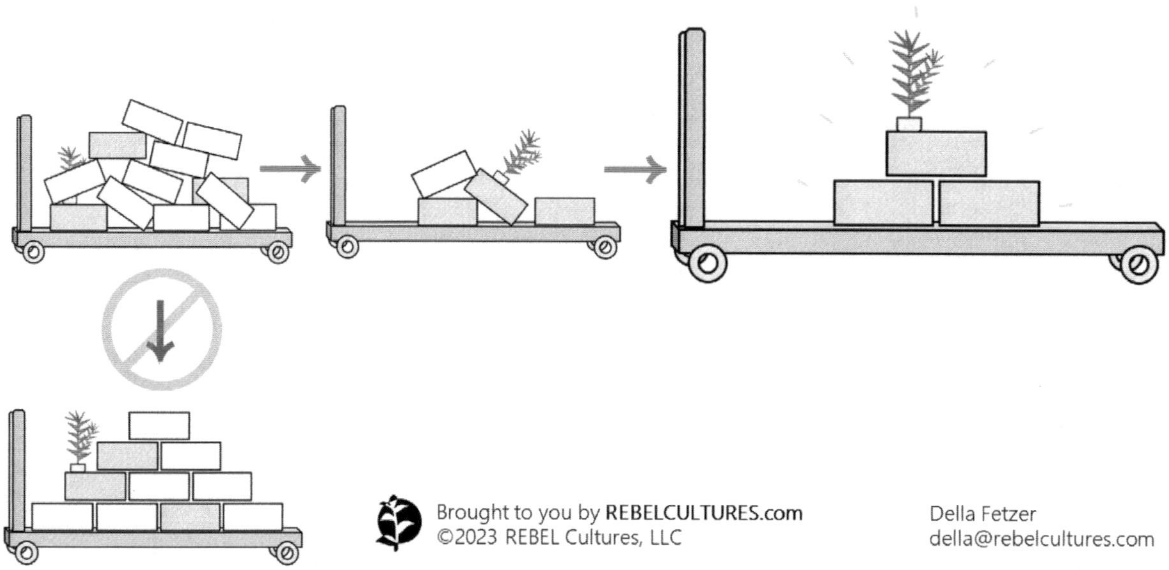

Figure 1. The result of organizing an overly complex process is an equally heavy process, which is difficult to maintain, in contrast to a dramatically simplified process that can continue to be improved and is more easily repeated and sustained.

Why does process improvement matter?

Every set of steps we do is a process, and processes are how we accomplish things. This means that without effective processes, we have no way of accomplishing key outcomes in our lives and careers; for our businesses; and within our industry.

Why does process improvement fail so often?

Process improvement fails when an implemented solution creates more problems than it solves. The likelihood of problems occurring increases as complexity is added to the process. Complexity is often disguised as features that appear useful, but are actually nonessential to the process. One example of this is evidenced by the gap between features developed by the software industry and used by end-users: Of all software features developed, launched, and purchased, 80% have been found to be rarely or never used by the consumer (Forbes, 2019). These features, often mislabeled as "improvements," actually create complexity instead of reducing it. This leads to less effective processes and failed process improvement efforts.

How can complexity be removed from processes?

Process complexity gradually accumulates as demands change and access to new tools grows over time. Very few processes remain exactly the same for long, even if people feel that a process has always been a certain way. In reality, every time that demands and tools change, the holistic process has also changed, even if some familiar process elements remain. The removal of complexity must start with the intention and readiness to simplify a process that has accumulated complexity over time. Once this readiness is achieved, the following progression of techniques may be used to guide a complexity reduction effort to improve a process **(Fig. 2)**:

First, understand the current state (without solving any problems). Leaders who invest adequate time in understanding the current state are guaranteed to solve more problems and avoid more pitfalls than those who rush to implement solutions. They should exist where the work is happening, and observe without judgment until it is understood what, when, and why things are happening. Leaders should also ask the team performing the work questions about the process and listen to responses without judgment. They should also individually map information, supplies and tools, and product flows of the current state. And then they should map the process of a hypothetical future state where only steps that add value occur.

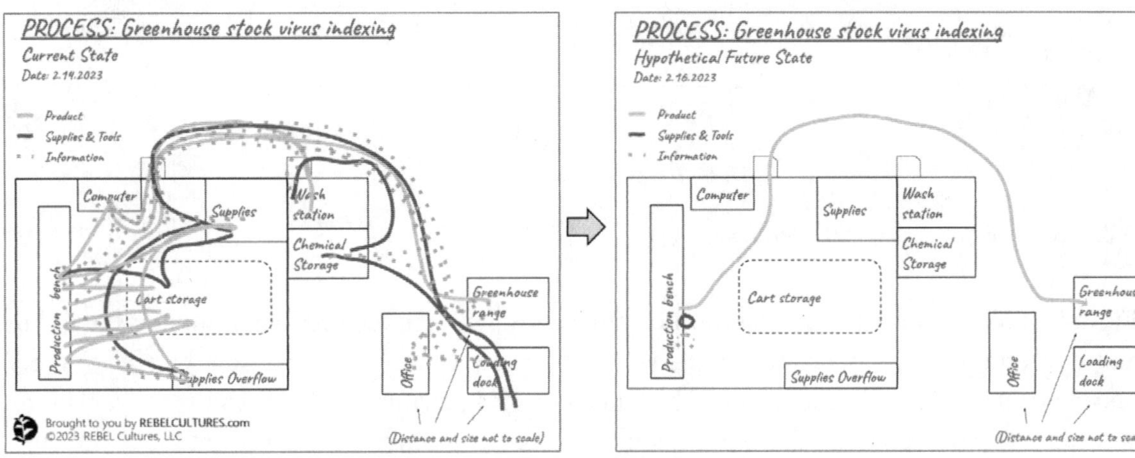

Figure 2. Diagrams of the current state of a process, including the flow of information, supplies and tools, and products in contrast with a hypothetical future state of the same process, focused exclusively on the highest value step(s).

Second, simplify the future state**.** This is where process change actually occurs. Leaders should discuss with process experts all requirements that would enable the hypothetical future state, recording all responses, even those which are not possible. All feasible requirements from this document can be pursued to reduce complexity and improve the process. Once a change is in place, check to validate that complexity was actually removed, rather than transferred elsewhere within or outside of the process. Efforts should then be made to provide necessary resources and support to all team members throughout changes, especially as unexpected complications arise.

Third, understand natural human ability. While every human is unique, understanding broad patterns in natural human ability can inform how to sustain less complex processes and where future automation can provide the highest value. Humans do *NOT* naturally excel in repetition or reading. Quickly repeating tasks with low variation, either in physical motion or mental patterns, such as simple calculations or counting. Optimize for comfort and variation wherever possible, and consider automation after the process has been simplified as much as possible. Intake of written information translated to an output of material is complex and requires translation. Visual instructions which detail exactly when a task is complete and correct are superior to written instructions **(Fig. 3)**.

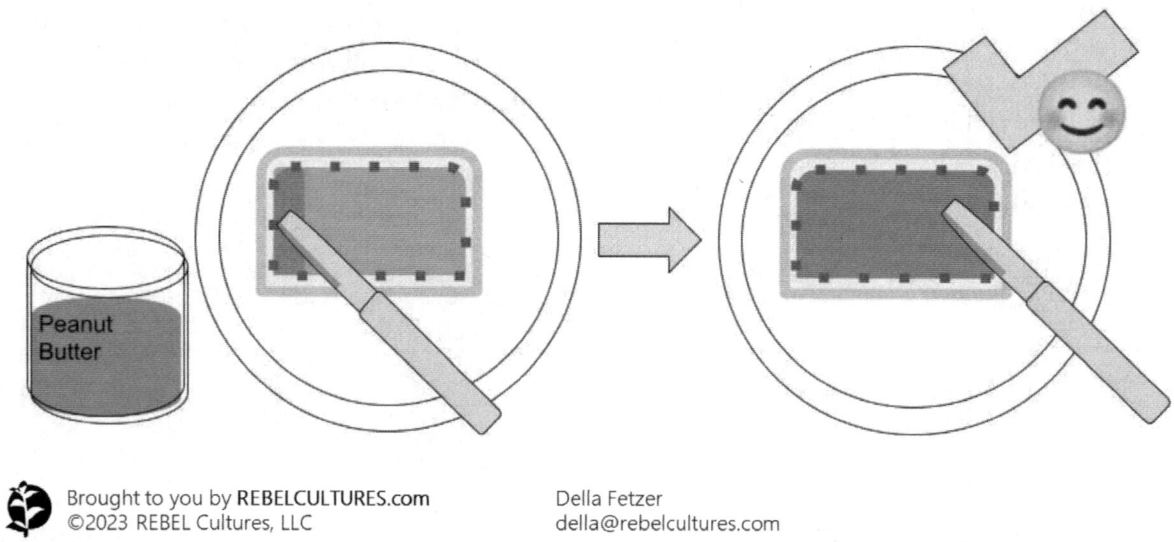

Figure 3: Visual instructions require no translation. They accelerate new employee training and cross-training. They are fast and easy for workers to reference, and are easy for others outside of the process to understand. This understanding enables higher-quality conversations and less misinterpretation with other managers, outside auditors, and even international suppliers and potential clients.

Humans *DO* naturally excel in adaptation and contribution. Like any living organism, humans adapt when changes in their process or environment occur. When teammates struggle to adapt, be sure to take extra time to understand the real source of resistance, which is often a fear that their effort in becoming equally proficient in a new method will not be adequately supported. Humans will adapt to a new process, but also have the power to inadvertently adapt any process to themselves. People are intrinsically motivated to contribute to solutions, which impact their work. Encourage mass participation and learn to value difficult conversations because difficult conversations lead to significant breakthroughs.

CONCLUSION

Effective process improvement requires a willingness to make many mistakes, so allow yourself to be wrong, but make it right, and allow others to do the same. A good team and good industry colleagues will be there to support you. The future of plant propagation depends on increasingly better practices. When we as leaders take steps toward making things better, more people who also care about making things better will want to work with us. This is the ultimate cycle of self-sustaining, continuous improvement. Lastly, think about what you want most out of your work. Let yourself be excited about simplifying and improving the process, and let your process make it real.

Combined Proceedings IPPS 73:404-409. 2023.

Biochar in Propagation Substrates: Sustainable solution or Impractical Idea?

Benjamin K. Hoover

Plant Sciences Department, California Polytechnic State University, San Luis Obispo, California 93407, USA

bkhoover@calpoly.edu

Keywords: biochar, substrates, media, sustainability, seed propagation, vegetative propagation, cutting

Summary

Identifying sustainable horticultural substrates is critical, but what does sustainability really mean? Biochar is perceived as sustainable in many settings, but does it deserve this status in plant propagation? I conducted experiments with coconut-shell biochar to assess its suitability in seed propagation and vegetative propagation substrates. Biochar performed well as a substrate amendment in my experiments. However, the costs associated with the biochar make classifying it as sustainable a nuanced discussion.

INTRODUCTION

Horticultural substrates need to be very consistent and high performing, especially in propagation settings (Davies et al., 2018). Traditional propagation substrate components, such as sphagnum peat and perlite, are increasing in cost and subject to supply-chain interruptions (Jackson, 2022). In addition, the sustainability of peat has become a controversial question.

Sustainability is not just an environmental issue; it is also economic and social (Purvis et al., 2019). Growers are looking for economically and environmentally sustainable substrates. Expanding the list of effective and proven substrates would allow growers to respond to supply issues and select sustainable options. Biochar, pyrogenic carbonaceous material that may be used as a growth medium for plants, has drawn attention as a promising candidate for the horticulture industry (Dumroese et al., 2011; Vaughn et al., 2013). Over the past 6 years, I have conducted seed- and vegetative-propagation experiments with herbaceous plants in substrates amended with coconut-shell biochar. Data from these experiments were analyzed with the appropriate test for the design, in most cases analysis of variance (ANOVA) with post-hoc Tukey HSD Testing. Test assumptions were checked, and the threshold for rejecting a null hypothesis was $P \leq 0.05$.

Biochar in These Experiments

The biochar used in my experiments came from a commercial supplier (Bay Area Biochar, Concord, CA). It was made from coconut-shell feedstock, which underwent fast pyrolysis at ~ 700 °C. The components were then ground and acidified to a pH of 6.4. The chemical and physical properties of the biochar was relatively consistent in the batches used in these experiments (Figure 1, Table 1). The raw material cost of the biochar was approximately three times that of coarse perlite or sphagnum peat by volume.

Table 1. Nutrient and particle-zize analyses of the coconut-shell biochar used in seed germination and vegetative propagation experiments in this report. Data are from a well-mixed sample submitted to a commercial laboratory (Waypoint Analytical, Anaheim, CA).

		Particle Size (dry weight)	
Nutrients (ppm)		*Screen (mm)*	*Passing (%)*
Nitrogen (N)	17	9.5	100.0
Phosphorus (P)	104	6.4	99.8
Potassium (K)	2831	4.8	99.4
Calcium (Ca)	248	2.4	97.7
Magnesium (Mg)	116	1.0	37.5
		0.5	15.4

Figure 1. A typical coconut-shell biochar sample, showing particle shapes and sizes (bar = 1 mm).

Seed- Propagation Experiments

A completely-randomized-design experiment was conducted to test biochar effects on seed germination in a laboratory setting. The experimental unit was a Petri dish, with a base substrate of sphagnum peat and fine perlite (1:1, v/v) amended with biochar (0, 10, 20, 40, 80% v/v), containing 20 seeds. *Coreopsis grandiflora*, *Eschscholzia californica*, *Lavandula angustifolia*, and *Rudbeckia fulgida* were the plant species tested. Each species was treated as a separate experiment (n = 10). Over a 14-day germination period, the only significant difference that occurred was a reduction of *Eschscholzia* germination in 80% biochar compared to the other treatments. The other species showed no significant germination differences related to biochar.

Following the lab study, two seed germination and growth trials were conducted in two germination rooms. The experiments were randomized complete block designs; each experimental unit was a 4-inch container with 10 seeds. The plant seeds were *Coreopsis grandiflora*, *Eschscholzia californica*, and *Leucanthemum* × *superbum*, with each species treated as a separate experiment (n = 10). The base substrate was a modified Cornell germination mix (sphagnum peat and fine vermiculite, 1:1, v/v) amended with biochar (0, 5, 10, 20, 40%, v/v). Root length and shoot length were measured. Growth of the seedlings increased slightly with added biochar content, but despite being statistically significant, the differences had little practical significance. The results from these experiments were published in 2018 (Hoover, 2018).

In my lab and germination room testing, coconut-shell biochar performed well as a propagation substrate component. Germination was mostly not affected by biochar incorporation, with just one instance of a slight negative effect at 80% biochar in one species. Growth rates were slightly increased in seedlings when biochar was added; however, this positive effect was minimal.

Vegetative propagation experiments

Herbaceous cuttings were stuck in a sphagnum peat and coarse perlite (1:2, v/v) substrate amended with biochar (0, 10, 20, 40, 80%, v/v, **Fig.2**), and then placed a mist house. Randomized complete block designs were used, with the species being treated as separate experiments (n = 15).

The experimental unit was a 2.5-inch rose pot containing one cutting. Most of the species were repeated in a second round of experiments (n = 10). The species tested were: *Achillea* hybrid, *Ajuga reptans*, *Coreopsis verticillata*, *Iberis sempervirens*, *Leucanthemum* × *superbum*, *Phlox subulate*, and *Salvia* × *sylvestris*. The cuttings were evaluated after recommended rooting periods (21 to 36 days, depending on the species).

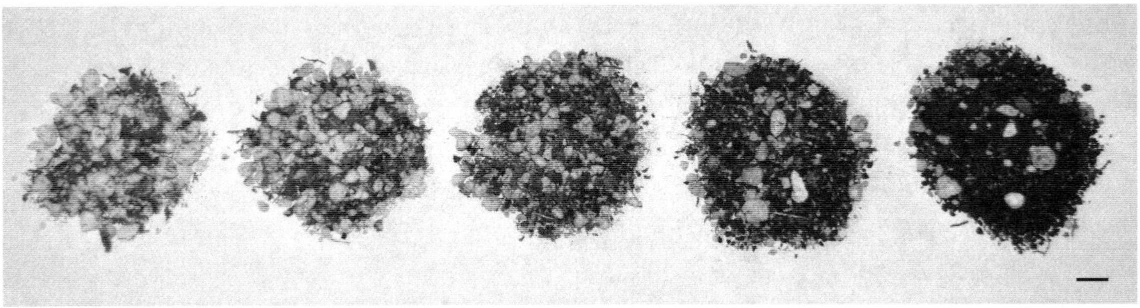

Figure 2. Sphagnum peat and coarse perlite (1:2, v/v) substrate amended with coconut-shell biochar at 0, 10, 20, 40, and 80% (left to right, v/v, bar = 1 cm).

I removed cuttings from containers, washed the roots, and assigned each cutting an adventitious rooting rating (0 = cutting dead, 1 = cutting alive but no root development, 2 = minimal root development, 3 = moderate root development but insufficient for transplanting, 4 = good root development and sufficient for transplanting, 5 = optimal root development). Roots were then excised and scanned (Epson Perfection V19). I used ImageJ to analyze root two-dimensional area, first-order root count, and primary root length in the scans. Biochar amendment of 0, 10, or 20% had no measured effect on root growth, though some slight positive trends were visible. Biochar at 40 or 80% either had no effect or a negative effect on root growth. The most pronounced effect was observed at 80% biochar, when many species had significantly more first order roots than the lower biochar treatments, yet those primary roots were shorter and less developed. Results from these experiments were shared via a presentation in 2017 (Hoover, 2017).

Follow-up experiments were conducted, matching the biochar particles with sand particles. Sand amendment did not affect rooting in the same fashion as biochar amendment, suggesting that the rooting difference was chemical or related to water and oxygen levels, rather than physical shape of the particles. Results from these experiments were presented via a poster in 2018 (Hoover and Mattlin, 2018).

I also conducted two experiments with herbaceous cuttings that involved biochar and drench treatment with indole-3-butyic acid with potassium salts (K-IBA). Cuttings were stuck in a sphagnum peat and coarse perlite (1:2, v/v) substrate amended with biochar (0, 10, 20, 40, 80%, v/v, Figure 2). I then applied 0, 1,000, or 3,000 IBA in either talc powder form or as a drench. Randomized complete block designs were used, with the species being treated as separate experiments (n = 20). The experimental unit was a 2.5-inch rose pot containing one cutting. The species tested were *Salvia* × *sylvestris* and *Scabiosa columbaria*. The cuttings were evaluated after 28 days. Biochar amendment at 80% negatively affected root development (**Table 2**), but low rates of biochar had no negative affect. Biochar presence did not influence K-IBA drench efficacy.

Table 2. Root measurement means of *Scabiosa* and *Salvia* cuttings in sphagnum peat and perlite substrate (1:2, v/v) amended with coconut shell biochar. Adventitious Root Rating (0 = cutting dead, 1 = cutting alive but no root development, 2 = minimal root development, 3 = moderate root development but insufficient for transplanting, 4 = good root development and sufficient for transplanting, 5 = optimal root development). Means in a column that do not share a letter are different according to a Tukey HSD test, $p \leq 0.05$, n = 20).

	Scabiosa		Salvia	
Biochar (%, v/v))	Rating	Root Area (cm^2)	Rating	Root Area (cm^2)
0	3.6 a	3.4 ab	3.0 a	3.6 a
10	3.5 a	3.4 ab	3.1 a	3.4 a
20	3.8 a	4.6 a	2.6 ab	2.7 a
40	2.8 b	2.3 bc	2.7 ab	2.8 a
80	2.0 c	0.8 c	2.2 b	0.9 b

In the vegetative propagation experiments, I saw favorable rooting responses when biochar was incorporated at rates of 20% or below. At 40% biochar, the response was either neutral or negative. The highest rate tested, 80% biochar, resulted in poorly developed roots.

CONCLUSION

The coconut-shell biochar in these studies had no negative effect on germination or rooting of cuttings when incorporated at rates up to 20%, and in some cases as high as 40%. While these outcomes are encouraging, they must be considered along with the financial cost associated with the biochar. At this time, the coconut-shell biochar used in my experiments could not be considered sustainable on an economic level. It is too expensive to displace other substrate options. On an environmental level the feedstock acquisition, processing, and shipping involved in making high quality biochar may also bring sustainability claims into question. While the biochar in these studies was very consistent, the product is expensive. Biochar may be created onsite or sourced from sellers, and as a general product it is very variable, spanning a large range of physical and chemical properties. This unpredictability means caution is required when predicting plant growth responses. Trials are recommended prior to large-scale adoption. If biochar production can become cost effective, with favorable porosity and chemical properties, it may have a future in propagation substrates.

Acknowledgments

Thank you to Chris Wassenberg, Pamela Crane-Hoover, Teresa Ray, Sara Do, Benjamin Taylor, and Jacob Mattlin for technical assistance in these studies. Thanks also to support from The California Association of Nurseries and Garden Centers (CANGC) CANERS Foundation, the Agricultural Research Institute (ARI), and the California Polytechnic State University College of Agriculture, Food and Environmental Sciences Summer Undergraduate Research Program (SURP).

LITERATURE CITED

Davies, F.T., Jr., Geneve R.L., and Wilson, S.B. (2018). Hartmann and Kester's plant propagation: Principles and practices. 9th ed. Pearson Education, New York, NY.

Dumroese, R.K., Heiskanen, J., Englund, K., and Tervahauta, A. (2011). Pelleted biochar: Chemical and physical properties show potential use as a substrate in container nurseries. Biomass Bioenergy *35*: 2018–2027.

Hoover, B.K. (2017). Adventitious rooting of herbaceous cuttings in biochar-amended substrates. The American Society for Horticultural Science Annual Conference, Waikoloa, HI. *52*(9): S148.

Hoover, B.K. (2018). Herbaceous perennial seed germination and seedling growth in biochar- amended propagation substrates. HortSci. *53*(2): 236-241.

Hoover, B.K. and Mattlin, J. (2018). Biochar and sand-amended cutting substrates: particle size effects. Combined Proc. Inter. Plant Prop. Soc. *68*: 221-222.

Jackson, B.E. (2022). Current and future state of the growing media industry. Greenhouse Grower. November 6, 2022. www.greenhousegrower.com/production/current-and-future- state-of-the-growing-media-industry-2/

Purvis, B., Mao, Y., and Robinson, D. (2019). Three pillars of sustainability: in search of conceptual origins. Sustainability Sci. *14*: 681-695.

Vaughn, S.F., Kenar, J.A., Thompson, A.R., and Peterson, S.C. (2013). Comparison of biochars derived from wood pellets and pelletized wheat straw as replacements for peat in potting substrates. Ind. Crops Prod. *51*(1): 437–443.

Combined Proceedings IPPS 73:410-412. 2023.

The Climate Ready Landscape Plant Trials

Lorence R. Oki

Department of Plant Sciences, MS6, Room 1110 Environmental Horticulture, University of California, One Shields Avenue, Davis, CA 95616-8780

lroki@ucdavis.edu

Keywords: climate, landscape, deficit irrigation

Summary

The Climate Ready Landscape Plant (CRLP) trials evaluates ornamentals at three levels of deficit irrigation to yield information about their performance and irrigation recommendations from a network of trial sites in Arizona, California, Oregon, Utah, and Washington. A weather-based irrigation protocol was used to apply deficit-irrigation treatments of 80%, 50%, or 20% of reference evapotranspiration. The frequency and quantity of irrigations varied between treatments while the volume applied was constant for all treatments. Thus, the 80% treatment was irrigated more frequently and received more water overall than the 20% treatment. Plant growth and aesthetic data were collected monthly from spring to fall when deficit treatments were imposed. Each site held at least one field day, inviting master gardeners and horticultural professionals to assess the aesthetics of the plants. This allowed researchers to contextualize the monthly aesthetic ratings with local norms and preferences while providing participants an opportunity to discover new plant material that they may use professionally. Using these data, researchers will develop irrigation recommendations for each of the 41 evaluated taxa, identifying the treatment where the least water was applied without compro-

IPPS Vol. 73 – 2023

mising aesthetics. With the results, the project seeks to leverage the wide distribution of sites across the western U.S. to identify taxa that might perform well across wide geographic ranges and inform plant selection in future locations as the climate changes to support nursery growers in selecting and offering resilient landscape plants.

INTRODUCTION

The Climate Ready Landscape Plant (CRLP) trials is a multi-state project evaluating ornamentals on three levels of deficit irrigation to yield information about their performance and irrigation recommendations. The project is an offshoot of the UC Landscape Plant Irrigation Trial (UCLPIT) program, started at UC Davis in 2004 by Lorence R. Oki and S. Karrie Reid. In 2020, funding from the Specialty Crops Multi-state Program allowed for the development of new tests sites located in five additional western states: University of Arizona (Tucson, AZ), Oregon State University (Aurora, OR), Utah State University (Logan, UT), and University of Washington (Seattle, WA), in addition to a previously developed site at the South Coast Research and Extension Center (Irvine, CA). The project seeks to leverage the wide distribution of sites across the West to evaluate new or recently introduced plant material to identify taxa that might perform well across wide geographic ranges and inform plant selection in future locations as the climate changes to support nursery growers in selecting and offering resilient landscape plants.

The project selected six taxa to grow at all sites: *Hibiscus* Purple Pillar®; *Philadelphus* 'Blizzard'; *Rosa* Petite Knockout®; *Vitex* Blue Diddley®; and experimental selections of *Hibiscus syriacus* and *Philadelphus*, which were developed by Dr. Ryan Contreras of Oregon State University. Cooperators at each site then selected 10 additional taxa for a total of 15 taxa per site. In some cases, the same taxa were planted at multiple sites to create a "regional" plant palette (e.g. the same taxon was grown at the Arizona, California, and Oregon sites). Plant selection was determined by industry cooperators submitting plants for evaluation or guided by suggestions from an advisory committee of industry professionals with members local to each site.

At each site, plants were installed in winter or spring of 2021, spaced 2 meters on center with rows spaced 2 meters apart, and covered with 5-8 cm of mulch. In 2021, supplemental water was regularly applied using a weather-based irrigation protocol to fully establish the plants. Starting in the spring of 2022, deficit irrigation treatments of 80%, 50%, or 20% of reference evapotranspiration (ET_o) were imposed with eight individuals per taxa assigned to each treatment, with the percentage functioning similarly to a crop coefficient. As a result, the frequency and quantity of irrigations varied between treatments while the volume applied was constant for all treatments. Consequently, the 80% treatment was irrigated more frequently and received more water overall than the 20% treatment. For all treatments, irrigations occurred when 50% of plant available water (PAW) was removed according to ET data collected from local agricultural weather station networks, e.g. AgriMet, CIMIS, and PAW information was derived from soil survey data. Treatments continued until Fall 2022.

Researchers at each site collected length, width, and height measurements monthly while treatments were imposed to assess if there was any growth difference amongst the treatments. Researchers also assessed aesthetics for each plant in the following categories: foliage quality, flowering quantity, pest/disease resistance, vigor, and overall appearance with a 1-5 point scale. Plants were rated in all categories on a monthly basis during the deficit season with additional mid-month flowering quantity and overall appearance recorded for blooming plants.

During the deficit season, each site held field day events, inviting master gardeners and horticultural professionals to rate the foliage quality, flowering abundance, and overall appearance of one individual per treatment for each taxon. This allows researchers to contextualize the monthly aesthetic data in local norms and preferences, while providing participants an opportunity to discover new plant material, which could be used professionally. Using these data, researchers will develop irrigation recommendations for the 41 evaluated taxa, identifying the treatments where the least water was applied without compromising aesthetics.

Combined Proceedings IPPS 73:413-417. 2023.

Initial Detection of Emerald Ash Borer in Oregon and Rapid Response

Max Ragozzino[1], Thomas Valente[1], and Alex Gorman[2]

[1]Oregon Department of Agriculture – Insect Pest Prevention and Management, 635 Capitol Street NE, Salem, OR 97301-2532, [2]Oregon State University, Corvallis, OR 97331-5704

Max.Ragozzino@oda.oregon.gov

Keywords: EAB, insect, invasive species, *Agrilus planipennis*

Summary

Emerald ash borer (EAB), a highly destructive invasive insect, has rapidly spread across North America since its introduction in the mid-1990s, causing extensive damage to ash trees. This paper details the detection and control efforts of EAB in northwest Oregon. The establishment of the Oregon EAB Response Task Force and implementation of intensive survey efforts revealed the extent of EAB infestation in the region. Acknowledging the challenges of eradicating EAB, the paper outlines a slow-ash-mortality (SLAM) strategy to limit the spread of EAB and to give affected parties sufficient time to make management decisions. Chemical control of EAB needs to occur before trees experience more than 20% canopy decline. Emamectin benzoate appears to be the most effective chemical to protect trees from succumbing to the invasive insect. Two additional strategies, including the creation of buffer zones and biological control, appear to worth further exploring as viable options to control EAB spread.

IPPS Vol. 73 – 2023

INTRODUCTION

Emerald ash borer (EAB), *Agrilus planipennis*, is one of the most devastating invasive insects in North America. EAB was first introduced to Michigan in the mid-1990s, but went undetected until 2002. By 2009, EAB was estimated to have killed 17 million trees, and caused over $25 billion in damage (Kovaks et al. 2009).

In the years since its introduction, it has spread to 36 states, reaching from Rhode Island to Oregon, and from Louisiana to Qubec. Its spread is not random. EAB adults fly relatively short distances, once per year. The primary mode of long-range dispersal is human-aided transport in infested wood material. We believe this is how EAB reached Oregon.

On 30 June 2022, a trained pest detector (and attentive citizen) detected EAB at Joseph Gale Elementary school in Forest Grove, Oregon. The elementary school parking lot had 16 ornamental ash trees planted when the school was built in 2012. All 16 trees showed the unmistakable signs of multi-year infestation, including canopy decline, branch dieback, split bark, D-shaped exit holes, and emerald ash borer adult beetles climbing up the bark. EAB larvae feed under the bark of the tree, eating the phloem tissue. This feeding damage girdles the tree from the inside, making early detection very difficult. Infestations are almost undetectable for at least two years.

Response

On the same day as the report, Oregon Department of Forestry officials were able to visit the site, and confirm the report. A sample was sent to Oregon Department of Agriculture (ODA) and the United States Department of Agriculture for taxonomic identification. Over the 4th of July weekend, Oregon Department of Agriculture officials contacted the elementary school, and requested their arborist destroy the infested trees. On 5 July 2022, ODA began scouting Forest Grove for the presence of EAB. Twenty-six green funnel traps, and two purple prism traps were immediately placed in the area around forest grove by USDA and ODA, respectively. ODA created and distributed a visual survey tool for detecting trees infested by EAB. By August, the Oregon EAB Response Task Force was formed, consisting of members from over 40 organizations.

A crew of four seasonal technicians were hired for the winter 2022-2023 season. Emerald ash borer damage is permenant. Bark splits, dead branches, and woodpecker damage are all visible after leaves drop. As of 8 January 2023, the visual survey has surveyed over 6000 ash trees in and around Forest Grove (**Fig. 1**). In doing so, they have helped to identify the current extent of EAB in Oregon.

Due to how easily emerald ash borer is spread unintentionally in wood, a quarantine was put into effect. Ash (*Fraxinus* spp.), white fringe tree (*Chionanthus virginicus*) and cultivated olive (*Olea europea*) are all viable hosts of emerald ash borer. Transporting woody material of EAB host plants risks spreading this destructive pest to uninfested areas, including Washington, California, Idaho, and the majority of Oregon. As of now, no woody material of EAB host plants can be transported out of Washington County, Oregon.

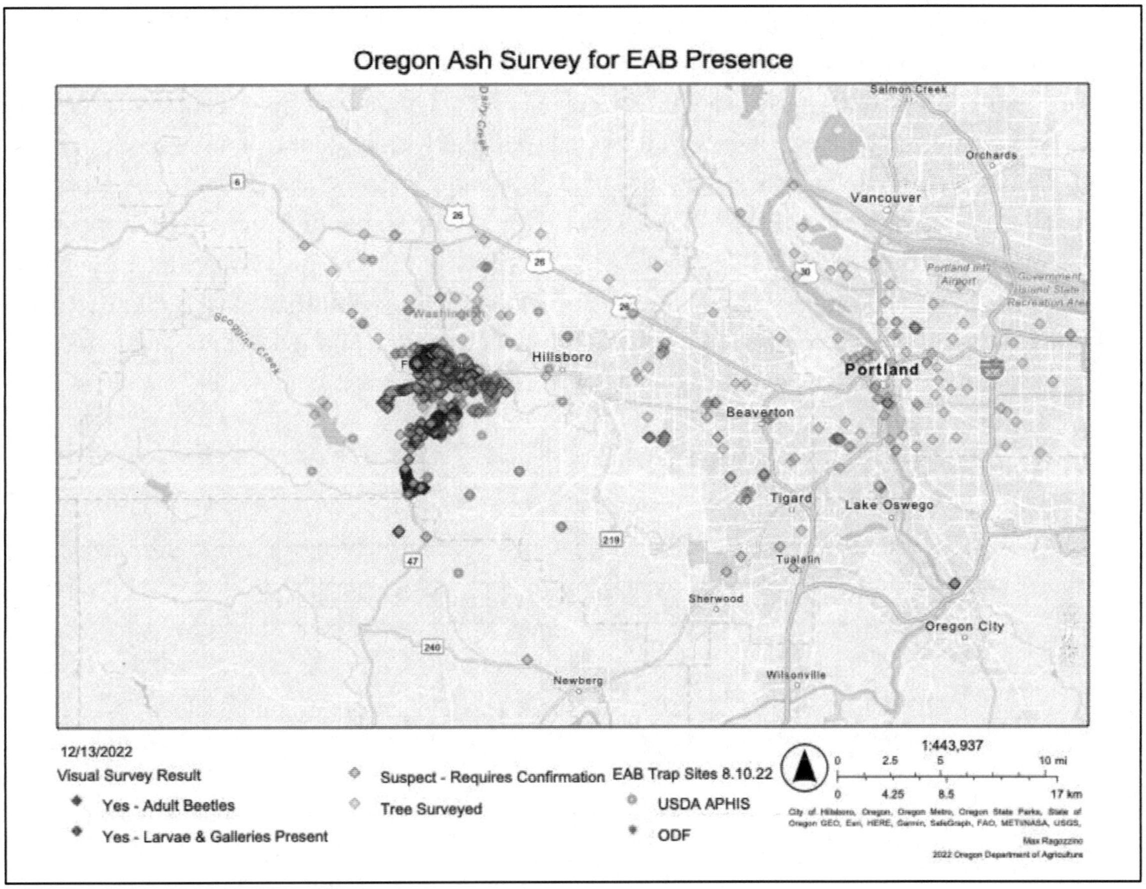

Figure 1. Visual summary of visual surveys for emerald ash borer in the greater Portland area of Oregon.

Planned Control Actions

Based on the response actions to EAB in other states and their related outcomes, we do not believe EAB can be eradicated from Oregon. We will be implementing slow-ash-mortality (SLAM) strategies to mitigate the spread of EAB from Forest Grove, and give growers, land managers, and home owners as much time as possible to make management decisions. SLAM strategies are intensive, and multifaceted. This includes conventional chemical control, creation of "buffer zone" of trap trees, and biological control using introduced natural enemies.

Chemical Control

Treatment applied correctly at the appropriate time of the year can protect trees with low level of infestation. Generally, trees with greater than 20% canopy decline are not considered good candidates for treatment due to the amount of damage already done. If a tree is going to be protected from EAB, treatment needs to occur **before** infestation within the tree has reached this level. Emamectin benzoate, applied via trunk-injection, has provided the highest efficacy of protection against EAB.

Buffer-zone strategies

Three strategies to create a "buffer zone" surrounding the known infested area (McCullough, et al. 2009; McCullough, et al. 2016). They include spring-girdling trees along a corridor to increase their attractiveness to ovipositing EAB. These "trap trees" will be removed and destroyed the following fall. In addition, selected trees in highly attractive positions within the spring-girdled corridor will be girdled and also trunk-injected with insecticide to create trap trees lethal to EAB larvae. Selected mature seed-producing trees will then be trunk-injected to maintain seed production and regeneration at the site.

Trunk-injection treatments are compatible with biocontrol release at nearby sites closer to Forest Grove. EAB parasitoids require healthy larvae for oviposition and will not oviposit in lethal trap trees. Treatments will utilize emamectin benzoate, which will be applied by independent certified contractors.

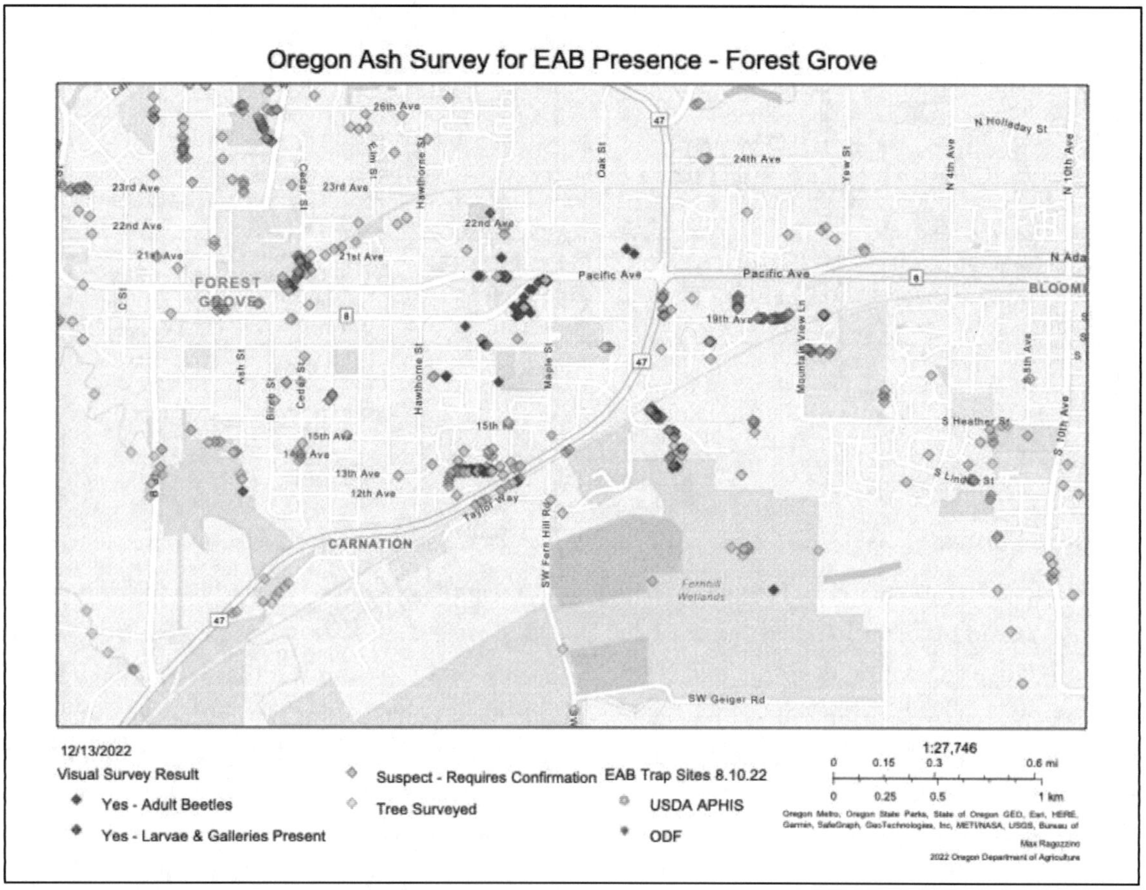

Figure 2. Visual summary of visual surveys for emerald ash borer presence in Forest Grove, Oregon.

Biological Control

Four species of introduced biological control agent have been introduced to North America against EAB. Three species which attack EAB larvae, *Spathius agrili, Spathius galinae*, and *Tetrastichus planipennisi*; and one species that attacks EAB eggs, *Oobius agrili*. Each of these species attacks EAB in a different niche. EAB infesting larger mature trees are heavily targeted by *Spathius* species, while EAB infesting saplings are more heavily targeted by *T. planipennisi*.

Both Gales Forest Grove and Fernhill Wetlands are planned release sites for the introduced biological control agents for EAB. Degree-day models suggest that *S. galinae*, *T. planipennisi*, and *O. agrili* will be suitable for release in Oregon. Within each site, we measured the diameter at breast height of ash trees to determine specific release locations for each species.

Combined Proceedings IPPS 73:418-420. 2023.

A Peek into the North Dakota State University Woody Plant Improvement Program

Todd P. West

North Dakota State University, Department of Plant Sciences, P.O. Box 6050, Fargo, ND 58108, USA

todd.p.west@ndsu.edu

Keywords: plant breeding, cold resistance, NDSU, plant introductions

Summary

North Dakota State University's Woody Plant Improvement Program (WPIP), initiated in 1971, has released 61 woody plant selections into the ornamental nursery trade since 1986. With a focus on the U.S. northern Great Plains, the program aims to evaluate and introduce new woody plants suitable for the region's challenging conditions. The evaluations, conducted at three primary research sites in North Dakota, have assessed over 200 genera and 3,000 species and cultivars of trees and shrubs. The extensive collection at the NDSU Horticulture Research Farm and Dale E. Herman Research Arboretum includes more than 11,000 accessions, making it the largest and most diverse woody ornamental plant collection in the northern Great Plains. The WPIP employs various germplasm collection methods and plant improvement techniques, including landscape observation, mass selection, and traditional and mutagenic breeding. Notable recent selections include cultivars of Japanese elm, dwarf Korean birch, Ohio buckeye, birch, mollis azalea, mugo pine, katsuratree, and mountain pine, with potential future introductions such as "Gumdrop" sugar maple and "Golden" littleleaf linden. The program's efforts contribute significantly to expanding the variety and resilience of woody plants for urban landscapes in challenging climates.

INTRODUCTION

Woody plant evaluations at North Dakota State University (NDSU) began in 1954. In 1971, Dr. Dale E. Herman initiated the NDSU Woody Plant Improvement Program (WPIP) with the first plant introductions beginning in 1986. To date, this program has released 61 woody plant selections into the ornamental nursery trade.

The NDSU WPIP has two primary goals. First, evaluate unreleased or released cultivars from the nursery trade, and second, increase diversity through selecting and/or breeding new woody plants suitable for the U.S. northern Great Plains. The NDSU WPIP program woody plant selections are ideal for urban climate conditions. North Dakota is the eighth driest state with respect to annual precipitation in the U.S. North Dakota soils are typically alkaline with a pH >8.0.

There are three primary research evaluation sites in North Dakota. The first is the NDSU Horticulture Research Farm (HRF) and Dale E. Herman Research Arboretum (DEHRA) (USDA hardiness zone 4a, Absaraka, ND Absaraka, ND, USA; Lat:46.9859, Long: -97.3549). The second are research plots in Fargo, ND (USDA hardiness zone 4a, Fargo, ND, USA; Lat: 46.918900, Long: -96.796681). The third is the NDSU Langdon Research Extension Center (USDA hardiness zone 3b, Langdon, ND; Lat: 48.7631, Long: -98.3713).

The NDSU WPIP has evaluated more than 200 genera and more than 3,000 species and cultivars of trees and shrubs. Over 11,000 accessions have been obtained and evaluated since planting began in 1974. The largest and most diverse woody ornamental plant collection in North Dakota and the northern Great Plains is located at the NDSU HRF and DEHRA with a total of 80 acres (~32 hectares).

The NDSU WPIP is involved with several woody plant evaluations, including cultivar comparisons done in conjunction with several industry cooperators and private breeders. For plant evaluation, selections and breeding, germplasm is collected using three different methods including foreign and domestic seed sources (growing out seedling populations and selection individuals with superior attributes), plant breeding (tradition breeding including F2 populations to observe segregation of traits, including hybridizing with both intra- and inter-specific hybridization), and *in vitro* tissue culture utilizing somaclonal variations, embryo rescue and mutagenesis. Three plant-improvement methods utilized are selections by landscape observation, mass selection (seed source and seed lot variation), and breeding (both traditional and mutagenic).

With the large germplasm collection located at the NDSU HRF and DEHRA, there are many accessions that have shown outstanding hardiness and make excellent parents for improvement through breeding efforts. These include Spring Welcome® magnolia (*Magnolia* x*loebneri* 'Ruth') and Fall Grandeur™ red maple (*Acer rubrum* 'Minnkota'). Magnolia breeding objectives focus on flower tepal color, introducing any color from *M. acuminata* hybrids coupled with the hardy Spring Welcome® selection (white flower color). Maple (*Acer* spp.) breeding objectives include utilizing known hardy and environmental tolerant selections to develop a better adapted Freeman maple (*A.* x*freemanii*). Current selections, such as Autumn Blaze®, do not have consistent performance in the northern Great Plains with respect to pH tolerance and winter hardiness. Utilizing a red maple selection that is known to be pH tolerant and have reliable winter hardiness would be better suited for a Freeman maple hybrid selection in colder climates.

Ornamental breeding research at NDSU includes developing freeze test procedures for earlier hardiness screening, traditional breeding efforts (making interspecific crosses with cold hardy species and hybrids), developing propagation protocols (micropropagation, grafting and vegetative cuttings), and polyploid induction for future sterility breeding.

Some outstanding recent selections that have come out of the NDSU WPIP include Northern Empress® Japanese elm (*Ulmus davidiana* var. *japonica* 'Burgundy Glow'), Cinnamon Curls® dwarf Korean birch (*Betula costata* 'CinDak'), Lavaburst® Ohio buckeye (*Aesculus glabra* 'LavaDak'), Emerald Flare® birch (*Betula tianshanica* 'EmerDak'), and Fireflare Orange® mollis azalea (*Rhododendron* x*kosteranum* 'FireDak'), Hyland Splendor® mugo pine (*Pinus mugo* 'HyDak'), KoolKat® katsuratree (*Cercidiphyllum japonicum* 'KoolDak'), and Hyland Guard™ mountain pine (*Pinus uncinata* 'GuarDak'. Potential future selections may include "Gumdrop" sugar maple and "Golden" littleleaf linden.

Combined Proceedings IPPS 73:42ろ42ゑ2023.

Effects of Cytokinin on Shoot and Rhizome Initiation in Leaf Cuttings of *Eucodonia* and *Achimenes*

Anna Baloh, Robert Geneve, Shari Dutton, and Marta Nosarzewski

Department of Horticulture, University of Kentucky, Lexington, KY 40506 USA

agba242@uky.edu

Keywords: propagation, gesneriads, BA, benzyladenine

Summary

The impact the cytokinin, benzyladenine (BA) on rhizome production in leaf cuttings of *Eucodonia* and *Achimenes* was determined. In *Eucodonia* 'Adele', untreated petiole cuttings were not statistically different from the untreated leaf lamina cuttings. Lamina leaf cuttings hereby defined as leaf segments without a petiole. Untreated lamina cuttings were found to have the highest adventitious shoot formation. BA had a negative effect in *Eucodonia* on rhizome formation in petiole leaf cuttings and reduced the number of shoots initiated in leaf cuttings with a cut lamina. Leaf cuttings in *Achimenes* only initiated rhizomes and roots. *Achimenes* 'Desiree' leaf cuttings produce a higher percentage and greater number of rhizomes compared to *Achimenes* 'Tarantella'. The effect of BA on *Achimenes* leaf cuttings was similar to *Eucodonia*. There was no statistical difference between the untreated and BA treated leaf cuttings in *Achimenes* 'Tarantella'. *Achimenes* 'Desiree' cuttings were observed to have negative effects from the BA treatment. Overall, a pretreatment with BA had no impact or a negative impact on rhizome initiation in *Eucodonia* and *Achimenes*.

IPPS Vol. 73 – 2023

INTRODUCTION

Eucodonia and *Achimenes* in the Gesneriaceae are native to Mexico and Central America. The diverse family consists of mostly tropical or subtropical herbaceous terrestrial or epiphytic plants. They are summer to fall flowering plants used as seasonal pot plants or for bedding. Both *Eucodonia* and *Achimenes* form scaly rhizomes (Zimmer et al. 2002). *Eucodonia* and *Achimenes* can be propagated using rhizomes or taking leaf cuttings (Dole and Wilkins 2005). New plants can form from the rhizome or pieces of the rhizome, or emerging shoots can be used as stock plants for future leaf cuttings (Geneve et al. 2022). Seasonal crop production requires an adequate supply of rhizomes as a starting material. Both species have the potential to become commercial seasonal potted plants but need a constant supply of rhizomes for stock plants. The objective of this research is to investigate the impact of cytokinin (BA) on rhizome initiation in leaf cuttings in *Eucodonia* and *Achimenes*.

METHODS AND MATERIALS

Plant Material, Growing Conditions, and Experimental Design**:** *Eucodonia* 'Adele', *Achimenes* 'Tarantella', and *Achimenes* 'Desiree' plants (**Fig.1**) were grown in the greenhouse from March through November during 2021 and 2022. The plants were allowed to dry seasonally and go dormant during the fall. Flats were placed into the greenhouse during mid-March of 2021 and then moved into the growth chambers under plastic domes during 2022. There were four replicate 6 packs per treatment for this study. The leaves were evaluated 60 days after the cuttings were taken.

Figure 1. A. *Eucodonia* 'Adele'; B. *Achimenes* 'Tarantella'; C. *Achimenes* 'Desiree'.

Shoot and Rhizome Initiation in Leaf Cuttings: *Eucodonia* leaf samples were harvested from dormant, stock plants and then cut with a razor 2mm from the petiole or across the lower portion of the lamina above the petiole. Leaves were either untreated or soaked in BA at 50 mg·L^{-1}. *Achimenes* leaf samples were either untreated or soaked in a BA (50 or 100 mg·L^{-1}). Treated leaves were placed in their solution upright with about 1 mm of the lamina in the treatment. Leaves were left in their treatments for 24 hours. Leaves were then placed into a 6-pack with vermiculite and placed into the growth chamber with light (~120 μmol·m^{-2}·s^{-1}) and constant 24°C.

RESULTS

Eucodonia leaf cuttings were able to produce roots, shoots, and rhizomes (**Fig. 2A**).

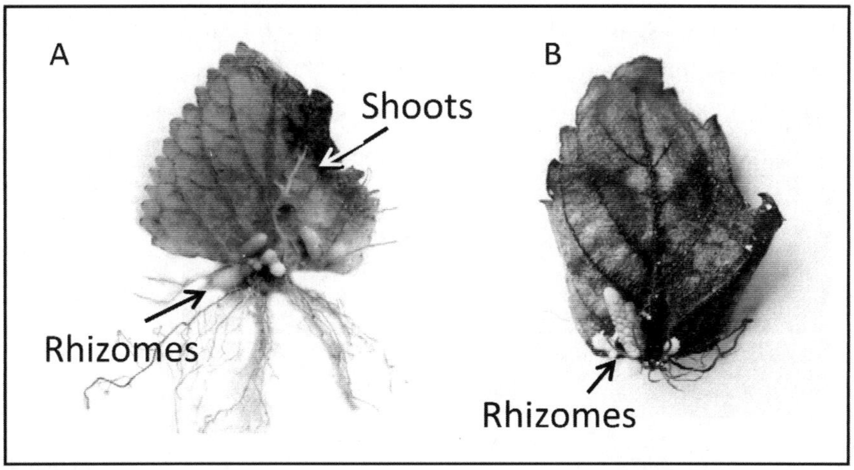

Figure 2. Rhizome initiation in A. *Eucodonia* 'Adele' and B. *Achimenes* 'Tarantella'.

There was no difference in the percentage of rhizomes initiated from untreated cut lamina and petiole leaf cuttings (**Fig. 3**).

Untreated cut lamina cuttings produced 3.5 rhizomes per cutting which was a slight increase compared to the 2.5 rhizomes per cutting in untreated cut petiole cuttings.

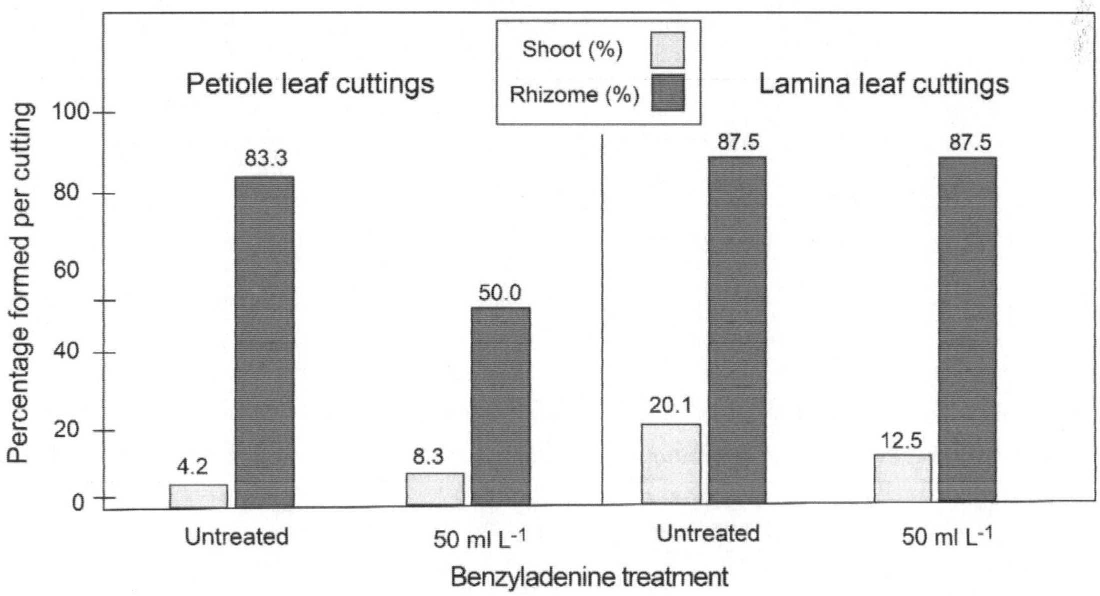

Figure 3. Percentage of shoot and rhizome production in *Eucodonia* 'Adele' petiole and lamina leaf cuttings treated with 0 or 50 mg L^{-1} benzyladenine (BA).

In contrast, untreated cut lamina leaf cuttings formed a higher percentage of adventitious shoots compared to petiole leaf cuttings (**Fig. 3**). BA had no impact on leaf cuttings with a cut lamina, but there was a reduction in the number of leaf cuttings forming shoots after a BA treatment.

Both *Achimenes* species only initiated roots and rhizomes from leaf cuttings (**Fig. 2B**). On average, leaf cuttings of *Achimenes* 'Desiree' produced a higher percentage and greater number of rhizomes compared to *Achimenes* 'Tarantella' (98% vs. 67% and 4.5 vs. 1.8 rhizomes per cutting, respectively). In *Achimenes* 'Tarantella', there was no difference between the untreated and BA treated leaf cuttings (**Fig. 4**). However, in *Achimenes* 'Desiree' there was a decrease in rhizome production for the cuttings treated with BA at 50 mg·L^{-1}.

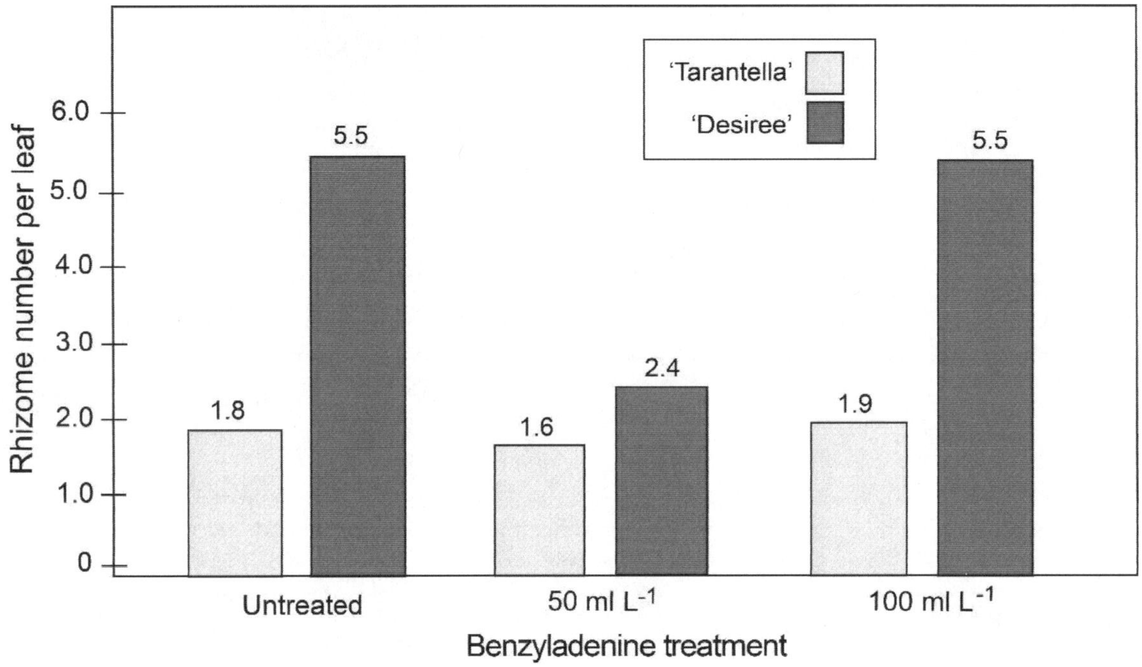

Figure 4. Rhizome production in *Achimenes* 'Tarantella' and 'Desiree' leaf cuttings treated with 0, 50, or 100 mg L^{-1} benzyladenine (BA).

DISCUSSION

A viable propagation method utilizing leaf cuttings requires adventitious organ formation. Selected members in the Gesneriaceae can form adventitious shoots, roots, and rhizomes from leaf cuttings (Miller and Brigden 2005). Cytokinins, including BA, have been used to increase shoot initiation in leaf sections for many crops cultured in tissue culture and a few species from traditional leaf cuttings (Davies et al. 2018). In *Eucodonia*, BA increased shoot initiation from rhizome segments (Geneve et al. 2023) but the impact of BA on rhizome initiation in leaf cuttings has not been evaluated. In the current study, treating leaf cuttings in *Eucodonia* with BA reduced the number of leaves forming rhizomes, but increased the number of rhizomes formed per petiole leaf cutting (**Fig. 3**). The addition of

BA to *Achimenes* leaf cuttings had no impact or reduced rhizome formation depending on the cultivar (**Fig. 4**). In this preliminary study, only a single concentration of cytokinin was used for a single exposure time. It is possible that a study varying concentration and exposure time could yield increased rhizome formation. A novel approach could be treating stock plants with a cytokinin foliar spray prior to taking leaf cuttings rather than post-severance cytokinin treatment.

LITERATURE CITED

Davies, F.T. Jr, Geneve, R.L. and Wilson, S.B. (2018). Hartmann and Kester's Plant Propagation: Principles and Practices (9th ed). Prentice-Hall, Boston, MA, USA.

Dole, J.M. and Wilkins, H. (2005). Floriculture: Principles and Species (2nd ed). Prentice-Hall, Boston, MA, USA.

Geneve, R.L., Dutton, S., Baloh, A.G. and Nosarzewski, M. (2023). Rhizome segments form shoots, whereas leaf cuttings form shoots and rhizomes in *Eucodonia* 'Adele' treated with benzyladenine. Hort Sci. *58*:785-791. https://journals.ashs.org/hortsci/view/journals/hortsci/58/7/article-p785.xml#B5

Miller, C.T. and Bridgen, M. (2005). Photoperiod and stock plant age effects on shoot, stolon, and rhizome formation response from leaf cuttings of *Achimenes.* Acta Hortic. *673*:349–353.

Parliman, B.J.and Stusnoff, C. (1979). Mutant induction through adventitious buds of *Kohleria*. Euphytica *28*:551-530.

Zimer, E.A., Roalson, E.H., Skog, L.E., B oggan, J.K. and Idnurm A. (2002). Phylogenetic relationships in the Gesnerioideae (Gesneriaceae) based on nrDNA ITS and cpDNA trnL-F and trnE-T spacer region sequences. Am J Bot. *89*:296311. https://doi.org/10.3732/ajb.89.2.296.

Combined Proceedings IPPS 73:42を43を2023.

Spring Meadow Nursery: Behind the Scenes – What it Takes to Become a Proven Winner ColorChoice® Shrub

Megan Mathey

Spring Meadow Nursery, 12601 120th Ave, Grand Haven, MI 49417

megan@springmeadownursery.com

Keywords: new product development, plant breeding, trialing program, tissue culture lab, mutation breeding, ploidy manipulation, seed collection, data collection

Summary

Spring Meadow Nursery was established in 1981 in Grand Haven, Michigan, and has emerged as a leader in the horticultural industry. Through collaboration with a global network of breeders as well as an internal breeding and trialing program, Spring Meadow has developed a wide array of woody ornamental plants that meet high standards for consumer success, disease resistance, and environmental friendliness. The nursery's commitment to innovation, quality, and sustainability has led to over 249 plant awards and a significant impact on the industry, underlining its role as a key player in advancing horticultural practices and ornamental plant varieties. A motto from the owner runs deep and if you know him and anything about the nursery you will know, “It’s a good start”.

IPPS Vol. 73 – 2023

INTRODUCTION

Spring Meadow Nursery was founded by Dale and Liz Deppe in 1981 in Grand Haven, Michigan, on the same site it exists today – what was formerly their backyard. Spring Meadow Nursery hired its first full-time employee in 1986, published its first catalog in 1988, and hired Tim Wood (as a Plant Hunter) in 1995 with the role of finding and trialing the newest and best in woody ornamentals from all over the world. The brand ColorChoice® was introduced in 1999, conveying the message that shrubs can be more than just a green blob or foundation plant; rather, they have a real place in the landscape. A partnership was formed between Proven Winners® and Spring Meadow in 2004, as both realized that they shared the common goal of introducing not only new plants but better plants to the industry. Noteworthy advancements include the addition of a tissue culture lab in 2011, overcoming an office fire incident in 2017 (which lead to an office expansion), and becoming the first company to reach $1,000,000 in donations to the Horticultural Reach Institute in 2021.

Spring Meadow first and foremost is a wholesale liner production nursery, offering three different sizes: 2 1/4", 4", and Quick Turn or 1-quart size liners (**Fig. 1**). Over the years, the nursery has seen substantial growth in its facilities, transitioning from what once was a mix of Quonset, Cravo, and Gutter Connect greenhouses to a more robust and extensive West Brook style (**Fig. 2**), which now encompasses almost 60 acres. The shift was partly due to the enhanced structural integrity offered by the West Brook designs.

Figure 1. (left) 2 ¼" liners, (middle) 4" liners, and (right) quick turn or 1-quart liners.

Figure 2. (left) Aerial photograph of Spring Meadow Nursery's facilities in 2007, and (right) aerial photograph of Spring Meadow Nursery's facilities in 2022.

TECHNOLOGIES

Spring Meadow Nursery emphasizes innovation and efficiency in its operations, marked by significant technological advancements. The adoption of a TTA Grading Machine (**Fig. 3**) in 2007 was the nursery's first major investment in automation, enhancing product quality. Subsequently, the integration of proprietary trimming machines in 2009, tailored for the new West Brook-style growing spaces, demonstrated a commitment to operational efficiency. This was further expanded with the introduction of spraying machines, streamlining the nursery's processes.

In 2016, the nursery introduced an ISO Robotic Sticking Machine (Fig. 4), leading to the addition of more machines over the years, facilitating uniform planting processes. In 2017, Spring Meadow began trialing broad scale biological pest control methods (**Fig. 4**), notably improving pest management and reducing the need for chemical controls. This successful trial led to a nursery-wide application of biological controls, including the use of beneficial organisms, starting in 2018, furthering the nursery's move towards more sustainable and efficient production methods.

Figure 3. (top left, top right) sorting machine at Spring Meadow Nursery and (bottom left, bottom right) trimming machines at Spring Meadow Nursery.

In 2016, the nursery introduced an ISO Robotic Sticking Machine (**Fig. 4**), leading to the addition of more machines over the years, facilitating uniform planting processes. In 2017, Spring Meadow began trialing broad scale biological pest control methods (**Fig. 4**), notably improving pest management and reducing the need for chemical controls. This successful trial led to a nursery-wide application of biological controls, including the use of beneficial organisms, starting in 2018, furthering the nursery's move towards more sustainable and efficient production methods.

Figure 4. (top left, top right) sticking machines at Spring Meadow Nursery and (bottom left, bottom right) integrated pest management at Spring Meadow Nursery.

WHERE DO THE NEW PLANTS COME FROM?

Spring Meadow Nursery's mission focuses on identifying, breeding, and evaluating superior plant varieties. To achieve this, Spring Meadow maintains a global network of over fifty active plant breeders, ranging from hobbyists to nursery growers and academic partners, including top university plant breeding programs. Plant Hunters Tim Wood and Dr. Judson LeCompte, who joined the team in 2020, manage these collaborations.

One of Spring Meadow Nursery's foremost partners is Dr. Tom Ranney (**Fig. 5**) with the Mountain Horticultural Crops Research & Extension Center at North Carolina State University. Ranney and his team developed the Invincible Spirit Line of smooth hydrangeas (**Fig. 5**). This initiative, launched in 2009, supports breast cancer research by donating a portion of the proceeds from each plant sold. Spring Meadow Nursery achieved a significant milestone of one million dollars in donations contributed to the Breast Cancer Research Foundation in 2018.

Figure 5. (left) Dr. Tom Ranney from North Carolina State University and (right) Invincibelle Spirit Hydrangea.

In addition to collaborating with an extensive network of external breeders, Spring Meadow Nursery has been breeding plants in house for 23 years. Initiated by Tim Wood, the success of this program has allowed for the expansion of the team, including hiring Megan Mathey in 2013 as well as Davis Harmon in 2022 as additional plant breeders, thus greatly expanding specialized breeding efforts.

BREEDING AND NEW PRODUCT DEVELOPMENT

At Spring Meadow Nursery, the starting point for any plant breeding project is having an idea and setting specific goals, such as making the plant more dwarf, increasing flower size, improving remontancy capabilities, or enhancing disease tolerance. Once an idea and goal are at hand, traditional plant breeding techniques are often instrumental to achieve them. This early phase of plant development involves selecting ideal parents, emasculating the female plant to prevent self-pollination (**Fig. 6**), and then introducing pollen from the chosen male plant to achieve the cross.

Spring Meadow Nursery has a small on-site tissue culture lab for tasks like mutation breeding, particularly through ploidy manipulation (**Fig. 7**). Actively growing meristems are exposed to a meiotic inhibitor, resulting in a doubling of chromosomes. Ploidy doubling is one avenue breeders often utilize as the first step in developing sterile plants and is a strong example of breeding for reduced fertility. Techniques such as flow cytometry (**Fig. 7**) are utilized for in-house confirmation of elevated ploidy. This technique is used to identify tetraploids, which can then be used as a parent to back cross with diploids in the same genera, resulting in triploid progeny.

Figure 6. Emasculating the female plant to prevent self-pollination.

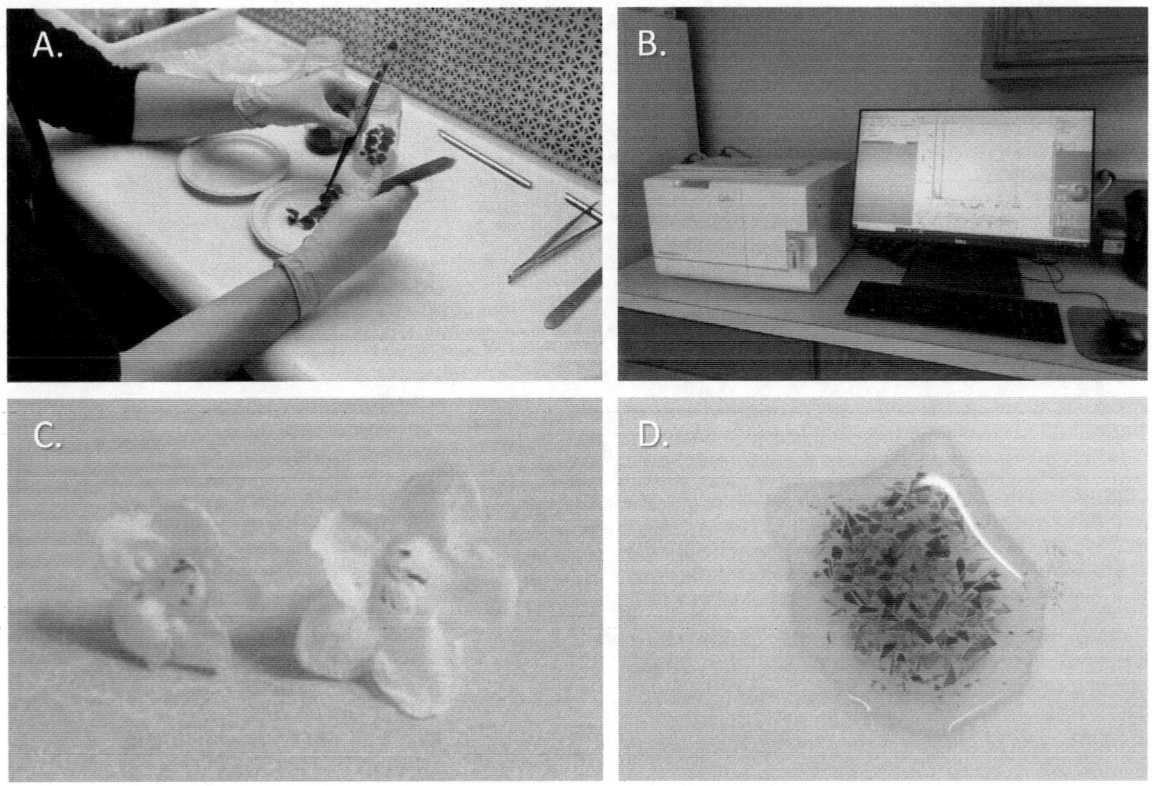

Figure 7. (A) Tissue culture lab for mutation breeding, ploidy manipulation, embryo rescue, and export, and (B-D) flow cytometry.

The breeding process continues with seed collection, cleaning, sowing, and, if necessary, stratification (**Fig. 8**). Tens of thousands of seedlings are produced annually from an average of 600 unique seed lots. About 10,000 - 15,000 seedlings are planted in the field, and another 10,000 are grown in the greenhouse. Sorting and rouging of seedlings begin early and can even start before the plants are transplanted out of their germination tray. Great efforts are placed on rouging seedlings even at the 4" stage, focusing on traits such as foliage color, habit, and sometimes even flower color. Rouging and selection continues, on average, for the next three to five years as we continue to turn greenhouse and research space over.

Figure 8. (A) seed collection, (B-C) seed processing and sowing, and (D-E) seed stratification.

An oftentimes overlooked means of generating new plants is finding sports/natural mutations. A few examples of recent instructions of plants found as sports include *Weigela florida* VINHO VERDE® 'SMNWFBGV' (**Fig. 9B**), arising as a sport on a seedling from a cross between two distinct varieties 'Briants Rubidor' (a yellow-leafed weigela) and 'Naomi Campbell' (a purple-leafed weigela). Weigela BUBBLY WINE® 'FLORACLAR' (**Fig. 9D**) is another Weigela sport discovered externally by plantsman Jeff Good of southeast Michigan from parent plant Weigela FINE WINE® 'Bramwell'. Similarly, Hibiscus RED PILLAR® 'GFNHSRP' (**Fig. 9F**) was found as a branch sport of 'Gandini van Aart' by Dan Baston of Green Forest Nursery out of Mississippi. This plant actually beat out another selection for the "Red Pillar" seat due to its superior flower color and dark red "eye", proving that, no matter the plant's origin, "the best plant always wins."

Figure 9. (A) Weigela Vinho Verde sport, (B) weigela Vinho Verde, (C) weigela Bubbly Wine sport, (D) weigela Bubbly Wine, (E) hibiscus Red Pillar sport, and (F) hibiscus Red Pillar.

SELECTION AND TRIALING

Spring Meadow Nursery's research and development facility was initially established in 2014, breaking ground for a 2-acre West Brook Greenhouse which has since expanded to encompass approximately 10 acres (**Fig. 10**). This facility centralizes all trials under one roof, ensuring that environmental conditions are consistent across trial groups.

The site is managed by the New Product Development team (**Fig. 11**), including the New Product and Stock Manager, Ginger Thurston, the Plant Breeding Manager, Megan Mathey, and Dr. Judson LeCompte, who oversees External Trials. The team is composed of individuals specializing in different areas, such as stock management (Alex Pott), R&D growing (Tori Boos and Leticia Pasqualino), plant breeding (Davis Harmon and Caleb Pritts), and tissue culture and intellectual property (Julie Crone).

Criteria for evaluating shrubs at the facility focus on multiple factors to enhance the likelihood of commercial success. These include consumer interest, foliage attractiveness, seasons of color, new uses, compact growth, long blooming or re-blooming properties, disease resistance, lower maintenance, retail appeal, and eco-friendliness.

Trialing involves weekly evaluation meetings starting in March, covering around 194 genera. A preliminary review conducted by a small team precedes the final presentation to Dale Deppe, President, Jeremy Deppe, General Manager, and Tim Wood, Product Development & Head of Marketing.

Figure 10. Aerial photograph of Spring Meadow Nursery's 10-acre Research and Development facility.

Figure 11. Spring Meadow Nursery's New Product Development team.

Trials are conducted in both container settings (**Fig. 12**), with each selection undergoing trials in 38 containers to assess uniformity and response to various treatments, and field settings (**Fig. 12**), where five plants are tested and in a landscape trial where three of each plant is grown.

Figure 12. (top) Spring Meadow Nursery's container trials, and (bottom) Spring Meadow Nursery's field trials.

Data collection extends beyond the onsite trials to include external evaluations (Fig. 13) in diverse climates such as Florida, Minnesota, and Texas, as well as feedback from growers, retailers, and consumers, ensuring a well-rounded and effective trial process for new plant developments. The entire process from breeding to product development typically spans six to eight years, though some plants may take longer.

Figure 13. Spring Meadow Nursery's external field trials.

CONCLUSION

Spring Meadow Nursery's comprehensive approach to plant breeding and new product development, combined with its thorough trialing process, has contributed to the nursery's success in the horticultural industry. With over 325 plant patents assigned or under license and 249 plant awards and counting, Spring Meadow Nursery demonstrates a high level of innovation and quality in its plant selections.

Because of its rigorous evaluation process, resulting in less than 2% of the varieties in trials being designated as Proven Winners®, the stringent criteria ensure that only the most outstanding, consumer-friendly, and garden-worthy plants reach the market. This selective approach underlines Spring Meadow Nursery's commitment to excellence and sustainability, ensuring that each new plant introduced has a significant impact on gardens and landscapes.

Combined Proceedings IPPS 73: 437-443. 2023.

Evaluating Remontancy and Rebloom in *Hydrangea macrophylla*

David Roberts

Bailey Innovations, 120 Walter Sams Rd., Winterville, GA 30683, USA

david.roberts@baileynursery.com

Keywords: plant breeding, plant trialing, genetics

Summary

In *Hydrangea macrophylla,* remontancy is a valuable trait that is relatively easy to incorporate into seedling lines, but determining which lines possess the strongest rebloom can be more challenging. Simply falling into the classification of "remontant" does not necessarily mean a hydrangea will be a strong rebloomer. Plant breeders at Bailey Innovations™ gauge hydrangea rebloom potential through seasonal cutback evaluations and annual rebloom trials. This paper details the procedures and protocols utilized by Bailey Innovations™, to evaluate selections of --*Hydrangea macrophylla* which have superior reblooming characteristics.

INTRODUCTION

Located in Winterville, GA., Bailey Innovations™ is the in-house, plant breeding and trialing division of Bailey Nurseries Inc. *Hydrangea macrophylla,* big leaf hydrangea as it's commonly known, is one of the most important species currently under development in our program. As more *H. macrophylla* cultivars enter the market every year, the need for not just remontant but strong reblooming selections has become an important target for the hydrangea breeding program.

Remontancy, often used synonymously with rebloom, is considered by many to be the ability to produce flowers on not just the previous season's growth (old wood) but on newly emerged, vegetative growth (new wood) as well (**Fig. 1**).

A.

B.

Figure 1. The difference between (A.) non-remontant, *H. macrophylla* varieties that only produce blooms on old wood and (B.) remontant *H. macrophylla* varieties that bloom on both old and new wood. Credit Bailey Nurseries Inc. 2015.

This phenomenon can lead to continuous flowering throughout the growing season (Adkins and Dirr, 2003) which is a highly desirable trait for a crop like *H. macrophylla*, whose main appeal lies in colorful, floral displays. Remontancy is widely believed to be a qualitative trait, controlled by a single or at least a small number of genetic loci (Wu and Alexander, 2020). Observations made over the last seven years of breeding efforts at Bailey Innovations, support this belief. In our hydrangea program, we treat remontancy as a trait controlled by a single, recessive gene and we have had some success in breeding reblooming hydrangea while operating under this assumption.

In *H. macrophylla*, remontancy is a highly desirable trait that allows certain varieties to continue flowering, or rebloom, well past the point when older, non-remontant, varieties would have finished flowering. The extension of a hydrangea's normal flowering period creates added value for consumers who desire season-long, floral displays. Remontant cultivars also provide more reliable blooming characteristics for homeowners in more northerly or frost-prone regions (Wu and Alexander, 2020). Remontancy is such a valuable trait that nearly all new *H. macrophylla* cultivars claim to have the ability to rebloom. However, our observations indicate that remontant cultivars can fall into a spectrum of potency when it comes to their ability to rebloom (**Fig. 2**), with some varieties performing better than others. We have observed purportedly remontant cultivars rarely produce flowers after their initial, Spring flush, while other cultivars can be aggressive to rebloom, even after a harsh cutback. From a breeding perspective, remontancy is a relatively easy trait to incorporate into *H. macrophylla* lines but determining which lines possess the strongest rebloom can be more challenging.

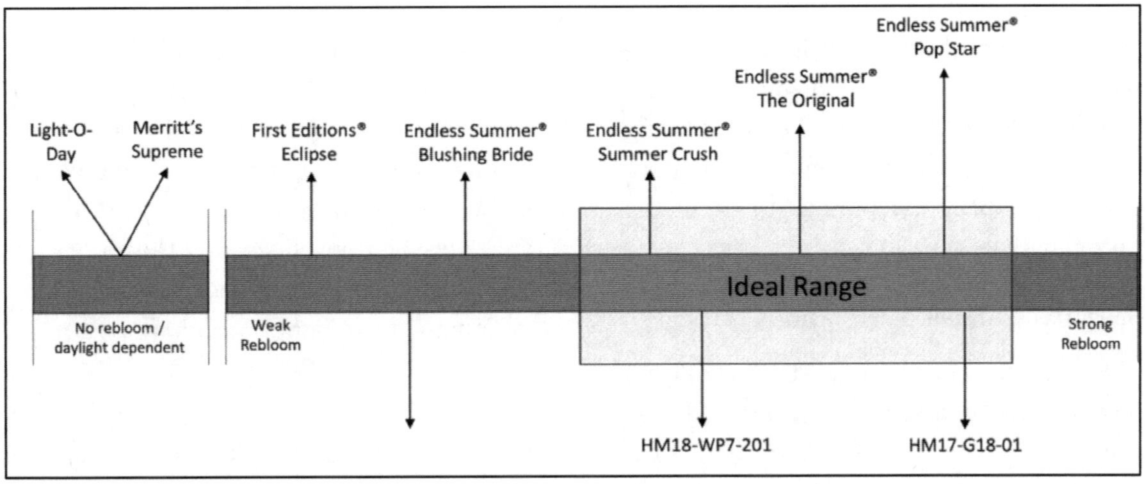

Figure 2. Theoretical scale of rebloom potency, based on 5 years of observations. Credit Justin Schulze, Bailey Innovations, 2020.

Great strides have been made in identifying the genetics mechanisms responsible for rebloom in *H. macrophylla.* Molecular testing may be able to positively identify hydrangea with remontancy but that does not guarantee a selection will rebloom to the breeder's expectation or level of standards. While remontancy appears to be a qualitative trait, the potential for strong rebloom may be more quantitative in nature, with many developmental factors contributing to whether a hydrangea will rebloom aggressively or weakly. The plant breeding team at Bailey Innovations believe the best way to accurately and confidently evaluate a hydrangea's reblooming potential, is through several years of trialing.

Evaluating for Strong Rebloom in *Hydrangea macrophylla*

First-Year Seedling Evaluations

Bailey Innovations™ typically grow between 10,000 - 15,000 *H. macrophylla* seedlings each year. Seed is germinated in late December and grown in a greenhouse until early to mid-March. Around mid-March, seedlings are transplanted into 3-gallon containers where they are fertilized with three tablespoons of Florikan 18-6-8, 270-day, control release fertilizer (CRF) and allowed to grow for several months while being observed for floral development. By June / July, seedling populations are approximately 6-7 months old and do not possess any growth from the previous season (old wood) for the plants to produce flowers on. The only growth these seedlings have is from the current season's, vegetative growth (new wood). According to the definition of remontancy, any seedling that is capable of flowering after only a few months of vegetative growth could be considered remontant. The hydrangea seedling populations at Bailey Innovations™ are frequently derived from parents which possess some degree of remontancy, so we often see very large sectors of seedling populations flower within a few months of germination. Since it is not realistic to keep every seedling that flowers in its first year, initial selections are based on unique aesthetics, quickness to flower, bloom load and overall biomass accumulation. These advanced selections are then moved to a holding area where they are observed for 2-3 years to gauge their rebloom potential. These selections are not pruned in their first year of observation, this allows the plants to build a strong body in year one, accommodating future propagation trials and allowing for seasonal cutback evaluations to occur the following year.

Seasonal Cutback Evaluations

The first step in quantifying strong-rebloom potential for advanced selections is a seasonal cutback evaluation that occurs twice, every year. In Winterville, GA, peak bloom for *H. macrophylla* typically occurs in early to mid-June. We allow our advanced selections to reach peak bloom before we perform an early-season cutback, usually around June 15th. A hydrangea in its first year of seasonal cutback evaluation will be pruned to about 3 - 4" above the rim of the container. With older selections, the cutback is less severe but still aims to eliminate at least half of the accumulated biomass. This practice is meant to eliminate preformed flower buds that may still be present on the previous season's growth. The pruned selections will re-flush vegetative growth from the areas that were cut back and we consider any flower buds that develop from this new growth to be rebloom. The strongest rebloomers are flagged for future propagation and production trials that will occur the following year.

The final aspect of the seasonal cutback evaluation comes in late-Fall, when all hydrangea selections are again, cut back to eliminate about half of the biomass which accumulated since the early-season cutback occurred.

A common frustration for some homeowners comes in not knowing when to properly prune their hydrangea in the landscape. Older varieties of *H. macrophylla* should not be pruned in the Fall as doing so could eliminate the next season's flower buds that have already developed on old wood. The late-season cutback evaluation is meant to address this concern, if an advanced selections does not flower in the Spring that follows a Fall cutback, it is eliminated from our program.

Rebloom Trials:

Once an advanced selection has made a positive impression through the seasonal cutback evaluations, it is flagged for our production and rebloom trials. The flagged hydrangea are propagated with a goal of having a minimum of 10 clones to trial for each selection. Two-node propagules are collected around mid-June, treated with 1,000 ppm K-IBA and stuck into 32 cell trays filled with Sunshine® Mix #4 Professional Growing Mix. Approximately two weeks after the cuttings are stuck, they are top dressed with a half teaspoon of Osmocote 15-9-12, 90 day CRF. The hydrangea cuttings are typically pruned or "tipped back" once or twice before they are upshifted into a different house. Approximately two months after the initial stick date, the hydrangea cuttings are upshifted into 3-gallon containers where they are top-dressed with 2 tablespoons of Osmocote 15-9-12, 90 day CRF. The cloned hydrangea selections are left to grow until Fall / Winter when they are consolidated so they can be protected from unseasonable weather. In the Spring of the following year, the hydrangea selections are distributed to a production area where the rebloom trials occur. Once buds begin to swell, the plants are lightly pruned for shape and to remove any stems that were damaged over the course of the winter.

The rebloom trial is divided into a year-one study and a year-two study. Selections in both studies are evaluated in replicates of 10 and grown in identical conditions. From April to mid-June, we collect initial bloom data on the same day of each week. The average number of blooms for each selection is determined by adding together the total number of floral buds counted in each clonal block and dividing by the number of clones. For our trial, we count any floral bud that is dime-sized or larger and once a bloom begins to decompose, it is no longer counted. Once we begin collecting data, we quantify how long it takes a selection to flower, the time to reach peak bloom, bloom quantity and bloom duration. As with the seasonal cutback evaluations, we allow the *H. macrophylla* selections in the rebloom trial to reach peak bloom (**Fig. 3A**) before they are pruned, which usually occurs around mid-June. All selections in the year-one rebloom trial are pruned to 3 - 4" above the rim of the container (**Fig. 3B**).

Figure 3. *H. macrophylla* selections in year-one rebloom trial. (A) pre-cutback in early June and (B) post cutback in mid-June. Winterville, GA. 2022.

Plants in the year-two rebloom trial receive a less severe cutback but we still aim to eliminate at least half of the accumulated biomass. Cutting the *H. macrophylla* selections back this harshly, in early Summer, results in a relatively quick flush of new, vegetative growth. When we observe floral buds developing from newly emerged, vegetative growth, we consider a selection to be reblooming. Once a single selection begins to rebloom, we resume collecting data for the entire trial. We make note of how long it takes a selection to rebloom, the time to reach peak rebloom, rebloom quantity and rebloom duration. On average, most known varieties take between 6-8 weeks to initiate floral buds after a hard pruning but we have observed internal selections that produce floral buds in as few as 3 weeks after the initial cutback (Fig. 3). The best performers in the year-one rebloom trial are saved for the next season's, year-two rebloom trials and are re-propagated so we can have the same selection represented in year-one and year-two rebloom trials at the same time. Any selections that do not perform to our expectations are eliminated from trial.

Selections are considered top performers if they rebloom quickly after a cutback and produce a relatively high number of floral buds throughout the post-cutback evaluation. Overall bloom load and density are valuable metrics but we also look at weekly bloom averages and seasonal averages (**Fig. 4**) to determine which selections are the most aggressive rebloomers. We have recorded instances of selections that produce so many blooms they fail to accumulate enough biomass to fill a pot by the end of the growing season. We believe this is due to the amount of energy invested in floral vs. vegetative development. Unfortunately, these selections do not offer production values high enough to warrant introduction.

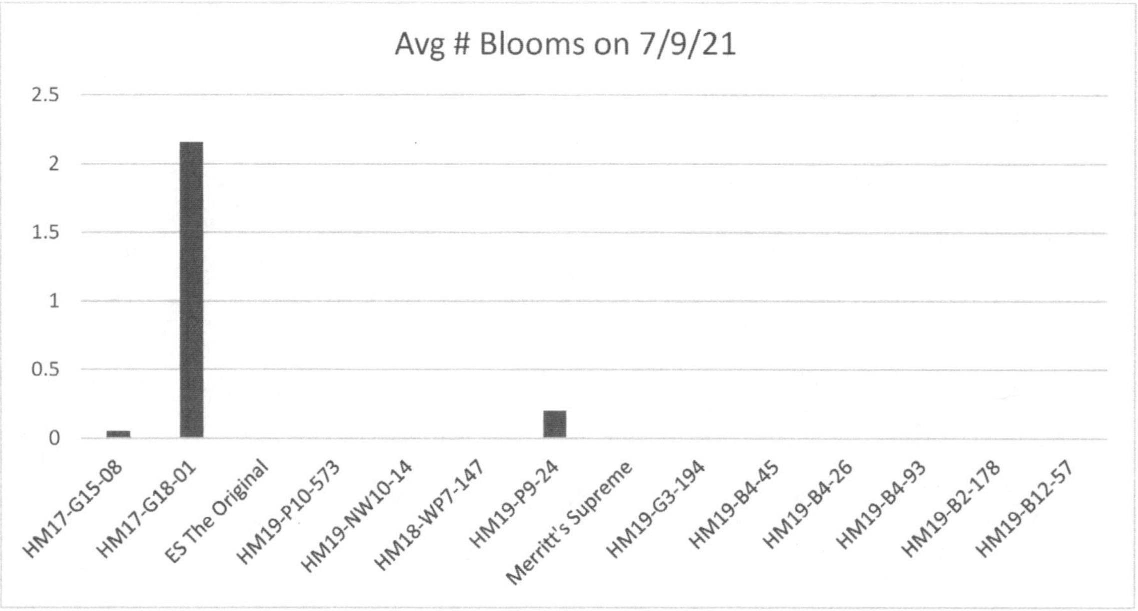

Figure 4. Average number of blooms recorded the week of July 9th, 2021, 24 days after cutback.

After two years of positive results from the rebloom trials, our top selections are flagged for additional trialing and distributed to various locations. Selections are entered into container trials at Bailey facilities located in GA, MN, IL and OR to gauge their production values and all selections planted in-ground, at Bailey facilities located in MN and IL to gauge cold tolerance. We often see a correlation between aggressive rebloomers and strong flowering performance in regions where Winter temperatures drop below a certain threshold. In Minnesota, where temperatures can often drop below - 20°F, all stems on *H. macrophylla* are killed back entirely, leaving only the crown and roots of a plant alive. Our strongest reblooming *H. macrophylla* selections will regrow from the crown and eventually flower from the newly emerged, vegetative growth. The higher a plant's rating in regard to rebloom strength, the stronger its flowering performance will often be in colder, Northern regions. Any non-remontant, *Hydrangea macrophylla* that experience total stem dieback will fail to flower the following Spring, further reflecting the limitations of varieties that are incapable of reblooming.

LITERATURE CITED

Adkins, J. A., and Dirr, M.A. (2003). Remontant flowering potential of ten *Hydrangea macrophylla* (Thunb.) Ser. cultivars. *HortScience 38*:1337-1340.

Wu, X. and Alexander, L.W. (2020). Genome-wide association studies for inflorescence type and remontancy in *Hydrangea macrophylla*. Horticulture research *7*:1-9.

Combined Proceedings IPPS 73: 444-450. 2023.

Solving Production Problems at Robinson Nurseries: Innovation is Part of Our Culture

Chris Robinson and Adam McClanahan

Robinson Nursery, Inc., P.O. Box 100, Amity, OR 9701 USA

Chris@robinsonnusery.com , Adammc@robinsonnusery.com

Keywords: container production, roots, air pruning, container design

Summary

Circling and girdling roots is a major problem for container nursery producers. Containers that permit air pruning can eliminate root circling and lead to better tree establishment in the landscape. The mount of root circling varies by genera and various container and air pruning systems were evaluated for popular tree species.

INTRODUCTION

Robinson Nursery Inc. of McMinnville, Oregon is a large grower of liner trees in a variety of sizes and forms. An important product line is container trees with sizes ranging from Grow Ready #3, Grow Ready #5, and Grow Ready #7. Product lines include *Acer, Betula, Celtis, Cercis, Crataegus, Gleditsia, Hibiscus, Hydrangea, Malus, Nyssa, Parrotia, Platanus, Prunus, Quercus, Salix,* and *Tilia* with some lesser known genera such as *Gymnocladus,*

IPPS Vol. 73 – 2023

Syringa and Ulmus varieties. It's a varied lot with significant requirements for specific species and varieties. *Quercus* do not always follow the requirements of *Salix* for instance.

Since we are a volume producer of container crops one of our biggest problem areas is circling and girdling roots in the containers (**Fig. 1**).

Figure 1. Circling and girdling roots are not optimal for good container production.

This condition results from root systems being too vigorous and presents several challenges such as difficulties in watering, upshift and establishment of the plants in fields or as part of a new landscape.

As a result, we started working with air root pruning and GRL insert. The utilization of these techniques resulted in eliminating circling roots which in turn lead to better transplant success and root establishment after planting. The fabric inserts enhance water retention in the pot which also makes for greater utilization of applied fertilizer. We can remove the bag and ship the established liner without the pot. This allows for a return of the pot to the production cycle and reduces weight for shipping. Removing the fabric bag at the final plant destination either for stepping up or field planting accomplishes a final root pruning at the time of planting. **Figure 2** shows the root ball without the fabric bag with no circling roots and with a layer of hazel nut shells as a container mulch which cuts down on weed seed germination and herbicide use.

Figure 2. Saleable container root ball with the fabric liner removed. Note the hazel nut crushed shells as a weed preventative mulch.

The air pruning journey starts with a container designed to prevent root circling coupled with multiple ports for extensive air pruning. In **Figure 3**, the four shots clearly show the positive effects of the container ribbing and air-ports. On the left is the fabric bag with minimal root emergence and to the right is a well grown ball with the fabric bag removed and hundreds of small feeder roots ready for a new environment.

Figure 3. The air-pruning journey begins with container ribs and air-ports.

In **Figure 4**, an ample liner crop in the air pruning containers with a close up of the root pruning container, a rootball with the container and fabric removed and a ball wrapped and secured with burlap for shipping and installation.

Figure 4. Transition from retainer pot to experimental burlaped plant ready for shipping or planting.

Another aspect of shipping and handling can be achieved with a ball sock which also can be removed upon planting (**Fig. 5**). The purpose of the sock is to ensure ball integrity for shipping and not allowing for a "joystick" movement of the trunk. Notice that there is a specialized machine to accomplish this which greatly makes the process efficient. In **Fig. 4**, the plant that has the burlap bag installed and is ready for installation without removing the bag. When the sock is used, it is imperative to remove it prior to planting.

Figure 5. The container ball sock insert.

We are trying a series of trials with a biodegradable liner that if successful would achieve the root pruning process and would not have to be removed at the time of final planting, thereby eliminating a labor step in the process, We are looking at both an Ellepot paper insert for 3, 6, 8, and 12 months and a biodegradable fabric for 12, 24, and 36 months. Again, using the *Quercus* vs *Salix* example, different plant crops will require different times of production and longevity of the liners in the containers (**Fig. 6**).

Figure 6. A. Biodegradable container inserts for root pruning. B. Close up of Ellepot paper liner.

It is a well-known fact that some production issues in larger containers and the field can be attributable to flaws in the propagation program. It is important to look at propagation to ascertain if potential problems can be corrected early on at a greatly reduced expense as compared to correction once the plant is moving up the production sequence. With that in mind we decided to look closely at production practices that could have an immediate impact on crop improvement further down the production cycle.

We discovered a ribbed propagation pot which appears to us to be a positive step in the right direction. Fig. 8/9 shows a close up of the resultant plant from a ribbed propagation pot.

Figure 7. Plant grown in a ribbed propagation pot.

Figure 8. *Quercus* seedlings being grown in a composite ribbed propagation tray.

Ellepot production for propagation is being evaluated with promising results.

Figure 9. The very good root system emerging from the Ellepot system for a *Quercus* seedling.

Barnes (2010) showed some time ago that rooted cuttings and seedlings have different root systems. He also showed that different containers exhibited different root systems for the same plant (Barnes, 1999). We decided to look at the potential root system response to the differences for seedlings and rooted cuttings. For this trial we selected Magnolia virginiana rooted cutting in a wood fiber insert, “Fertil”. As can be seen in **Figure 10**, a well-balanced root system is emerging which should lead to a quality liner in the future.

Figure 10. *Magnolia virginiana* rooted cutting in a wood fiber Fertil pot.

Light quality and quantity are two factors that can significantly impact liner quality . The greenhouse industry has used colored films for a number of years. Only recently has colored films made a show in the woody production systems. One of our trials was with a rose pink colored shade cloth. The plants see it as a form of red. The principal involved is altering the red to blue ratios which promotes stem elongation a d stretch. The direct result is a faster growing stem which is often much straighter than that which can be found under ordinary ambient light. Also by influencing the stem to grow taller the formation of axillary shoots is diminished and the resultant plant requires less corrective pruning. Proof in the pudding can be readily seen in **Figure 11** with white poly vs. red shade cloth.

Figure 11. Stem elongation via red shade cloth.

Figure 12. Plant on left is with white poly, plant on the right is with red shade cloth.

Here's a thought. What about biodegradable paper as a mulch for bareroot production? The initial plan looked good on paper, (no pun intended) well maybe.

Well maybe not. Too much light transmission triggered a massive weed seed germination in such a nice, protected environment.

Figure. 13. A. Biodegradable paper mulch. B. Weed growth under the mulch.

Our mission: We grow people and plants to change the world. Innovation is part of our culture. As with the Japanese concept of Kiazen: the breakdown and build up. The breakdown is to look at our processes, look for areas where we can be more efficient and to turn out a better product. The build-up is to expand on what we as well as others know to achieve that goal.

Acknowledgement

Special thanks to Redi Root, Discount Nusery Supplies, and Blackmore.

LITERATURE CITED

Barnes, H. W., jr. (1999). Affects of container size and shape on the rooting of cuttings. Combined Proc. Intern. Plant Prop. Soc. *49*:304-307.

Barnes, H. W. (2010). Some fundamental differences separating rooted cuttings and seedlings. Combined Proc. Intern. Plant Prop. Soc. *60*:416-421.

Combined Proceedings IPPS 73: 451-453. 2023.

The Effects of IBA Treatment and Surfactant on Root Development during Vegetative Propagation of *Hibiscus grandiflorus*

Jack Schaefer

University of Cincinnati, 6210 DAAP P.O. Box 210016, Cincinnati, OH 45221 USA

Jack.Schaefer@cincinnatizoo.org

Keywords: cuttings, swamp rose mallow

Summary

Preliminary rooting trails were conducted in Swamp rose mallow (*Hibiscus grandiflorus* Michx.). Cuttings were treated with K-IBA with or without a surfactant. There is no significant impact of the rooting hormone or surfactant on rooting success or plant survival.

INTRODUCTION

Swamp rose mallow (*Hibiscus grandiflorus* Michx.) is a beautiful plant that has unique traits within the genus yet is as approachable to the public as the popular species *H. moscheutos*. *H. grandiflorus* has a pinwheel flower silhouette and a luscious velvet leaf that adds foliar interest throughout the growing season (**Fig. 1**). The species has flowers of variable color, from almost pure white on some plants to soft pink petals and a magenta throat on others. Culturally, *H. moscheutos* and *H. grandiflorus* are very similar, both preferring to be grown in partial to full sun with moist to wet soils. This plant could be sold at garden centers as an Eastern United States native and as a rain-garden plant. Breeding programs can incorporate the unique floral shape and foliar textures into *Hibiscus moscheutos* genetics, as demonstrated by the F1 hybrid *Hibiscus* 'Moy Grande' (Yu et al., 2016). Key objectives of this project were to identify effective propagation techniques, introduce an under-represented species to the region, and share this research with the horticultural community.

Figure 1. Flower and leaf characterization in swamp rose mallow.

MATERIALS AND METHODS

Softwood cuttings were collected on June 30th. A large cooler was lined with ice. Sharpened Felco #2 pruners were used to cut 24" lengths of stem from established *Hibiscus grandiflorus* plants at Boone County Arboretum. The cuttings were placed in a damp plastic bag and secured in the prepared cooler with frequent misting as they were transported to the propagation location. Stems were cut into two node cuttings and re-bagged to keep the cuttings cool, as well as to randomize the cuttings for the sticking process.

Media was constructed of 50% Fafard Professional Potting Mix and 50% perlite. Four treatments with 23 specimens per treatment were tested: hormone/surfactant, no hormone/surfactant, hormone/no surfactant, and no hormone/no surfactant. Cuttings were selected at random and given one of the four treatments. The hormone treatment consisted of a three-second quick-dip in Dip'N Grow liquid hormone at a 10x concentration. The surfactant was a spray bottle foliar application of Liquid Harvest Non-Ionic Surfactant three times on a seven-day interval with the first application at the time of sticking. Water and Peters Professional 20-20-20 General Purpose Fertilizer were applied as needed.

RESULTS and DISCUSSION

Plants were evaluated after 30 days using a 0-4 method with 0 being dead, 1 being alive with no callus, 2 having callus but no roots, 3 having young roots, and 4 with vigorous root development (**Fig. 2**).

Figure 2. Rooting rating system for evaluating cutting success.

The highest percentage of cuttings that were fully rooted was seen in the IBA alone treatment (56%, but there is no significant need for either rooting hormone or surfactant (**Fig. 3**). This plant is not able to be purchased readily via seed or from Eastern nurseries, so large-scale propagation of desired plant material was not possible. The specimens generated from this experiment can be used to run a larger scale experiment and generate plant material for future propagation.

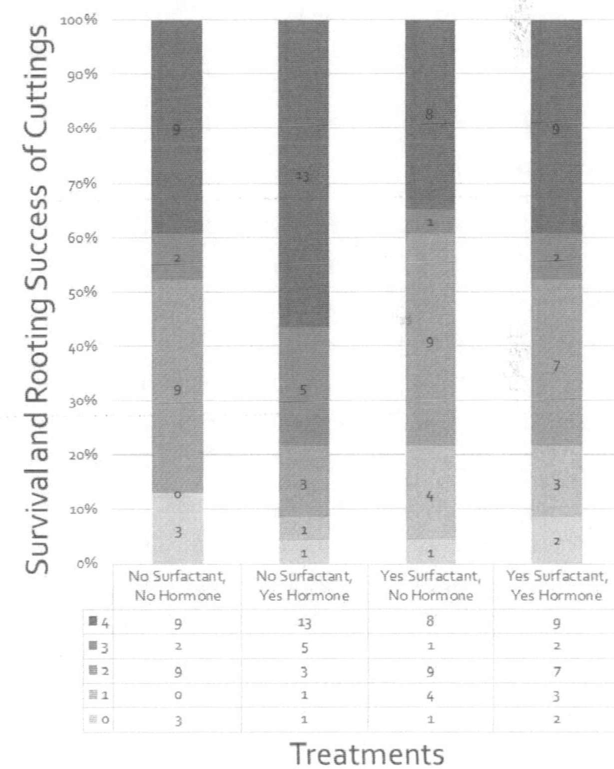

Figure 3. Survival and rooting success in swamp rose mallow cuttings.

LITERATURE CITED

Hibiscus grandiflorus. North Carolina Extension Gardener Plant Toolbox. (n.d.). https://plants.ces.ncsu.edu/plants/hibiscus-grandiflorus/

Yu, C., Yin, Y., Creech, D. L., Lu, Z., and Xu, J. (2016). Morphological characters and SRAP analysis of two hybrids between *Hibiscus dasycalyx* and *Hibiscus* 'Moy Grande.' Scientia Horticulturae *198*:118–124.

Combined Proceedings IPPS 73: 454- 460. 2023.

The Horticulture Club's Native Plant Program at the University of Kentucky

Katie Taliaferro, Anna Baloh, Shari Dutton, Richard Durham and Robert Geneve

Department of Horticulture, University of Kentucky, Lexington, KY 40506 USA

klta242@uky.edu

Keywords: education, internship, leadership

Summary

For over 40 years, the Horticulture Club at the University of Kentucky has provided students with outside-the-classroom educational and social experiences. Over the years, the members' interests have changed along with horticultural trends. Primary interest in landscape management has shifted to a current emphasis in production and use of native species, especially as pollinator and wildlife plants. A major goal of the club is to provide opportunities to expand their members' horticultural knowledge beyond the classroom including national and international study tour experiences. The current Vice President and former Native Plant Nursery Intern for the Horticulture Club, Katie Taliaferro, shares her unique insight into the inner workings of one of the most successful clubs in the Martin-Gatton College of Agriculture, Food, and Environment.

IPPS Vol. 73 – 2023

INTRODUCTION

The Horticulture Club at the University of Kentucky provides student members with leadership opportunities, expanded extra-curricular plant-related education, and life-long friendships. Currently, the Horticulture Club is one of the most popular clubs in the Martin-Gatton College of Agriculture, Food, and Environment, approaching 100 members from various departments on campus.

The club's mission is to:

1. Foster a spirit of friendship between students mutually interested in plants, horticulture, and related areas.
2. Provide a supportive environment where students can learn and work with plants.
3. Provide opportunities for students to broaden personal and interpersonal perspectives through travel and other club supported activities.

Members meet weekly to socialize and learn various aspects of horticulture and related subjects. The most prominent activity is propagating and producing plants for the club's Native Plant Nursery managed by horticulturist, Shari Dutton. However, the club's influence extends further into each member's growth and the larger community on campus. UK's Horticulture Club accomplishes this mission by providing students with weekly opportunities to dive deeper into horticulture through workshops, guest speakers, and trips to horticultural events. In addition, native plant sales, campus projects, and international trips enhance the campus experience by bringing people in for a common cause: a profound love of nature.

LEADERSHIP

The Horticulture Club has three committees, each with their own committee head: Plant Production, Sales, and Outreach. While the officers and committee heads are first faces of leadership students meet, the club relies on the thoughtful participation of each member to guide newcomers in the right direction.

Senior member mentorship is a large part of the environment and success of the club. As members grow with the club, they share their knowledge and experiences with newcomers, deepening their understanding of the techniques and procedures that keep plant production activities safe and the plants healthy (**Fig. 1**).

Figure 1. Plant Production Committee workdays give members the opportunity to practice their skills with plants native to the eastern region of the United States.

Additionally, tasks like watering the tropical greenhouse and maintaining the club's carnivorous plant collections are given to new students with a desire to learn; these individuals will then go on to pass down their tips and tricks to the next member. This not only creates a stronger community, but also exemplifies how members learn the communication and technical skills necessary to thrive in their future endeavors by participating in club activities.

However, this community does not end when a student graduates. Several club alumni stay in contact, providing materials, workshops, and their unique skill set with current members throughout the year. Alumni specializing in carnivorous plant care and the native cut flower business have been key participants when showing students the range of opportunities the horticulture degree can offer them (**Fig. 2**).

Figure 2. The carnivorous plant collection was started by a previous Horticulture Club president who frequently comes by to help teach new members how to maintain the plants.

OPPORTUNITIES

The club is key for getting students the exposure necessary to discover and pursue a career in horticulture and its adjacent fields. Advisors and alumni provide invaluable contributions to the club, providing students with professional connections and guidance. Another large role of club advisors is bringing awareness to scholarship and internship opportunities. Club meetings regularly include guest speakers and hands-on workshops, so students can get a glimpse into an expert's professional work.

In the past couple years, the Horticulture Club has used its connections to curate workshops to members' interests. In just the Spring 2024 semester, the club has eight workshops and guest speakers lined up in between workdays, sales, and tabling events. These workshops range from wildflower walks at a local nature preserve to hands-on work with bonsai and mushrooms. Utilizing these workshops, the club combines education with community to teach people of all majors how to enjoy and appreciate the natural world. Take, for example, a workshop done in the fall of 2022. Shari Dutton, one of the club's beloved advisors, shared her love of natural dyes by using the plants of the dye garden she had designed the year prior at the UK Arboretum. After giving a brief history of natural dyes and their uses, Shari taught members how to dye cloth with indigo leaves. An instant favorite among students.

UK's Horticulture Club also provides members with the chance to travel abroad. Over the past 25 years, the club has traveled on study tours to numerous locations in North, South and Central America, Europe, Asia, and New Zealand (**Fig. 3**). Traveling with fellow horticulture students and professors offers a deeper learning experience that encourages members to expand how they view horticulture with a global perspective. The club is designed to make these kinds of opportunities accessible to students. In 2023, four students were able to use their volunteer hours to offset the cost of a club trip to the Canary Islands. Anna Baloh, a senior horticulture major and the current club President, enthusiastically encourages her fellow members to join these club trips. She has been able to use her volunteer hours to offset the costs of attending two club trips abroad and four horticulture conferences. She frequently describes these trips as some of the highlights of her time at UK.

Figure 3: In 2023, the Horticulture club took a group of mentors and students to the Canary Islands where they learned production techniques used in a semitropical environment.

NATIVE PLANT NURSERY

The Native Plant Nursery is the backbone of the Horticulture Club, growing and housing 123 herbaceous and 45 woody species native to the eastern region of the United States during the 2023 season (**Fig. 4**). The club's Native Plant Sales program brings plant enthusiasts from around campus to purchase plants they know will support their local environment. These sales have become widely popular within Lexington's native plant circles. Students are given the opportunity to answer questions concerning the care and use of the plants.

Figure 4. A blooming passion vine (*Passiflora incarnata*) surrounded by Virginia creeper's (*Parthenocissus quinquefolia*) vibrant autumn colors in the Horticulture Club's hoop house.

In fact, many of the plants sold had been sown and maintained by these same students, providing a full-circle experience they can be proud of. The club's main source of funds comes from the Native Plant Sales program. Students exchange their volunteer hours for club funds to help pay for their spot on the club's annual trip abroad or to cover costs at educational opportunities like conferences and symposia (**Fig. 5**).

Figure 5: Katie Taliaferro making a presentation at the 2023 IPPS-ER conference in Hamilton, Ontario supported by the region and partially funded by Horticulture Club funds.

Many of the plants grown as part of the Native Plant Nursery are used for planting projects on campus, the Horticulture Club has been able to provide professors and landscape architects with native plants to fulfill the campus's many needs. This can include for use in classes or to create large bioswales. The current project, The Diverse Landscapes Nature Rx, intends to beautify UK's medical campus using large blocks of colorful flowers (**Fig.6**). While the project does include many native plants, the Diverse Landscapes Nature Rx project puts an emphasis on a plant's ecological functionality. This project aims to utilize nature's calming effect to reduce the stress of students and medical patients on campus. Approximately 4,000 plants were grown by the Horticulture Club with a substantial portion coming from the Native Plant Nursery. This is only the most recent example of the long history of the Horticulture Club supporting its community.

Figure 6. Katie Taliaferro tending to the plants of The Diverse Landscapes Nature Rx Project during her summer internship funded by the Horticulture Club.

CONCLUSION

The Horticulture Club at the University of Kentucky is a great resource for anyone interested in plants. As an industry that continues to grow, horticulture needs to encourage that interest in the field grows with it. The club is not major-specific. It is approachable for those with even the slightest affinity for nature and draws them in. Many are not aware what horticulture even is until they get the chance to experience it themselves. The club has been a turning point for some students, helping them realize that horticulture is their passion. Others may decide to add a plants and soil science minor. The Horticulture Club offers many ways to interact with plant production, giving members an opportunity to see what aspect appeals to them and their career aspirations. They can then use the club and its resources to gain the experience and skills they need. Nevertheless, the students in chemical engineering and political science still have a place in the club. As stated in its mission statement, the Horticulture Club aims to create an environment where students can learn about plants. Education about nature is a necessity in a world that continues to be covered in concrete. Through their involvement with the club, members foster a thoughtful perspective of their surroundings. Something as approachable as a workshop on plant identification can make all the difference in someone's ability to pause and consider their impact on their environment.

My time as the native plant intern for the Horticulture Club was a truly enriching experience. Going into it, I had very little knowledge of or exposure to horticultural production. I had only recently switched from a biology to a horticulture major, having taken only one related class. I am incredibly lucky that the club advisors saw my potential and dedication despite this.

Over the next seven months, I worked intimately with nursery director Shari Dutton, learning the foundation of her work: thoughtful observation. It is simple, but much more difficult than I thought. That concept alone made each of my everyday tasks intentional. As I continue as the Vice President of the club, I continue to learn through my work with Shari and my peers. I try to show people the value of thoughtful observation. I know that I could not possibly explain everything I learned to my peers with other various majors. However, I also know that this intentionality will gradually strengthen anyone's ability to effectively take part in our workdays and sales.

Thanks to the Horticulture Club, I was able to have this opportunity and get paid during the process. Two years ago the club made the decision to use club funds to sponsor the Native Plant Nursery Intern position to support passionate students studying horticulture. Everything about this internship was incredibly impactful. It gave me a foundation to confidently pursue my future in horticulture.

Find out more about the Horticulture Club at UK at: Instagram & Facebook: @UKHorticultureClub
or HortClubUK@gmail.com

Combined Proceedings IPPS 73: 461- 468. 2023.

Back to Basics, Setting Standards for Success with Seed

Jaime Manlove

388 North Creek Rd, Landenberg, Pa 19350

jaime@northcreeknuseries.com

Keywords: germination, dormancy, native plants, nursery, collection

Summary

Plug and plant production of native plants is described for North Creek Nursery including seed collection, dormancy treatments, germination and seedling growth. Standard plugs are produced and finished in a variety of sizes including transplant ready Landscape plugs.

INTRODUCTION

North Creek Nurseries was founded in 1989 by C. Dale Hendricks and Steve Castorani. Dale Hendricks retired from the business in 2008. Steve Castorani is the current President and CEO. North Creek employs approximately 85 full time and 5 part time employees, plus a few interns. Annual liner production is approximately 8 million starter plugs. Facilities are spread between two locations, comprising a total growing range of 320,000 square feet. Modern gutter

IPPS Vol. 73 – 2023

connected greenhouses have a full complement of the latest in production and growing technology. Plants continue to be propagated in a series of older Quonset style houses as well. North Creek offers plugs of perennials, grasses, *Carex*, ferns, woody vines, and a few woody shrubs (**Fig. 1**).

Figure 1. shows the product mix by plant collection and items offered in multiple sizes. All plants are grown as plugs in standard plug trays configurations.

Plugs are grown in four sizes that fit an industry standard 1020 tray. Horticultural sizes, which are produced for finished growers are 72's (72/tray) and 50's (50/tray); 72's are a square plug of 3.60 cu inches and the 50's are round with 6.77 cu inches. Landscape plugs ^{Tm} are LP 32's, round, with a soil volume of 10.07 cu inches and LP 50's, square and tapered with a soil volume of 11.90 cu inches (**Fig. 2**). Landscape Plugs are often grown for use in landscape projects and can also be employed by nurseries to pot up for the occasional quick finish in containers.

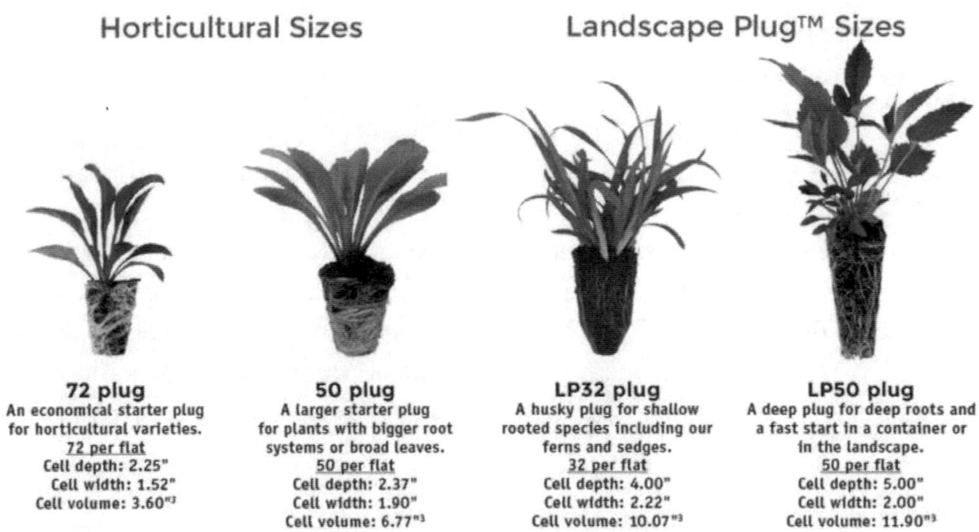

Figure 2. Plug sizes produced at North Creek Nurseries.

Production Efforts

Various production methods are utilized at North Creek based on a history devoted to the production of native plants and native plant selections. Seed production figures prominently in these efforts. In many cases, as dictated by projects or customer type, vegetatively produced native plants offered cuttings may be in short supply or not available (**Fig. 3**). Also, certain projects demand genetically diverse, open pollinated, seed sourced material for revegetation and establishment in natural environments to increase biodiversity.

Other production methods include vegetative cuttings from stock or purchased in, divisions of stock plants, tissue culture, and fern plugs purchased from other propagators and grown on into larger plugs. (example: 288 plugs to LP32).

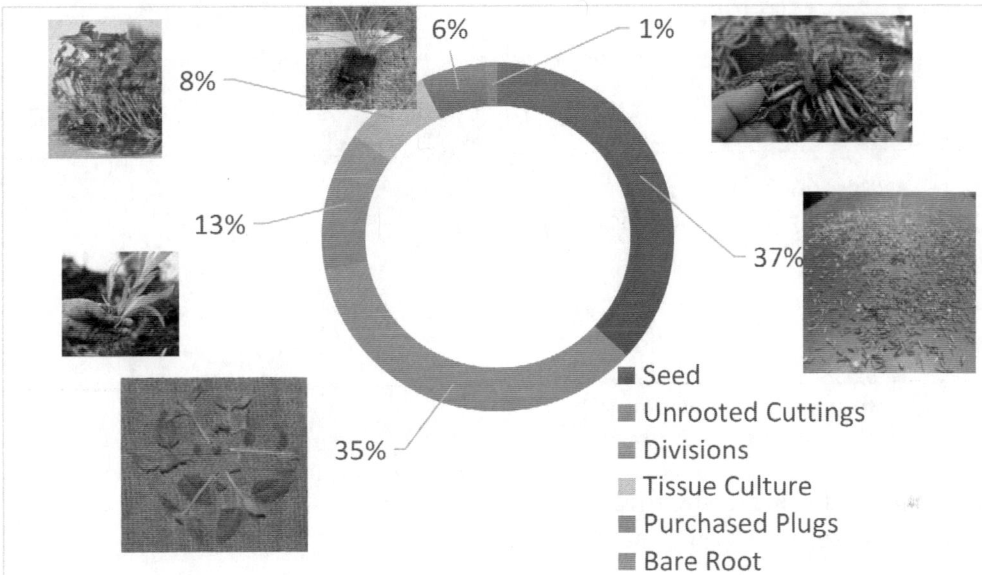

Figure 3. Percentage of production by type.

Seedling Production Data

Seedling production details include:

- 426 total product SKUs
- 135 SKUs from seed
- 23 SKUs collected in house
- 17% seed collected in house
- 83% seed purchased from supplier
- 25% SKUs directly sown on main production line
- 17% SKUs directly sown on seed production line
- 57% SKUs transplanted from 288 trays

In some cases, seed is not obtainable via outside sources and there are situations where seed quality is an issue. In those cases, it is preferable and a priority for the company to produce their own seed. About 23 SKU's (varieties) are grown and collected in-house, which represents 17% of seed needs. It is important to note that growing and producing seed in house also eliminates the possibility of extraneous hybridization with undesirable genetic relatives. Genetically pure seed may not be obtainable on a commercial level. An example would be *Echinacea tennesseensis*, where much of the commercially obtained material is cross bred with *Echinacea purpurea*.

On site seed production is an important aspect of North Creek's production including 20+ species like:

Allium	*Tiarella*	*Heuchera*
Delphinium	*Solidago*	*Hibiscus*
Rudbeckia	*Acorus*	*Monarda*
Zizia	*Amsonia*	*Clematis*
Aster	*Chasmanthium*	*Scirpus*

The company devotes a significant amount of effort to ensure the highest quality seed available is obtained and ensures a systematic approach that encompasses a full growing season which is aptly laid out in **Figure 4**.

A sophisticated system of maps is employed for positive identification on our growing plots, Full color pictures of the specific plant including what the seed looks like (**Fig. 5**). It cannot be stressed enough that proper labelling insures accountability.

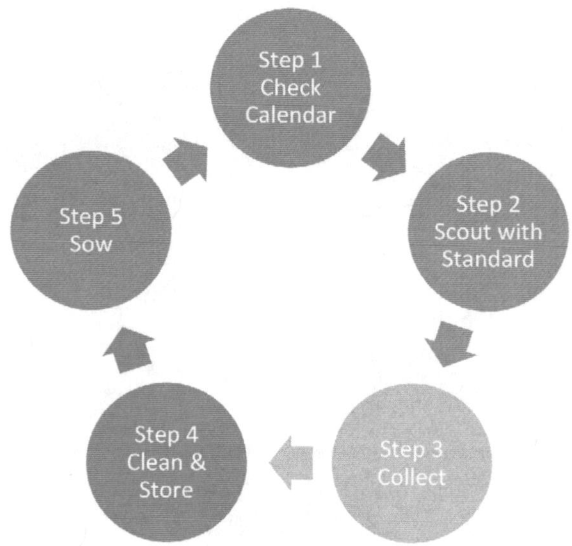

Figure 4. Steps in quality seed production.

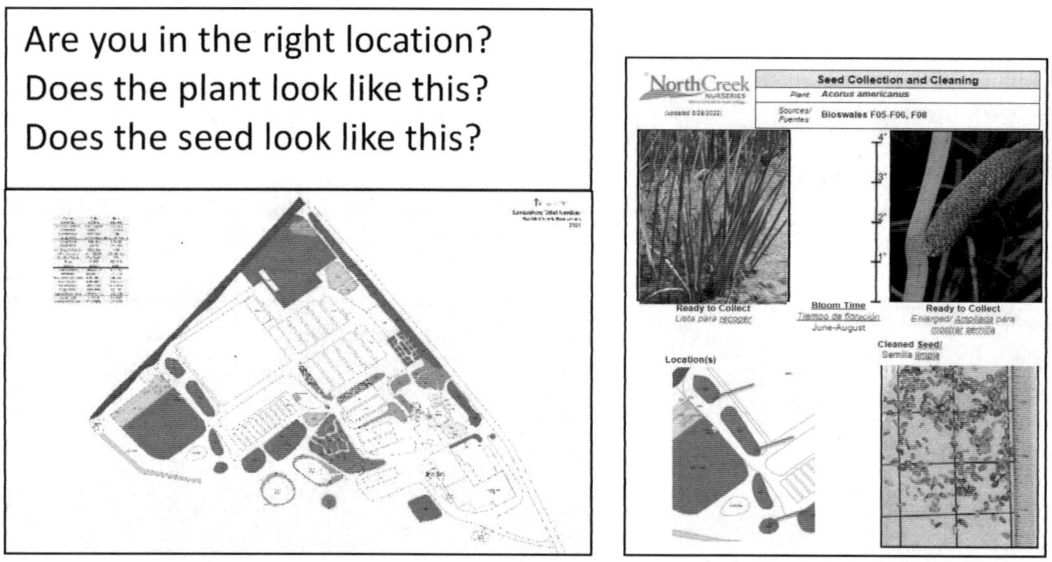

Figure 5. Outlines the process that insures reliable and accurate seed collection.

Figure 6. A target plant in the growing area circle in yellow for positive identification.

Cleaning Seed

Cleaning seed is a critical element in the process. Cleaning is a double check to assure we have the correct seed. Many plants can readily be identified based solely on the seed if one knows what to look for. Cleaning also allows for an immediate status of seed quality so that hollow or malformed seed can be discarded, plus this allows for visual inspection for insects that invariably hitch a ride. Also cleaning allows for easier handling of seed and increases the potential that such seed can be utilized via a seed sowing machine. Cleaning is critical to our success. We hold regular workshops for key employees who work with seeds so that they stay up to date on our processes and needs (**Fig. 7**).

Figure 7. In-house seed cleaning tool use training secession.

Seed Storage

Seed storage is an important aspect of our operation. Some seeds, such as grasses do not necessarily need cold storage and there are instances where a two-year-old grass seed actually germinates better than one year old seed. Other seeds have a short life expectancy unless they are held in dry cold from 38 F to 42F. In all cases it is vital to store seed in a dry environment that is fully rodent proof. Storage in metal containers is critical (**Fig. 8**).

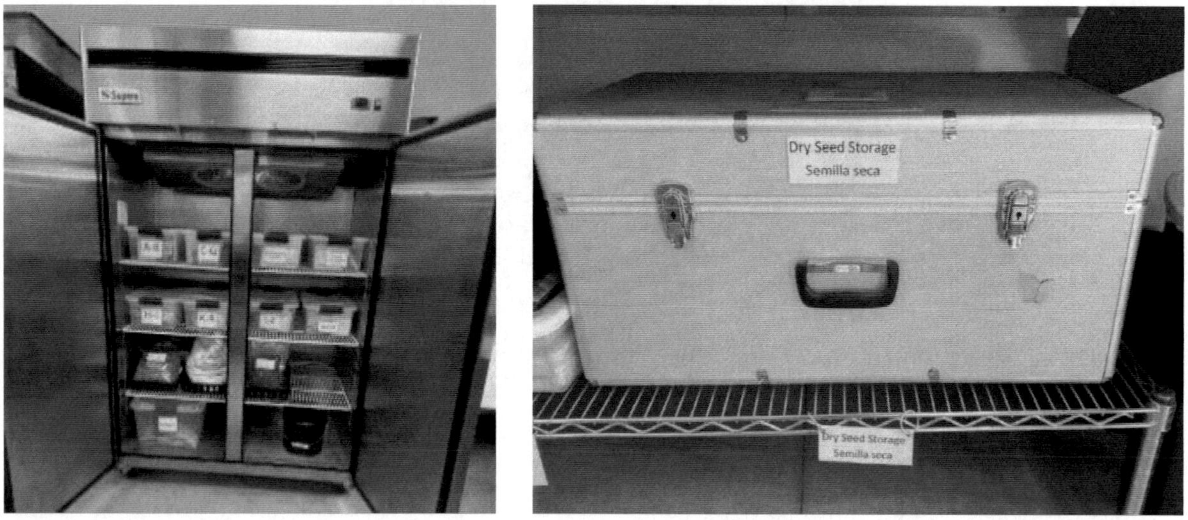

Figure 8. Left - refrigerated cold dry storage at 38 – 42°F. seed packets are stored alphabetical for easy sorting. Right – dry storage.

Curious Facts

No two seed varieties are the same. There would be obvious differences between Helleborus and Asclepias. Two different genus and species, two different families, different natural growing environments, and different cultural requirements, both in the production sequence and in the natural environment. Each individual species has its own brand if you will, and this must be taken into account in the production sequences.

But how different is often determined by practice and experience. In the contest of *Helleborus* vs. *Asclepias* the differences are clear.

Helleborus 'Brandywine' has a grow time of 37 weeks from start to finish, *Asclepias incarnata* can do a turnaround in 8 weeks. The economics of this is significant. When to start the process, anticipate bench requirements for "x" number of weeks. Plug size is also very important. Market demands come into play too. Above all, how is final pricing affected by the production cycle?

The whole production effort is governed by available space, timing, weekly work orders, patching need (filling gaps), cheat sheets (efficiency), and readily available tools to do the job. Certain seeds need a lot of human intervention.

Machines are an important part of the production process. For high volume flat fillers coupled with seed sowing machines are a must (**Fig. 9**). A one-yard mixer can be used for soil incorporation with slow release fertilizer, biologicals, etc. It can be used to can fill ~180 72s/50s cell flats per hour or ~200-gallon pots per hour. We still fill flats the old way by hand for sowing some seeds.

Figure 9. High volume flat filler and seed sowing machine.

Refrigeration is a general requirement for both seed storage and cold moist stratification. Even small operations need several refrigeration units (**Fig. 10**). Cold storage has many applications including seed prep, cold moist stratification, beneficial insect storage and in a separate specialty chemical storage area.

Figure 10. Cold storage units vary from walk-in storage to something as common as a kitchen refrigerator.

Ideally a small sample of seed should be given a water soak to make a judgement to the percentage of sound seed in the sample. Seed that floats are generally considered to be defective and are usually discarded. There are occasions where good seed floats, but it is not common. If adequate supplies are available, then it is probably best to discard the floaters so there are no skips in a plug tray.

Cold Moist Stratification (CMS) can best be described as seed dispersed in a friable (high air content) substrate that is moist and subjected to storage between 32 and 40F. The length of time for each seed lot is carefully monitored, for instance *Helleborus* vs. *Asclepias* is a good example.

Some seeds require **Warm Moist Stratification (WMS)** prior to cold moist stratification. The process is like CMS, but WMS depends on a much warmer 70F storage condition. Caution, some seed will inadvertently germinate during either of the two cycles. Seed lots should be examined in a timely manner to detect this pre-germination. Explanations as to why this occurs are variable and are not generally identifiable prior to the stratification process. Sometimes the entire sequence must be repeated a second time for adequate germination to occur. Finally, it should be concluded that no stratification will occur if the seed is not moist during the process. Moisture is critical for both warm and cold stratification.

Scarification can be described as applying mild physical damage to seed so that the seed coat or shell becomes permeable to water and oxygen. Tools to achieve this vary but can include rock tumblers, hot water soak, and acid soak (caution, dangerous and might harm the seed),

Seed can sometimes be coaxed into germinating by the use of Gibberellic Acid (GA) at 500ppm for GA3, absolutely no more than that. Too much GA will promote germination, but the resultant seedlings will stretch abnormally. Other chemicals include a potassium nitrate soak at 5000ppm. Still others might use a urea solution at 250 ppm. Usually, a 24-hour soak is required, then seed is removed, washed, allowed to dry a little to facilitate handling at sowing.

Some seeds will germinate readily if given a simple water soak, since tap water can vary, it is probably best to use distilled water or rainwater for this process unless it is known that the tap water is acceptable. Lupines for instance don't care, whereas Sarracenia are quite particular about the water used. Rainwater is best.

CONCLUSION

As illustrated here, seed propagation on a grand scale takes many steps to insure a successful crop. Since all crops are affected by production timing, as well as the needs of customers, crops can be in various stages of development at any given time. Certain plants can be tricky, such as Helleborus seed, which has very narrow requirements. Should environmental conditions fluctuate, and higher than anticipated temperatures prevail, Helleborus seed will cease to germinate, and the only recourse is to start the process over again. The Helleborus seed will not die, it just reverts to a new dormancy window. Alternately, *Asclepias incarnata* will not do that. These are just two examples of the many variations and challenges seed production pose. Time, familiarity, experience, and experimentation all play a part in the ongoing success with seed propagation.

Combined Proceedings IPPS 73: 469- 478. 2023.

Perspectives, Trends and Sustainability Initiatives in the Soilless Growing Media Industry

Brian E. Jackson

Department. of Horticulture, North Carolina State University, Raleigh, NC

Brian_jackson@ncsu.edu

Keywords: Substrate, peat, wood fiber, nursery, greenhouse

Summary

Growing media is fundamental to greenhouse and nursery production. Sustainability issues as well as competition for substrate from a growing controlled environment food production sector is making it important to reevaluate existing media resources and examine viable alternatives.

INTRODUCTION

The growing media industry today is being altered on a continual basis due to a number of factors that have a worldwide impact and what was once a local phenomenon no longer has a geographic center. Factors such as projected global demand are influenced by a variety of driving forces behind this demand. Potting media is no longer potting soil, along those lines the original one gallon nursery pot is no longer a discarded food can and it is no longer a true one gallon. Factors such as this change the focus of what it means to project the coming needs of the nursery industry. But it does not stop there. The industry has sprouted branches into many different aspects of horticulture which often requires tailor made media substrates for specific crops and that in turn affects component availability and supply. How do we keep abreast of these changes and what does it mean for our future endeavors?

By 2050 the projected future demand for soilless substrates in all aspects of horticultural production is expected to quadruple. Currently, supply and availability issues of many substate products will extend well into 2023 and perhaps beyond. The immediate take home lesson is to act now and do not get cut short.

Table 1. Total estimated substrate market in 2050 based on the expected market increase and a more realistic estimate of the potentially available materials.

	2017 (Mm^3 $year^{-1}$)	2050 (Mm^3 $year^{-1}$)	Increase (%)
Peat	40	80	200
Coir	11	46	418
Wood fiber	3	30	1000
Bark	2	10	500
Compost	1	5	500
Perlite	1.5	10	667
Stone wool	0.9	4	433
Soils/tuffs	8	33	413
New		65	
Total	**67**	**283**	

The future demand for growing media materials can be clearly seen in the trend towards 2050 as a massive increase in the use of materials **(Table 1)**. Notice that while peat moss is still being used the increase in need for peat moss is 200% while that of wood fiber is 1000% followed by significant increases in other materials. This is important to know because there is a worldwide effort to limit peat moss production, it is not a matter of availability so much as it is a response of many countries looking at the actual and hypothetical environmental damage that the peat moss industry is causing. Issues such as water quality, land degradation and potential release of greenhouse gases is causing concern. The increase in peat moss usage from 2017 to 2050 is 2x but the total increase for all materials is closer to 4.5X. Most notably coir, wood fiber, bark, and compost. This increase in demand for alternative substrates as compared to peat moss will inevitably lead to both shortages and higher prices as time goes by.

While the trend has been increasing for years the amount of materials being allocated to alternative crops on a worldwide basis has become substantially significant. One of the biggest "new crops" is the tremendous increase in growing herbs, vegetables and fruits in greenhouse environments.

The amount of acreage of such systems is increasing steadily and is being driven in part due to the high costs of shipping and a greater understanding of the use of fossil fuels. Systems that can produce edible crops more locally are gaining great favor over those that are produced via conventional means. Less environmental damage, cleaner environment of the production facilities, less over the road fuel usage all contribute to a rise in this type of production. This results in an accelerated need for significant amounts of substrate that used to be destined for the ornamental plant industry, but quite frankly not anymore.

Figure 1. New crops in soilless substrates under greenhouse conditions.

Typical vegetable production in closed greenhouse systems include cucumber, peppers and tomato (**Fig. 2**). Production of vegetables and fruits on a local scale eliminates costly shipping and promises better product quality and availability. "Organic" suddenly becomes a prominent reality. The downside of course to the ornamental plant industry is that there is increased competition not for sales but for starting materials such as substrates.

Figure 2. Closed greenhouse systems for vegetables and fruits.

Production of vegetables and fruits on a local scale eliminates costly shipping and promises better product quality and availability. "Organic" suddenly becomes a prominent reality. The downside of course to the ornamental plant industry is that there is increased competition not for sales but for starting materials such as substrates.

The prominence of the marijuana and hemp production activities presents a relatively new venue for the usage of substate materials as well as talented horticultural professionals. That trend is not likely to decrease anytime soon.

In the future there could well be other "medicinal" crops other than Cannabis that will also occupy significant greenhouse space and materials. The growing field of "nutraceuticals" is on a steady march as is a trend towards more natural products for food flavoring and dyes instead of the artificial coloring agents so commonly used presently.

There simply no end to the possibilities of new crops coming on line in the future. In Holland efforts are underway to produce vanilla from greenhouse grown orchids to offset the problems facing the natural vanilla production from Madagascar.

Soft fruits meaning berries of all sorts including specialty product lines such as petite grapes or Zante currents (a type of grape) are coming into prominence and will increase as time progresses (**Fig. 3**).

Figure 3. Soft fruit (berry) crops being grown under controlled environment systems. A. Strawberry. B. Blueberry. C. Raspberry.

Not only are soilless media used they are necessary to produce both the propagules as well as the finished production plants. Soilless media are fundamental in production due to the greater control over nutrient applications, water usage and in maintaining a high degree of sanitation and disease prevention.

In **Table 1**, it was detailed that wood fiber would increase 1000% from 2017 to 2050. Such an increase obviously means an increase in raw wood production either by seed, cuttings or tissue culture. Efforts are underway to produce woody plants tailored to provide an optimum wood product for the revolution of greater use of wood fiber as opposed to peat-based substrates. These "new" targeted woody trees will in turn require new production facilities to meet the increasing demand (**Fig. 4**).

Figure 4. Woody conifer seedlings destined for the production of wood fiber.

These seedlings are not just for Christmas trees anymore. Wood fiber for substrates is an absolute necessity for the soilless media industry. Efforts have been underway for a couple of decades to produce softer hardwoods such as clonal *Liquidambar* to meet such demands. It should be noted that wood fiber also has a significant application in the alternative fuel industry which could potentially affect substrate raw material availability.

Challenges facing the Growing Media Industry

One of the most significant concerns to the growing media industry is the availability of peat and peat products. While it is true that the overall supply of sphagnum peat is not in jeopardy this does not mean that availability is guaranteed. **Figure 5** highlights some of the current situation on concern to the world supply of peat. We in North American cannot arbitrarily consider the peat issue a local issue, much like petroleum resources and due to the one world concept of international trade what happens in country A thousands of miles from country B still has an effect.

Figure 5. Peat availability concerns in Europe and North America.

The headline" Ireland Stopped Peat Harvest" pretty well sums it up (**Fig. 5**). It should be said this is not an isolated case.

Regulatory threats to the peat industry are alive and well and not just in the Northern Hemisphere. Worldwide there is greater and greater emphasis on doing the "right" thing when it comes to environmental issues. Chile and other South American countries are struggling with an increasing awareness of the population on environmental issues and how the voting public is making their concerns known even if it means dumping an entire industry. The horticulture industry needs to be fully aware of and not take for granted that the future will be business as usual. **Table 1** illustrates the point with respect to alternatives to peat entering the field of soilless media and it should be seriously considered that more alternatives will be needed in the future. The industry as a whole needs to embrace this and endorse more research into new technologies for soilless media and techniques for growing in those media.

Coconut coir is one such alternative substrate constituent with the center of world production being in India and Sri Lanka. Granted production is a reality but with the increased awareness of the people of the world to the many facets of commercial activities what we have traditionally taken for granted may or may not be the same as it has always been.

Other New or Ongoing Challenges

- Geopolitical crises – pending and yet unknown
- Increasingly global weather patterns affecting peat extraction
- Spread of misinformation and use of "cherry picked" data to drive corporate and political agendas (i.e. Sustainability Impact)
- Lack of trained professionals in growing media (substrate science)
- Increasing global demand/reliance for MORE growing media!
- War in Ukraine and closing of RUS and BLR.
- Continued sea freight costs and Labor Shortages
- Fuel prices (material processing and ground transport)
- Discovery of "Issues and Threats" of peat alternatives

Figure 6. Other new and /or ongoing challenges facing substrate and soilless media availability.

In the headlines of late, hardly a day goes by that the war between Russian and the Ukraine does not enter the picture. Such geopolitical crises both actual and potential (even the threat of a conflict has massive ramifications on the world market). Picture a conflict between India and Sri Lanka, how that might affect the coir industry and in turn affect the production of strawberries in Holland or Society Garlic in Florida.

One subject that is prominent and yet seems innocuous is the high cost of shipping, alas, it is rising to the top of major concerns. But there is another equally innocuous and perhaps more deleterious element in the mix and that is labor shortages, not just in North America but in Europe, China of all places and a whole mix of countries throughout the world. This is especially exacerbated by the lack of trained and knowledgeable workers. Coupled with the need for more directed research to solve some of the substrate issues as presented here could lead to a decrease again in substrate availability.

While there are some who discount the effects of global warming the reality is that the impact of climate change on substrate materials can and will be significant. To address this consequently demands more talented workers and researchers to find new methods to produce the vast array of horticultural crops.

As given in the **Table 1**, peat presently is the #1 component of all soilless substrates worldwide. This will change. Unprecedented demand is taxing the system so that in short order it will no longer be sustainable to use peat and peat products.

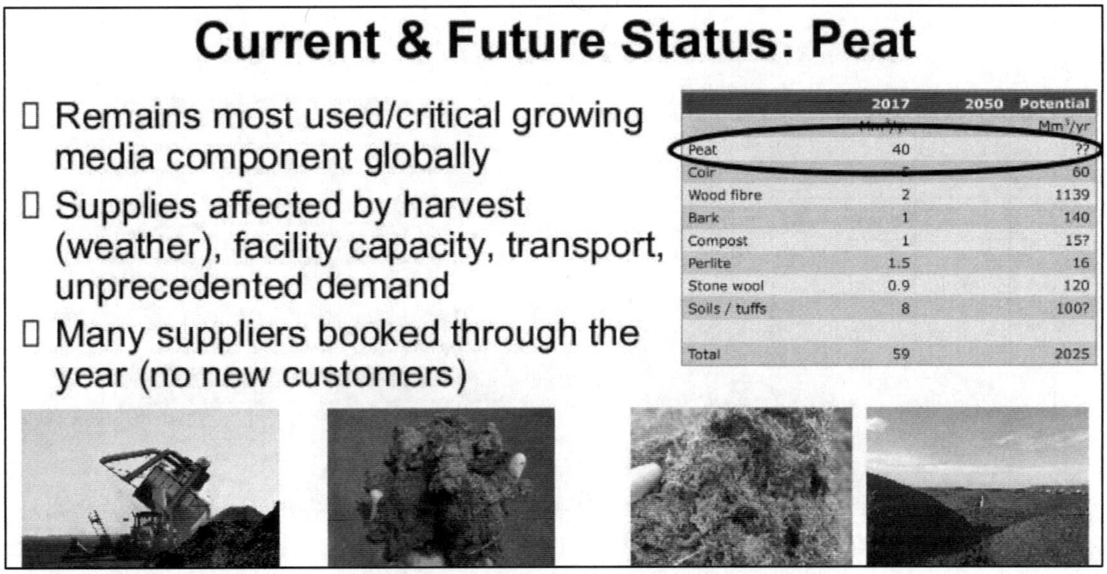

Figure 7. Just how sustainable is peat for soilless substrates.

Table 1 also indicates that wood and bark products and compost (often made from wood by products such as sawdust will take a greater center stage in the soilless media industry. Engineered wood fiber and other forest products are thought to have the greatest overall potential for sourcing, development and utilization in soilless media. Many peat companies are proactive in substituting 20-30% peat with wood materials (sawdust or pulverized wood fiber). In the past 3 years over a dozen wood processing facilities have been built and are operating to supply this type of materials.

Perlite can be considered to be a "filler" in the soilless substrates. It allows for porosity, some water retention, and provides consistent avenues for air into the substrate . It is sterile, and generally non-reactive although for certain tropical plants it presents problems with excess fluoride. It should be noted that while useful in a soil-less media it has limitations and too much perlite in a given mix often results in a poor root system for the intended crop.

On a global scale most of the world's production is from Greece. Production does occur in the United States limited quantities due to industry consolidation with resources being stretched thin. It is an expensive production due to the extraction methods and the high costs of fossil fuels or electricity. It unusual for perlite production to be near adequate energy supplies thereby requiring expensive transportation of fuels or ores to the production facilities.

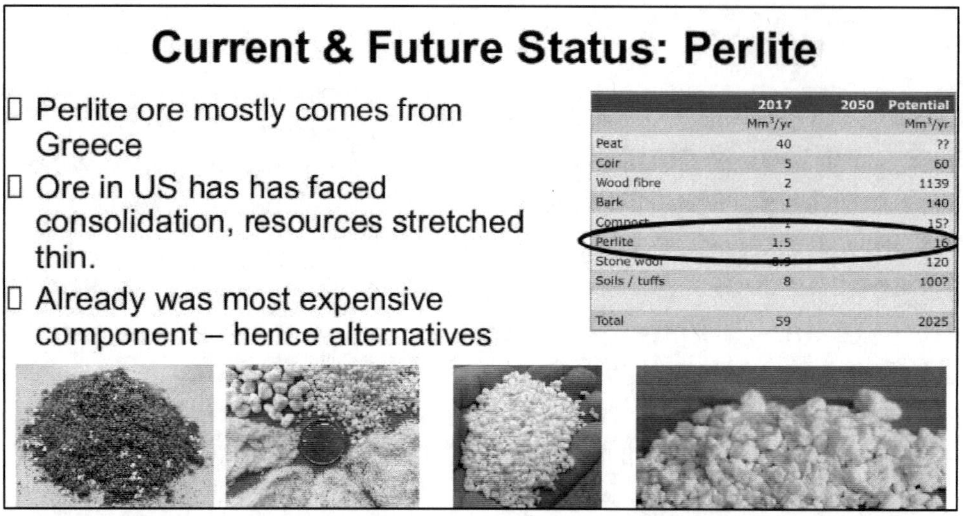

Figure 8. Status of perlite.

The nursery/greenhouse and allied industries are now being confronted with having to make noteworthy changes in their production methods. It is important to note that while Canadian peat is still widely available that might change in the future due to some of the factors mentioned in this thesis.

Can new growing techniques be implemented in a strategic manner that allows for a weaning from peat products to more renewable resources. Perhaps a thorough study of what is happening in Europe is a good starting point. In Germany, the use of peat is not allowed. Decades ago in the United States, some growers were using what is known as sedge peat, not a sphagnum peat at all but decayed organic matter dredged from ponds and lakes perhaps there could be a resurgence of this activity. New suppliers and alternative products seem to be some of the possible solutions to the coming shortage of sphagnum peat supplies. Euro peat, Coconut products (coir, fiber, chunks, etc) and of course wood products derived from a variety of sources, not just tree trunks. With the advent of other new industries such as hemp the number of new possibilities is increasing (**Fig. 9**).

Figure 9. New and developing substrate products.

A Lot Goes into Alternative Substrate Evaluation

There are many factors in evaluating a new product for commercial availability. Specific areas of interest are:

- Business Performance (will this thing fly?)
- Social Performance, (what are the societal consequences good or bad)
- Environmental Performance (water use, carbon footprint, ecosystems)

From a research endeavor to bring a new product into the market these questions have to be asked and acceptable answers provided. It is not merely an operation involving just horticulture. Granted that is a big component of the process but the consequences of the social and environmental aspects also have an important presence in the process.

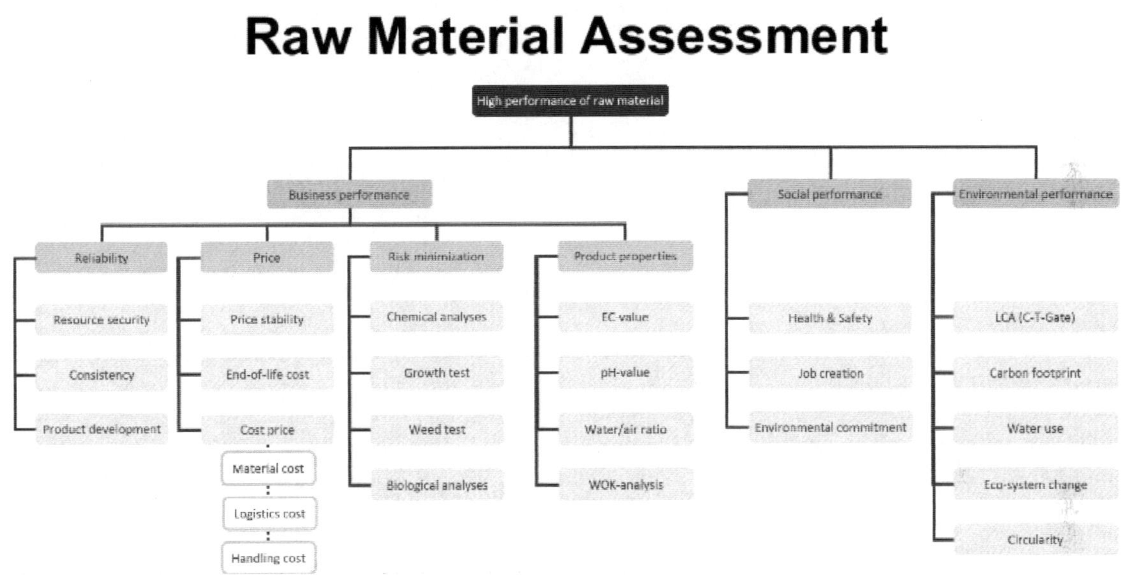

Figure 10. Raw Material Assessment.

In today's world, growers have a myriad of resources available to them. Educational webinars, substrate supplier online tutorials (many questions can be answered by these tutorials), publications, advice services by many companies. They are there to help if it means that the activity helps create a market. One thing to know is that when searching add "PDF or PPT "to the verbiage as that makes the search more concentrated to the subject matter.

There is lots of materials available for information. One caveat - don't necessarily stick with what you already know. Branch out to other possibilities. For instance, the *Cannabis* growers are perhaps some of the most innovative horticulturists going. New techniques, new ideas are often on the front burner. Just because you grow pine seedlings or eucalyptus does not mean you might not learn something from a hemp grower. The internet is a vast resource, take advantage of it.

Looking Ahead to the Future

It is important to remember that our industry is on the forefront of global change and that "substrate security" is a food, health, and National security issue. To meet these challenges there is a need for creative, collaborative, and innovative solutions. Global trends based on perceptions, social media, political platforms, etc. are helping to steer the future of our industry. We will need to stay proactive and reactive.

For the immediate future, order your substrate components early – very early. Potentially expect limited volumes even if once "guaranteed". Soon, 100% peat will be almost impossible to get.

Combined Proceedings IPPS 73: 479-482. 2023.

New Plant Forum 2023 – Eastern Region IPPS

Elizabeth (Dunham) Erickson Moderator

Knight Hollow Nursery, 7911 Forsythia Court, Middleton, WI 53562 USA

Liz@Knighthollownursery.com

Keywords: breeding, plant introduction, genetics

Summary

New plants for 2023 are highlighted and described. This year six IPPS-ER breeders presented herbaceous and woody perennial plants.

IPPS Vol. 73 – 2023

PRESENTER

Dr. Mark Bridgen
Cornell University
Long Island Horticulture and Research Extension Center, 3059 Sound Ave
Riverhead, New York USA
Mpb27@cornell.edu

Alstroemeria 'Coral Chaos'

Coral Chaos Inca lily shows continuous flowering from late May until killing frost.
Propagate by rhizome divisions.
Winter-hardy to USDA Zone 5

Figure 1. Flowering plant (left) closeup of flowers (right).

PRESENTER

Connor Hagemeyer
North Dakota State University, Dept 7670,
PO Box 6050, Fargo, North Dakota 58108 U.S.A.
connor.Hagemeyer@ndsu.edu

***Betula costata* 'CinDak'**
Cinnamon Curls® Dwarf Korean Birch is a distinctive dwarf selection of Korean birch, which grows in a diminutive, compact, single stemmed form. Attractive creamy white exfoliating bark with cinnamon-colored undersides curling in strips. Cold hardy and tolerant of higher pH soils. Originated from seed collected from the U.S. National Arboretum germplasm (NA39939). NA39939 originated from a collection trip in 1966 to the Republic of Korea along a trail on Mt. Deogyusan.
Hardiness: USDA hardiness zones 3b to 6; **Mature Size:** Height: 9', Spread 9'; **Form (Shape):** Rounded compact growth habit.

Figure 2. Cinnamon Curls® Dwarf Korean Birch throughout the seasons.

PRESENTER

Dr. Yi Li
Dept. of Plant Sciences and Landscape Architecture
Transgenic Plant Facility
105 Ahern Lane, Unit 5082
University of Connecticut, Storrs, CT 06269 USA
yi.li@uconn.edu

A New Seedless form of *Euonymus alatus*. invasive, sterile varieties of Burning Bush (Euonymus alatus 'Compactus') using gamma ray-mediated mutation and other breeding techniques. These non-invasive varieties have undergone over 12 years of field evaluations. Many of them produce less than 2% of the seeds compared to currently available cultivars on the market.

Among these sterile varieties, one stands out as completely sterile, with no single seeds produced from the original plant in 20 years, and from a large number of vegetatively propagated plants in 12 years. Both the original plant and its offspring derived from stem cuttings exhibit indistinguishable characteristics such as uniform growth and a brilliant red color in fall. This non-invasive, sterile variety of Burning Bush is now ready for release

Figure 3. *Euonymus alatus* selections.